Internet of Things

Technology, Communications and Computing

The series Internet of Things—Technologies, Communications and Computing publishes new developments and advances in the various areas of the different facets of the Internet of Things. The intent is to cover technology (smart devices, wireless sensors, systems), communications (networks and protocols) and computing (theory, middleware and applications) of the Internet of Things, as embedded in the fields of engineering, computer science, life sciences, as well as the methodologies behind them. The series contains monographs, lecture notes and edited volumes in the Internet of Things research and development area, spanning the areas of wireless sensor networks, autonomic networking, network protocol, agent-based computing, artificial intelligence, self organizing systems, multi-sensor data fusion, smart objects, and hybrid intelligent systems.

Indexing: *Internet of Things* is covered by Scopus and Ei-Compendex **

Sohail Saif · Suparna Biswas · Seemanti Saha ·
Samit Kumar Ghosh · Nauman Aslam
Editors

Secure Intelligent Sustainable Applications Using IoT for Society 6.0

A Holistic Approach to Protocols to Prototypes

Springer

Editors
Sohail Saif
Maulana Abul Kalam Azad University of Technology
Haringhata, West Bengal, India

Suparna Biswas
Maulana Abul Kalam Azad University of Technology
Haringhata, West Bengal, India

Seemanti Saha
National Institute of Technology Patna
Patna, Bihar, India

Samit Kumar Ghosh
Khalifa University
Abu Dhabi, United Arab Emirates

Nauman Aslam
Northumbria University
Newcastle Upon Tyne, UK

ISSN 2199-1073 ISSN 2199-1081 (electronic)
Internet of Things
ISBN 978-3-032-27247-8 ISBN 978-3-032-27248-5 (eBook)
https://doi.org/10.1007/978-3-032-27248-5

This Springer imprint is published by the registered company Springer Nature Switzerland AG
The registered company address is: Gewerbestrasse 11, 6330 Cham, Switzerland

Preface

IoT-based smart applications ranging from smart home to smart city to smart industry, etc., are being used in every sphere of life for quality living, societal benefit with automatic, affordable, usable and sustainable solutions. Society 6.0 focuses on integrating advanced, intelligent technologies with human life in more efficient, deeper way to develop fully sustainable, intelligent, adaptive, and ethically governed ecosystem. Here in this edited book, we have focused on convergence of secure, intelligent, and sustainable technologies through the Internet of Things (IoT) technology. IoT serves as backbone to integrate data collection, transmission, processing, and automatic decision-making for immediate actions to be taken in time-critical domains. The targeted application areas are Healthcare, Agriculture, Environmental monitoring, disaster prediction, Weather forecast, etc., for better and enhance quality of living. But this advantage comes with a drawback of privacy sensitive data sharing over insecure heterogeneous wireless links and thus the data are vulnerable to security attacks and threats. Hence, appropriate prevention mechanisms must be applied and in parallel, the solutions must be energy efficient, less carbon footprint emission should be ensures etc. to make the solutions green and sustainable. Our edited book is aimed to cover all such issues, challenges and opportunities of utilization of IoT to develop smart, secure, sustainable solutions for resilient, future-ready applications suitable for Society 6.0.

Haringhata, India — Sohail Saif
Haringhata, India — Suparna Biswas
Patna, India — Seemanti Saha
Abu Dhabi, UAE — Samit Kumar Ghosh
Newcastle Upon Tyne, UK — Nauman Aslam

Acknowledgements

We would like to express our sincere gratitude to everyone who contributed to the successful completion of this edited book project. We deeply acknowledge our students, research scholars, and collaborators for their significant contributions support throughout this journey. We also extend our heartfelt thanks to our university, Maulana Abul kalam Azad University of Technology, India, for providing a conducive ecosystem for inception, planning, execution of such academic and research-related project. We are especially grateful to our co-editors for their commitment and collaborative spirit. Our sincere appreciation goes to the publisher, Springer, and their entire editorial and production team for their professional guidance and continuous support in bringing this volume to fruition. We also acknowledge all reviewers and contributing authors for their scholarly efforts and meaningful contributions that have enriched the quality and scope of this book.

Above all, we humbly acknowledge the divine power that guided us and gave us the strength and wisdom for the successful execution and completion of this edited volume.

Contents

Modern Approaches of Secure and Intelligent IoT

Amalendu Si, Ayan Ghosh, and Sushovan Khatua

1 Internet of Things (IoT)

The Internet of Things (IoT) is a network of physical devices or "things," embedded with sensors, software, and other technologies that connect and exchange data over a network [1]. The IoT evolved a revolutionary paradigm in which everyday physical objects are ingrained with sensors, communication modules, and computing capabilities to collect and exchange data over the internet. The pipeline system architecture of the IoT model is shown in Fig. 1a. This interconnected ecosystem enables seamless interaction between devices, environments, and users, supporting intelligent automation in different aspects like healthcare, smart cities, industrial automation, agriculture, transportation, and home environments [2]. The exponential rise of connected devices expected to reach tens of billions globally has significantly enhanced operational efficiency, real-time decision-making, and service personalization. The primary structure of an IoT system consists of four layers [3] and those layers are fully dedicated to their functionalities which is shown in Fig. 1b, four layers of IoT system are defined as follows:

Sensing Layer: The sensing layer forms the foundation of any IoT system. Its primary function is to collect data from the physical environment. The functionalities of the sensing layer are achieved through the use of various sensors and actuators that

A. Si (✉) · A. Ghosh
School of Information Science and Technology, Maulana Abul Kalam Azad University of Technology, West Bengal, Nadia, India
e-mail: siamalendu@gmail.com

A. Ghosh
e-mail: send.info.2.ayan@gmail.com

S. Khatua
Department of Computer Science & Engineering, SRM University -AP, Andhra Pradesh, India
e-mail: sushovankhatua79@gmail.com

S. Saif et al. (eds.), *Secure Intelligent Sustainable Applications Using IoT for Society 6.0*, Internet of Things, https://doi.org/10.1007/978-3-032-27248-5_1

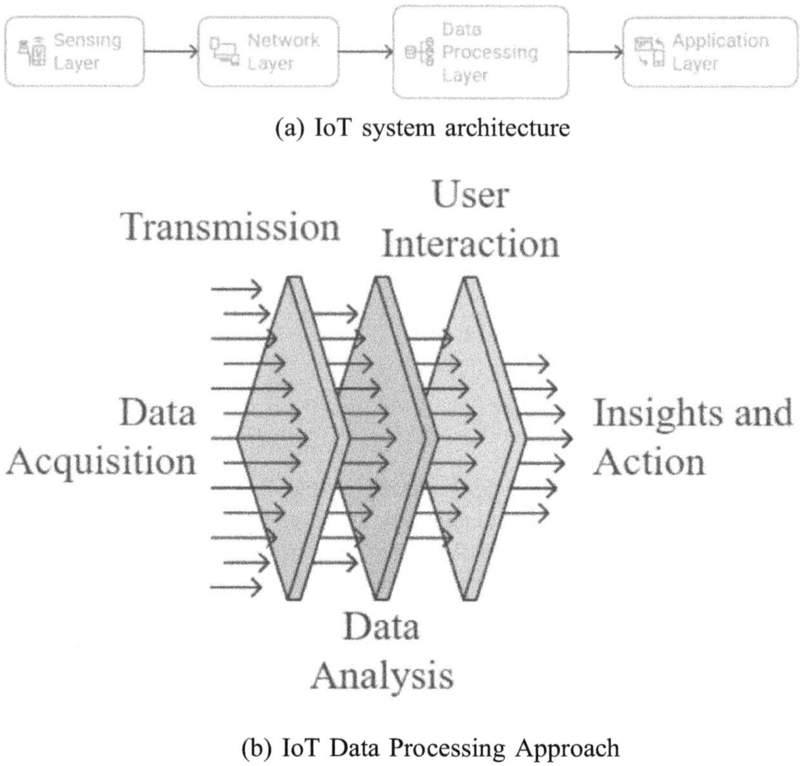

Fig. 1 IoT system architecture funnel

are deployed to monitor and measure different parameters. The sensors are the digital devices or hardware components that use to detect and measure the physical quantities such as temperature, humidity, pressure, light, motion, and more [4]. Similarly, the actuators are the electro-mechanical devices that perform defined action based on the data received from the sensors or from the data processing layer. Some of the well-known actuators are relays, motors, valves, LED, and so on [5]. The sensing layer is responsible to acquire data from the environment accurately and reliably. In this regard, this layer performs some specific operations such as aggregation, calibration, and modification. Data aggregation is used to combine data from multiple sensors to provide a more comprehensive view of the environment. Calibrate data to ensure the sensors provide accurate readings [6]. Modify the sensing data through amplifying, filtering, and converting into a format that can be processed by the next layer [7].

Network Layer: The network layer is responsible for facilitating communication between the devices in the sensing layer and the internet or other networks. It provides the necessary infrastructure for secure data transmission and reliable connectivity. Network layer uses the various types of communication protocols and network topologies to smooth and accurate data transmission. The network layer utilizes various communication protocols to enable local area as well as wide-area data transmission. Some common features of IoT-based network connections are low power, variation of range, and conflict-less transmission. The used protocols are Wi-Fi, Bluetooth, Zigbee, Cellular (4G/5G), low-power wide-area network (LPWAN), Narrowband IoT, and so forth. During the data transmission, the network layer is responsible for secure data transmission from the sensing layer to the data processing layer. So this layer involves data encoding to converting the data into a well-formed format, routing technique to find the optimal path for data transmission, error detection, and correction technique for accurate data transmission, and protecting data from unauthorized access.

Data Processing Layer: This layer is responsible for processing the raw data collected from the devices in the sensing layer. It transforms the data into meaningful information that can be used to make decisions and take actions. The layer uses storage component to store the raw and processed data and various data analytics techniques to extract insights from the data. The data processing layer often includes data visualization tools to present the data in a clear and understandable way. The modern cutting-edge technique: Edge computing processes the data at the edge of the network, closer to the devices that are generating the data. This modification tries to reduce latency and improve the overall performance.

Application Layer: The application layer is the topmost level of an IoT system, which serves as the critical interface between the system and its end users. The responsibility of this layer is delivering processed data and enabling meaningful interaction with IoT devices. The application layer is basically a user-friendly interfaces such as mobile applications, web portals, web platforms, or interactive dashboards where the users engage to monitor and control their IoT ecosystem. These interfaces transform complex information into clear, visual representations through charts and graphs, making it easier for users to understand actual facts across their system at a glance. The application layer empowers users for interacting with the devices in real time. The users without sufficient technical knowledge can remotely manage devices from a single interface through intuitive controls such as adjusting thermostats, toggling lights on and off, or securing doors. Informing users is a crucial functionality of the application layer. The system delivers timely alerts and notifications to ensure that users do not miss critical events, detect anomalies promptly, and perform necessary maintenance activities. The modern IoT applications are boldly interconnected with other systems for better performance. They are integrated with various platforms such as Customer Relationship Management systems, Enterprise Resource Planning systems, or social media networks. Then, the application layer creates bridges to remove the difficulties among the IoT infrastructure and broader business operations, which creates cohesive, robust, and responsive digital environment.

The IoT systems consist of interconnected layers, and each layer with its own distinct security vulnerabilities and challenges [8]. One challenge interacts with another and is amplified. After that, those are collectively affecting the reliability and credibility of the whole system. These are addressing effectively requires moving beyond universal security approaches and instead understanding what makes each layer unique and where it's most exposed to attack.

The *sensing layer* has serious vulnerabilities that are due to its nature of constraint. By nature, sensing devices have to run on small computational costs and are often deployed in geographically unprotected locations. These machines are therefore faced by real-world exposure which is limited in other systems. This provides a wonderful environment to enemies: they could seize gadgets, create spoofed sensor values, or directly alter data. These are not just technical issues but they undermine the accuracy of the data, endanger the operational safety and in some cases lose the privacy of the data owners.

After architecture development *network layer* faces its own set of critical problems. Weak authentication schemes, inadequate cryptographic services, and the fact that sending of communications without encryption remains common give the enemy of the state an opportunity to use the loopholes. In addition, the risks are compounded by the existence of non-standardized protocols that do not pass through a strict security examination. Networks are therefore vulnerable to distributed denial-of-service (*DDoS*) attacks, loss of data to third parties, and hijacking of devices—threats that often exploit poorly set up routing protocols or devices that are left at default mode of operation [9, 10].

Data processing layer that includes edge computing, fog, and cloud platforms presents a different range of security concerns. Most of these systems are hampered by susceptible software and firmware implementations, feeble or missing access controls that do not suffice to constrain user and process privileges and interfaces which were not initially designed with security in mind. By adversaries finding out these vulnerabilities, they can gain unlawful access to confidential information, intercept communications in a man-in-the-middle, cripple comprehensive services, or otherwise interfere with the normal operations used by Internet of Things (IoT).

The *application layer*, which contains the user interfaces that direct contact with authentication mechanisms, is often wanting or lacking altogether. Encryption measures are sometimes not applied, update procedures are seldom executed and managed securely, and device management practices are typically informal and carried out haphazardly. The combination of these deficiencies leads to system vulnerabilities, like account impersonation, privilege escalation by attackers, service disruptions, and prolonged compromises due to undetected vulnerabilities that could remain unresolved for months or even years

Overseeing the intricacy of an IoT system is challenging and requires careful, multifaceted security. The key understanding is that IoT systems are highly interconnected: a vulnerability in one layer can compromise the entire ecosystem. This interrelation indicates that neglecting security at any one layer can pose a significant threat to the entire system. For this reason, organizations must implement coordinated security measures across all four layers of the system. Based on this security concern, integrated security strategies build *IoT* deployments that are truly resilient, trustworthy, and safe.

2 Secure Internet of Things

The secure and intelligent IoT, often referred to as *AIoT*, where the artificial intelligence is combined with *IoT* technology. This hybridization allows sensing devices to do far more than simply collect data. Then, the devices can operate and interact with minimal human during the execution, process large-scale datasets in real time, and make intelligent decisions on their own. This highly efficient and automated systems span everything from the smart home on your nightstand to sophisticated industrial installations.

Based on this intelligence capability, the *IoT* systems are built on a layered architecture where each layer handles a specific responsibility. Some layers responsible for transmitting data reliably, some for processing that data, and the remaining for users to interact with the system (as illustrated in Fig. 1b). This modular design makes it easier for developers to build systems that actually fulfilled the requirement of the end users. To fulfill the requirement for a significant number of more devices to connect to IoT systems, security becomes increasingly difficult to manage. The propagation of interconnected devices introduces vulnerabilities at every occasion. The privacy concerns grow alongside every new device added to the system, and begin to realize the importance of security due to these systems are no longer safe. It has become an absolutely critical system.

IoT has develop from its inception, and the game-changer has been the integration of artificial intelligence and machine learning technologies. The relatively passive sensors have evolved into something far more sophisticated—intelligent *IoT* systems, or *AIoT*, this special capabilities can think and adapt. These systems passively collect sensor readings, actively process vast amounts of data, identify patterns that humans might miss, and continuously adjust their behavior to optimize performance. All these functionalities are performed automatically without any human instructions.

These improved smart systems facilitate the features that could not be possible before. They help to predict failures in advance, which allows true predictive maintenance. The contextual services are adjusted to the preferences and needs of the individual users. Distribution of the resources becomes more dynamic as real-time demand governs the allocation of the resources as opposed to the unchanging rules. Abnormalities are immediately identified and resolved without the need to manually discover them. As a result, the implementation of IoT becomes significantly more stable and inherently more efficient.

The high rate of growth in the field of the so-called IOT presents a disturbing fact that is not easily to be brushed off in the form of major security and privacy issues. The reasons behind this are due to the inherent replicability of the IoT systems. Devices have been more prevalent both in design and in capacity. Owing to harsh resource limitations that are inherent in the hardware, implementing effective security mechanisms is sometimes not possible. Besides, networks are explicitly open and interconnected, and the system architecture is decentralized, thus, there is no single bastion to protect. This meeting place creates a space in which security becomes broken and hard to govern in a holistic manner.

Attacks that include unauthorized access to the IoT environment are diverse and enduring. The opponents can bring in malware, launch denial-of-service attacks that disable the systems, or control the devices to act in a way they never wanted to. Conventional security methods intended for desktop computers and data centers are inadequate for the IoT, necessitating an entirely different approach. This has resulted in organizations employing lightweight cryptographic techniques that require minimal power and computational resources. Authentication systems need to be secure and effective. The systems require intrusion detection solutions that can function with restricted hardware. Emerging technologies, such as blockchain-enabled trust management and privacy-preserving data analysis, present fresh possibilities for realizing extensive IoT deployments.

The Intelligent and Secure IoT model revolutionizes the perspective on connected systems. The combination of AI-driven features with effective cybersecurity measures tackles security and intelligence challenges. This synthesis fosters more robust ecosystems that are resilient, dependable, and self-sufficient, as they possess the ability to reason, learn, and protect themselves. To realize this vision, it is crucial to be devoted at every level; security must not be regarded as an additional feature confined to distinct layers. Organizations must safeguard four essential areas: the device, the network it connects to, the cloud space where the data is handled, and the applications utilized by users for communication. This multilayered security approach fosters user confidence and enables the secure implementation of large-scale IoT in mission-critical infrastructures.

In this context, researchers are actively engaged in exploring innovative concepts that transcend conventional methods. These new intelligent security frameworks signify a shift away from reactive, post-event protection. They improve early threat identification that recognizes risks before they lead to damage. The modern architectures with self-healing functionalities can automatically recover from attacks without human intervention. Adaptive risk-aware decision-making allows systems to adjust their security posture dynamically based on changing threat conditions. These innovations provide the unique demands of upcoming smart environments with billions of connected devices operate autonomously.

3 Rise of Intelligent IoT (AIoT)

The familiarity of the IoT system spread significantly for its affective utility [11]. Then the large number of multi-functional devices becomes connected, the potential for cybersecurity risks and data breaches increases significantly. Most of the interconnected IoT devices are designed and built with weak security, making them vulnerable targets for attackers.

The integration of artificial intelligence with IoT, known as AIoT, enhances the capabilities of standard IoT networks. While IoT devices historically only collected and shared data, AIoT devices can process and analyze that data to make autonomous

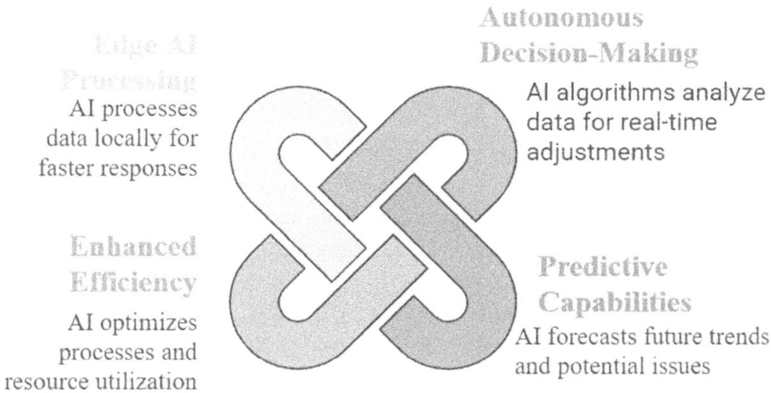

Fig. 2 Composite outcomes of AIoT

decisions. Figure 2 shows the effects due to integration of AI techniques in the IoT system. This integration leads to the following capabilities:

Autonomous Decision-Making: AI algorithms analyze data from sensors in real time, allowing devices to adapt to changing conditions and automate processes without human intervention [12]. For a smart home system, an AIoT system automatically adjusts the thermostat based on occupancy and weather conditions, optimizing energy consumption without requiring manual adjustments. To set the industrial setup, AIoT systems monitor the performance of equipment and, based on this observation, automatically adjust operating parameters to prevent failures and optimize efficiency.

Predictive Capabilities: Predict the future using historical data by utilizing the transformative abilities of *AIoT* systems. These systems analyze patterns that show the coming scenario, whether the equipment failure or the ideal time for intervention. In manufacturing settings, this results in an invaluable aspect known as predictive maintenance that ensures production operates seamlessly. The *AIoT* systems continuously track sensor data being streamed and recorded from equipment. Simultaneously, they analyze historical patterns and accumulate insights to recognize subtle indicators that a component could fail in the upcoming weeks or months. This understanding enables maintenance crews to plan work ahead of time, swapping out components before they fail and resulting in costly downtime. The outcome is convincing: fewer unforeseen failures, extended equipment longevity, and considerably lower operational expenses [13].

A markedly different area where these forecasting capabilities are applied is agriculture. Enacted through the *AIoT* systems, the agricultural sector can estimate crop yields by examining the complex interplay of weather conditions, soil quality, water levels, temperature fluctuations, and numerous other environmental factors. Armed with these forecasts, farmers can improve their choices regarding when to begin planting, how many seeds to apply, and when to gather the crops. This knowledge aids in optimizing the coordination of planting and harvesting operations in ways that were previously not feasible. These consequences encompass all farming activities, enabling farmers to enhance output, minimize waste, and allocate resources more effectively

Better efficiency: The ability to streamline the use of resources and automate repetitive, routine activities is one of the most important contributions of the use of the technology of AIoT. Companies implement the use of the AIoT systems with smart features, which have enabled them to operate in virtual industries with remarkable effectiveness and efficiency in their operations, which have been dramatic in their performance in this domain, as we know it today, due to these smart features (liu2023enabling). A good example is the logistics: *AIoT* systems can monitor the real-time traffic situation, weather conditions, and shipping requirements and optimize the routes, thus saving time and fuel. The drivers are directed to more efficient routes, idle time of vehicles is reduced as well as delivery times of packages are minimized. Likewise, in energy management, the change is also the most incumbent. *AIoT* systems also learn the occupancy habits of buildings and optimize lighting, heating and cooling without human intervention. It will not be necessary to manually switch switches or adjust a thermostat. The sensors will sense when space is empty and, as such, will decrease energy use; when people come back, the systems will respond instantly. The overall impact on the organization is huge, resulting in a decrease in energy use, decreased environmental effects, and increased occupant comfort.

Edge AI: The solution to a fundamental challenge in distributed systems is in transmitting massive amounts of data across networks to distant cloud servers. Edge computing brings processing power to the edge of the network and closed between the sensors and devices. This proximity creates several advantages that are important for time-sensitive applications. Here, data is locally processed and sent to the cloud, which reduces the response time and reducing latency that could be critical for safety or performance. This approach eliminates the need for continuous transmission of every sensor reading over potentially expensive network connections, thereby utilizing channel bandwidth more efficiently. In the autonomous vehicles ecosystem, edge *AI* proves essential for survival. The vehicle's onboard processors rapidly assess sensor data, enabling swift responses to unexpected obstacles, pedestrians, or alterations in road conditions. A delay in receiving a response from a cloud server could be critical. In medical environments, edge *AI* revolutionizes diagnostic processes Medical imaging systems can analyze scans locally; physician gets the insights immediately rather than after round-trip delays to remote servers. Patients get faster diagnoses, doctors make quicker treatment decisions, and improve the outcomes [14].

4 The Importance of Secure IoT

In the recent scenario, the relevance and usefulness of IoT is growing exponentially. Then, the IoT requirement links billions of intelligent devices that amass, analyze, and share data to facilitate automation, intelligent decision-making, and smart real-time services in different fields, including healthcare, industrial automation, smart cities, transportation, and agriculture [15]. With the further development of the IoT systems, their safety has become of paramount importance as the information sent over them is sensitive, the number of devices interconnected is growing, and the crimes of cyberattack are becoming more frequent. Secure IoT is needed to ensure the functionality of the systems, privacy of users and to remain trusted in the related technologies.

4.1 Expanded Attack Surface

The implementation of IoT devices in smart houses, industrial automation and health-care, transport and critical infrastructure grows fast, and the total attack surface of contemporary digital systems grows exponentially [16]. Attack surface is the sum of all possible entry points of malicious individuals to use security vulnerabilities. Since IoT systems involve a wide variety of heterogeneous devices, communication protocols, and cloud services, they present numerous layers of vulnerabilities that were used to cause cyberattacks, interfere with the operations of systems or even steal sensitive data. The vast number and the variety of IoT devices complicate the process of implementing and enforcing uniform security standards within the whole ecosystem.

4.2 Critical Infrastructure

IoT system integrated into the critical infrastructure systems such as power grids, water treatment plants, and transportation networks. This integration provides such type of benefits, including improved efficiency, automation, and real-time monitoring. However, this hybridization makes the IoT-related infrastructures more vulnerable to security breaches [17].

A successful cyberattack on the critical infrastructure system has devastating consequences and great impact in the community, including widespread power outages, water contamination, and transportation disruptions. The potential for physical damage and loss of life makes it imperative to secure IoT devices and networks that are used in critical infrastructure.

4.3 Data Protection and Privacy

IoT devices often collect sensitive user data, including personal information, location data, and health data. This data is used to track user behavior, create detailed profiles, and even make predictions about future actions. Protecting this data from unauthorized access, data theft, and privacy violations is essential for maintaining user trust and complying with privacy regulations [18].

Attackers may establish illegal access to sensitive user information when the IoT devices are not secured appropriately. This information may be subjected to identity thefts, financial frauds or other ill motives. IoT devices or cloud storage systems could also be stolen by the attackers. This information may be sold in the black market or blackmailed against people or groups. The IoT devices even include security concerns in terms of privacy, even though the information is not stolen or accessed without authorization, the collection and use of personal data by these devices is also questionable. The users have no idea what information is being recorded, how it is utilized or who can access it.

4.4 Business Impact

Cyberattack on mission-critical IoT assets has a significant business impact, including financial damage, loss of productivity, and reputational harm. Cyberattack causes financial damage in several ways. Companies have to pay for incident response, data recovery, and legal fees. They also lose revenue due to system downtime or loss of customer trust [19]. Cyberattack disrupts business operations and leads to a loss of productivity. Employees are unable to access critical systems or data, and production lines may be shut down.

A cyberattack damages a company's reputation and leads to a loss of customer trust. Customers are reluctant to do business with a company that has been the victim of a cyberattack, especially if their personal data has been compromised.

5 Common Security challenges

The common security challenges that contribute to the vulnerabilities of IoT devices. These challenges range from weak authentication mechanisms and lack of encryption to vulnerable software and firmware, network security issues, and supply chain vulnerabilities [20]. Addressing these challenges is crucial for enhancing the security posture of IoT ecosystems and protecting sensitive data. The fundamental vulnerability of the security threads among the IoT devices logically defined in the Fig. 3.

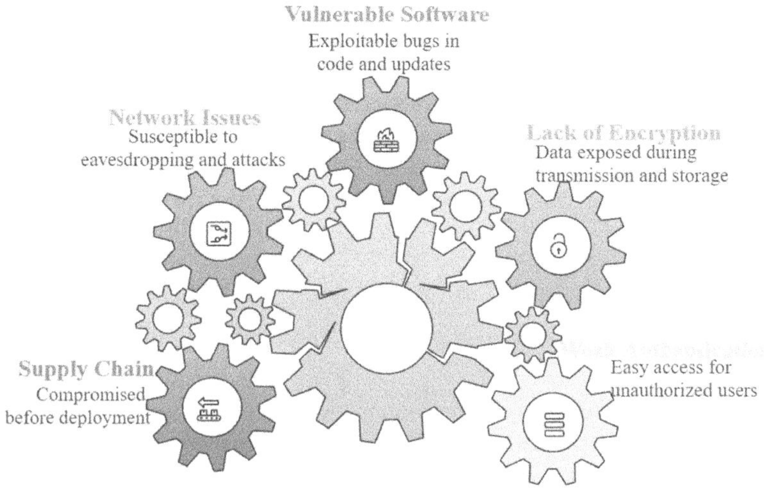

Fig. 3 IoT devices vulnerable to security threats

5.1 Weak Authentication

The main widespread security issue in IoT devices is poor authentication. Numerous devices come with simple, guessable default passwords that users frequently neglect to alter. This renders them simple targets for aggressors who can swiftly obtain unauthorized access [21]. Even today, certain devices employ obsolete authentication methods, including basic username/password pairs without multi-factor authentication (MFA) or robust password guidelines. The absence of strong authentication methods allows attackers to more easily breach devices via brute-force attacks, dictionary attacks, or credential stuffing. To address this issue, it is necessary to establish robust authentication measures.

The major security requirement posed by the Internet of Things (IoT) devices is the vulnerability of authentication procedures. Many devices come with default passwords that are easy to guess and standard gestures that users often do not change, making them easy targets by attackers who can easily access them without authorization and change them to their preference within a short period of time [21]. Many devices are supplied with default passwords that are easy to crack, and common gestures that are often not changed by the users, which makes them a perfect target of attackers who can easily access them without authorization and transform them to suit their preferences in a very short time frame. In addition, a lot of equipment still uses outdated authentication methods, including a basic username/password combination with no multi-factor authentication (MFA) or a strong password policy. Lack of strong authentication systems encourages the activities of attackers to weaken the devices through brute-force, dictionary attacks, or credential stuffing. This in turn requires the use of strict authentication systems in order to curb this challenge. Some of such mechanism as follows:

- Users need to change default passwords through a specific time interval.
- Enforcing strong password policies with complex and uncommon combinations.
- Implementing multi-factor authentication (MFA).
- Introduced certificate-based authentication for multi-device communication.

5.2 Lack of Encryption

Another significant security challenge among the IoT devices is the lack of sufficient encryption during data transmission. Many devices in IoT system transmit sensitive information and confidential data, such as personal data, financial information, or health records, without proper encryption, leaving it vulnerable to interception by attackers [22]. Attackers use different techniques such as man-in-the-middle (MITM) attacks to intercept unencrypted data transmitted over the network. They intercepted and used this data for malicious purposes, such as identity theft, financial fraud, or espionage. In this consequence, strong encryption protocols are required such as:

- Transport Layer Security (TLS) or Secure Sockets Layer (SSL) for encrypting data during transmission from devices to devices or from devices to servers.
- Advanced Encryption Standard (AES) or other strong encryption algorithms for encrypting data at rest on devices.
- Use secure communication protocols, such as HTTPS, SSH, and VPNs, to protect data transmitted over the network.

5.3 Vulnerable Software and Firmware

The dedicated IoT devices are plagued by vulnerable software and firmware to prioritize speed over security, leaving flaws in their devices. The difficulty of patching these devices leaves them exposed for extended periods [23]. These vulnerabilities exploited by attackers to gain unauthorized access to devices, execute malicious code, or launch denial-of-service (DoS) attacks. The lack of timely security updates and patches further exacerbates this problem, leaving devices vulnerable to known exploits.

To mitigate this challenge in the following ways:

- Given prioritize the security during the software and firmware development process.
- Check identity on every occasion and fix vulnerabilities before devices are released.
- Update security and patches to address known vulnerabilities.

- Prevent unauthorized firmware from being loaded on devices through secure boot mechanisms.
- Mechanisms are provided for users to update the firmware on their devices easily.

5.4 Network Security issues

The challenge begins at the network level with complex configurations, mixed protocol environments, and poor segmentation. Those are created vulnerabilities that attackers actively exploit. The connected *IoT* devices lack proper security controls, creating an open invitation for those with malicious intent. The consequences are serious. Attackers can infiltrate devices, eavesdrop on data transmissions, or pivot to launch attacks against other devices on the same network. Even worse, insufficient network segmentation allows attackers to move sideways through the network after gaining initial access, progressively compromising additional systems and sensitive data. Organizations need a systematic approach to address these interconnected challenges:

- Isolate *IoT* devices from other critical systems through proper network segmentation, preventing lateral movement by attackers.
- The introduced firewalls and intrusion detection/prevention systems are actively monitoring network traffic and automatically block suspicious activity.
- Establish strict network access controls that ensure only authorized users and devices can communicate with *IoT* systems and their data.
- Monitor network traffic continuously for signs of suspicious activity, creating a defense-in-depth strategy.
- Maintain rigorous patch management schedules for all network devices and software, closing security gaps before attackers can find them.

5.5 Supply Chain Vulnerabilities

There is a more hidden yet equally harmful threat in the form of the supply chain, i.e., the making and distribution of devices. Weakened hardware or software components may be supplied during the manufacturing or distribution process, by bad actors exploiting components, or by unintentional security processes. The vulnerabilities are especially sneaky because they are hard to detect once a device has been put in place. Attacker that has compromised a component in the manufacturing process has successfully implanted a backdoor within your system before you have even gotten it. The impact is devastating: attackers can gain unauthorized access to devices, steal sensitive data, or disrupt critical operations. Supply chain security should not be a secondary objective but an activity, and it has to be taken as an initial step by the organizations:

- Establish comprehensive supply chain security practices that vet all suppliers and components before integration.
- Conduct thorough security assessments of suppliers and manufacturers to ensure they maintain adequate security standards.
- The authenticity and integrity of all hardware and software components are checked before deployment through cryptographic verification and testing.
- Secure manufacturing processes are employed to prevent the unauthorized access to components and equipment during production.
- Develop and deploy mechanisms for rapid detection and response to supply chain attacks, enabling quick containment if compromises occur.

6 Reducing Mitigating Risks and Securing *IoT*

Effective risk mitigation requires a comprehensive multilayered strategy that addresses every aspect of design, deployment, and operation of the *IoT* systems. This strategy rests on several foundational principles such as security must be considered from the design phase onward, network infrastructure must be architected defensively, data must be protected throughout its lifecycle, and users must understand their role in maintaining security. The organizations are systematically combined security-by-design philosophy, robust network architecture, comprehensive data protection protocols, user awareness training, and intelligent security systems powered by AI before implementation of these principles. Then, they can dramatically improve the security posture of their *IoT* implementations and build resilience against emerging threats. Figure 4 shows these mitigation strategies work in practice within operational *IoT* systems.

6.1 Device Security

The establishment of the secure IoT ecosystems is established on the device level. The security of the device refers to the implementation of security features in both hardware and software from the beginning. This is referred to as security-by-design principle that provides assurance that security concerns are taken into account in all the choices of the development, recently added this core requirement. These are the key measures of that should be targeted at the organizations the device security:

- Frequent Security Patches: IoT devices should be provided with the security patch regularly around the lifetime of their operations. Patches vulnerabilities known and react to emerging threats before an attacker can take advantage of it. Manufacturers

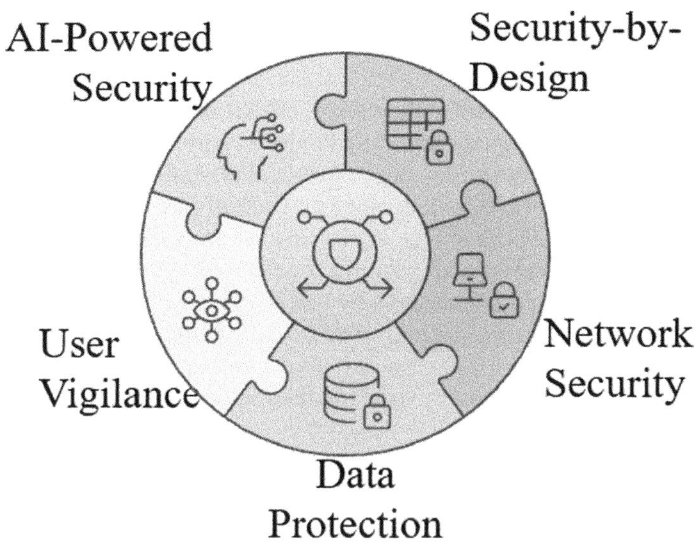

Fig. 4 Comprehensive framework of *IoT* security

ought to develop effective update systems that provide patches effectively and efficiently reliably. OTA updates are the most viable solution and provide flexibility to the latter updates its remote device automatically and without needing a physical or user intervention.

- Effective Authentication: Authentication devices must have high-quality authentication that access is granted to authorized users and devices only. This is more than mere passwords. Best practices involve the use of strong and complex passwords with certificate-based authentication and multi-factor authentication (MFA). Manufacturers provide default credentials which have to be altered as soon as possible deployment, because attackers constantly compromise devices that have not been customized.
- Physical Tamper Resistance: When devices are positioned in public or unattended locations, physical security becomes a point of issue. The security measures that manufacturers should include are tamper-evident enclosures with signs of unauthorized access access, Hardware security modules which defend cryptographic keys and secure boot mechanisms that deter any attempts of firmware alteration.

6.2 Network Security

Level-wise protection of the network infrastructure and the associated IoT devices is needed vigilance. Connection between the devices, gateways, and the cloud systems is the channel of traffic of passing data. Compromising the network often is to have access to all that is related to it. A robust network security These main elements should be reflected in the strategy:

- **Firewalls**: Firewalls are intelligent gatekeepers between the IoT network and the external world. This computer hardware device monitors outgoing and incoming traffic, and blocks unauthorized connections and permitting needed communication. It is essential to set firewall equipment, firewalls are to be adjusted to allow traffic that only IoT devices in fact require and lowering the attack surface.
- **Access Controls**: Not everybody requires access to everything and not everything to everything have to communicate between each other. The use of strict access control policies enforces that authorizations are in compliance with the real needs. Role-based access control Job responsibilities are matched with permissions in (RBAC) which enables the user to perform their duties access and limiting unwarranted access.
- **Network Segmentation**: Dividing the *IoT* network into smaller, separate areas limits the effects of security incidents. If a single segment is breached, the attacker's influence is limited to that separate segment only. They are unable to shift sideways to different areas of the network, safeguarding untouched systems and information.
- **Zero-Trust Architecture**: Conventional security believed that all elements within the network could be trusted. Zero-trust adopts a contrasting viewpoint: treat all entities as untrustworthy, irrespective of their location within or beyond the network. All access requests are subjected to verification and authentication prior to approval. This mentality greatly minimizes the chances of unauthorized access and the spread of breaches.

6.3 Data Security

The primary focus of an organization is to safeguard the sensitive data created and processed by *IoT* devices. Data consists primarily of user information, operations, and business activities, serving as a crucial asset transmitted via IoT systems for processing. An effective data security plan must encompass data protection for its entire lifecycle, rather than only in separate parts. The subsequent steps are the cornerstone of a successful data protection strategy:

- **Encryption for Stored and Transmitted Data**: Data needs safeguarding in two conditions: while it is saved on devices or servers (at rest) and while it is being transferred over networks (in transit). Encryption serves as a protective measure that blocks unauthorized access even if threats breach a device or capture network data. This two-layer strategy is crucial for confidential data. Organizations ought to utilize robust encryption methods like AES for stored data and TLS for data being transmitted across networks.
- **Verification of Data Integrity**: In addition to safeguarding data confidentiality, organizations need to confirm that the data has not been covertly modified. Detec-

tion mechanisms ensure that data stays intact from sender to receiver. Digital signatures, cryptographic hash values, and hash algorithms fulfill this role, enabling systems to identify alterations instantaneously.

- **Access Control Policies**: No individual or application should have access to all data. Access control policies implement the least privilege principle, providing individuals and systems solely the permissions required for their designated tasks. This separation restricts harm if an account is breached or a user's access is exploited

6.4 User Vigilance

Organizations frequently allocate substantial resources to technical security controls, yet these initiatives can only succeed if users serve as the ultimate line of defense. Users need to be aware of the security environment related to *IoT* devices and take intentional measures to safeguard themselves and their systems. Security awareness turns users from passive entities into engaged participants in security. Key user habits consist of:

- **Modifying Default Credentials**: Producers send out devices with preset usernames and passwords to aid in the initial configuration. Regrettably, these default credentials are well-documented and commonly recognized. Attackers commonly search for devices that retain their default configurations and take advantage of them right away. Users need to modify these credentials to robust, distinct passwords during the initial setup, prior to linking the device to any network.
- **Using Strong Passwords and Multi-factor Authentication**: IoT systems should also assign various devices and accounts that are unique and different complex password that cannot be guessed easily and through force. Nonetheless, passwords alone are not sufficient in the present-day threat environment. Multi-factor authentication (MFA) adds an additional security measure requiring the need verifying a password on top of, such as a one-time password that is sent to a security key, a biometric scan or a mobile phone. Such a mixture makes it very difficult unauthorized entry.
- **Keeping Devices Up to Date**: Manufacturers declare firmware and security patches to address vulnerabilities that are discovered as threats. These updates repeatedly apply patches to critical security weaknesses that hackers are stealing advantage of. Delay of updates or failure to do so exposes the devices. Ideally, organizations should develop specification of schedules and processes of updates using automatic updates wherever possible, to ensure that devices are kept safe.

6.5 AI-Powered Security

The introduction of the AI and ML is a major transformation in the organizations approach of IoT security. Instead of using the set of rules and the manual reactions, AI-based security systems are able to learn constantly based on the work they are operated drills, detect new threats as they occur and coordinate the necessary protective actions on a more automated basis. This intelligence-based practice makes security more of a transformation. to turn a proactive position into a reactive one. The most significant AI uses in the internet of things security include:

- **Anomaly Detection**: The machine learning algorithms are trained on what each one of them looks like as normal of the device by tracking its time patterns of behavior. The system is capable of signaling the deviations instantaneously in case the suggested security breach is beyond the limit of setup baseline behavior. To illustrate, in case a device abruptly sends high data contents the system identifies such an abnormal pattern and alarms to an unforeseen destination an alert. Many attacks are identified in this anomaly-based approach, which are not detected by signature-based. protections are absent, such as zero-day attacks and hacked hardware exfiltrating data.
- **Threat Intelligence**: The AI systems never stop analyzing new threat intelligence acquired through various sources to determine attacks and vulnerabilities that are specific target IoT ecosystems. Organizations can employ force in advance of being attacked and this visionary wisdom to fortify in advance. These insights assign priorities to patch management, network segmentation strategies, and user awareness training, which enables the teams to pre-equip them in the face of threats.
- **Automated Security Responses**: AI systems have the capability of automatic protective measures without having to wait to be detected depending on the detection threats. In case any signals of compromise are noted in a device then a system may instantly isolate the device off the network to avoid malware propagation. It can throttle suspicious traffic, suspend suspicious credentials, or elicit more authentication challenges. This automatic reply adds a lot of time to the interim between danger identification and control, reduction of harm.

7 The Intelligent Side of *IoT*

The development of the Internet of Things (IoT) technology has been experiencing a radical change as developers started to introduce artificial intelligence (AI) and machine learning (ML) capabilities directly into interconnected systems and devices. This intersection provides opportunities which are beyond those which can be achieved by traditional IoT architecture. The current deployments consider security and intelligence as different but mutually supportive issues; the assumption is made that the safe functioning of intelligent behaviors is impossible without a strong security base. To achieve such a secure base, a number of key elements are required; devices should authenticate trusted and systems before providing access; sensitive data should be put under protection throughout the entire lifecycle; networks should be segmented to be able to confine possible attacks; firmware should be updated on time to address a vulnerability; and systems should be continuously monitored to detect anomalous behavior.

The combination of AI and ML with the IoT technologies paves the way to the groundbreaking change in the functioning of the systems. Such convergence practically superimposes intelligence on the current IoT systems, which essentially reinvents the scope of interconnected systems. Combinations of AI and ML and data created by connected devices can open possibilities hitherto unavailable in organizations. These smart systems evolve on the basis of historical trends, adjust to changing circumstances and make independent resolutions that maximize output and minimize risks. Instead of following strict programmed conventions, smart IoT devices change how they behave according to the actual circumstances of operation. This flexibility opens up opportunities in a range of industries, by predictive maintenance systems that read the impending failure of equipment, personalized experiences that are sensitive to individual user preferences, intelligent automation that optimizes resource usage, real-time analytics that reveal patterns unnoticed, and edge computing processing information where it originates. All these developments demonstrate the disruptive nature of smart and safe IoT systems at scale.

7.1 Predictive Maintenance

Predictive maintenance is one of the applications of AIoT that have the most significant effect. With industrial sensors installed on the machine, one is able to monitor any slight wear and tear that would otherwise have remained unseen [24]. These sensors constantly monitor the different parameters, including temperature, vibration, and pressure. This data is then compared using AI/ML algorithms to determine patterns and anomalies that may be indicative of a failure.

The most important advantage of predictive maintenance is the capability to automatically plan the maintenance prior to the expensive losses. This is a proactive strategy that saves time in the event of downtime, lowers repair expenses and also prolongs equipment life. As an illustration, predictive maintenance applied in a manufacturing plant to detect a damaged pump even before it shuts down a production line and enters the company into losing a lot of money.

7.2 Personalized Experiences

AI is also an important part of developing personalized and adaptive user experience in smart homes [25]. The IoT is abundant in behavioral data, such as speakers, thermostats, and lighting systems. This data is studied with the help of AI algorithms in order to comprehend the preferences, habits, and routines of users.

According to this analysis, the system automatically sets the settings to produce a specialized environment. As an example, a smart thermostat programmed to know the types of temperature an individual likes at various times of the day and turns the temperature down and up automatically. On the same note, an intelligent speaker understands the music preferences of the user and generates custom playlists. This form of personalization improves the comfort and convenience of a user as well as the use of energy. The system is user-friendly in that it is flexible to user needs.

7.3 Smart Automation

The field of AI can use data collected by linked sensors to automate complicated processes in any application, including agriculture and manufacturing. This automation saves on the aspect of having human intervention all the time as the human workers are able to concentrate on more strategic activities [26].

In farming, as an example, the sensors are able to detect the moisture content, temperature, and nutrients in the soil. This data can be then processed by AI algorithms to fit the irrigation, fertilization, and pest control automatically, maximizing crop yields and reducing resource wastage.

In the production industry, AI can be used to automate activities which may include quality control, inventory, and optimization of processes. This results in high efficiency, cost reduction, and quality of products.

7.4 Real-Time Insights

A huge amount of data created by the IoT devices is an excellent chance in deriving real-time knowledge. This data could be processed using AI algorithms to find patterns and trends and anomalies that would be hard or impossible to find by humans.

These applications can be applied to make informed decisions and optimality in operations using the insights in real time. As an illustration, in smart cities, AI may be used to analyze the feedback of traffic sensors, which may help to optimize the traffic flow, shorten it, and improve the quality of air. In the retail sector, AI has the ability to study point-of-sale data, customer behavior and inventory levels to maximize pricing, promotions, and inventory controls.

7.5 Edge AI

In critical applications with a high rate of response, including autonomous vehicles and industrial robots, local AI processing on the device, at the "edge" of the network, is needed. This is referred to as Edge AI and does not cause the latency that is caused by transmitting data to a centralized cloud to process it.

Edge AI allows making decisions in real time even in conditions when network connectivity is low or unstable. As an example, autonomous vehicles can apply Edge AI to receive information through sensors and make instant steering, acceleration, and braking decisions, which assures the efficient and safe work. Another aspect that makes Edge AI more privacy and security conscious is the ability to store sensitive information on the device instead of sending it to the cloud. This eliminates the danger of data breach and unauthorized access.

8 The Convergence of Security and intelligence

The full potential of IoT is realized when intelligence and security are integrated, creating a more trustworthy, resilient, and effective system[27]. The present scenario of tradeoff between security and intelligence is shown in Fig. 5. This convergence occurs in the IoT system in different ways. AI-powered anomaly detection establishes a "baseline" of normal device behavior. Any deviation from this baseline can be flagged as a potential security threat, allowing for automated and proactive mitigation. Decentralized security models use for security purpose, like blockchain, can remove single points of failure, making IoT systems more resilient and secure. Edge security performs as the threat detector and mitigation directly on the device; edge computing can respond to threats instantly without relying on a central server, which is crucial

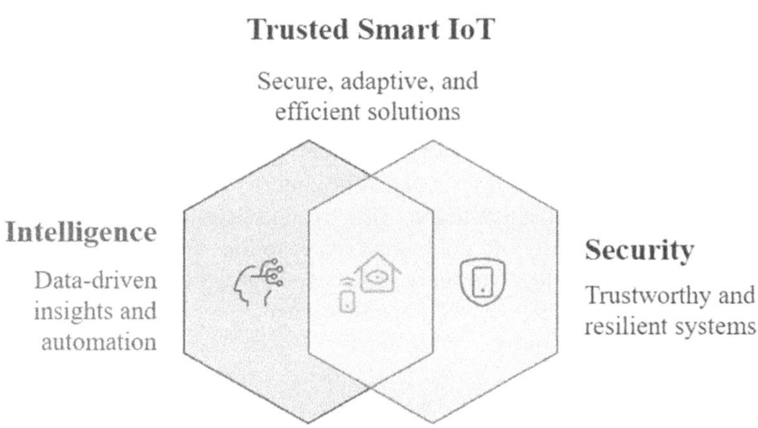

Fig. 5 Unlocking IoT potential through convergence

for applications like industrial control systems [28]. Regulatory push to increase the awareness of security and privacy risks has spurred governments and industry groups to create new regulations and standards to drive the adoption of more secure devices.

9 The Future of Secure and Intelligent IoT

The imminent direction of AIoT, with a closer approach to the interiority of AI and IoT devices and systems. It also identifies some of the most critical advancements in terms of enhanced AI applications across all industries, more security systems enabled by AI and blockchain, the development of edge computing, and the development of smart cities and infrastructure with integrated AIoT systems.

The next stage of AIoT development is premised on the idea of combining artificial intelligence with the society of IoT devices and systems [30]. This integration is not only on the simple automation but also on the more advanced and proactive functions.

Predictive Maintenance: AI software processing information is an input that is used by IoT sensors installed on machines and equipment to preempt the occurrence of failure before it occurs. This predictive maintenance allows companies to do their maintenance ahead of time and therefore reduce downtime and lower cost of operations.

Automation of the Production Process: AI plays a crucial role in the automation of the manufacturing process by examining real-time data of the IoT sensors mounted in the production line. This information can be used in the identification of a bottleneck, efficiency enhancement as well as reduction of wastes.

Data-Driven Decision-Making: Data-driven decision-making AI allows organizations to make more informed and data-driven decisions due to the processing of massive amounts of data generated by IoT devices. This assists businesses gain a more insight on their businesses, their consumers and their markets.

Cybersecurity with AI-powered cybersecurity is an important development in the way organizations identify and react to threats. Instead of being limited to human analysis or a set of established detection guidelines, AI-driven systems are self-operating systems that learn as they track traffic patterns on the network and discover anomalies that indicate an active threat. This live intelligence allows the automatic protection measures even before the human intervention. The technologies such as blockchain are complementary to these defensive strategies because they establish entirely new strategies of data protection and storage.

AI-Powered Cybersecurity: The automatic detection and mitigation of threats in real time through machine learning algorithms help to increase the capabilities of cybersecurity. The systems will constantly observe the trends within the network traffic, detect possible suspicious activity, which does not align with the typical operation of the network, and automatically initiate the necessary defensive mechanisms. This is contrary to rule-based systems, which use security teams to predict the trends of the attacks before they happen; AI-based methods get to know what normalcy looks like in each network and instantly indicate the abnormalities. Upon detection of threats, the system would also automatically block malicious traffic, isolate the compromised devices on the network to prevent the transmission, and inform security personnel of what has been done, and what systems to pay attention to. This automatic reaction decreases significantly the time lag between the detection of the threat and its competent control.

Decentralized Data Storage Using Blockchain: Blockchain technology is a technology that develops tamper-proof, decentralized data storage services that entirely transform the process of securing IoT data. Instead of storing all the information in centralized repositories that can be compromised, blockchain disseminates information in more than one node and is cryptographically authenticated. The security and data integrity are enhanced by this distributed architecture. Since the coding of information must be done on a large scale to change it, i.e., attack most of the nodes at once—a near impossibility, there is no way that a hacker can secretly change the history. Also, the audit trail offered by blockchain offers verifiable evidence on the authenticity and provenance of data, so the information has not been manipulated and substituted with fake data under the carpet.

Ubiquitous Edge Computing: Edge computing has recently become more of a standard in the contemporary IoT implementation, and it has completely altered the location and processing of data. Instead of sending all the information to remote cloud repositories to analyze it, edge processing information at the edge of the network, much nearer to the place of origin of data. This architecture change comes with a significant gain in various dimensions of system functioning and security.

Less Latency: Edge computing ensures much less time is taken between collecting data and making decisions since the data is processed locally instead of being sent to remote cloud servers. Applications which need response in the split-second are advantageous and especially out of this latency reduction. Self-driving cars are required to respond immediately to traffic incidents without the need to have cloud computers reply to them. The automation systems used in industries need to stop the hazardous machinery the moment they detect any unsafe environment. The remote healthcare applications should be able to handle vital patient information within a short time. Such time-sensitive applications can only be viable when it is processed at the edge.

Improved Security: Edge computing retains sensitive information within the premises instead of sending it to cloud storage via the networks. Such an architectural decision minimizes the surface of attack as it removes redundant data transmission. Information that does not pass through the network cannot be intercepted by hackers. The localized approach is highly useful in applications involving personal information or medical records as well as confidential business data. Organizations are able to encrypt sensitive data at rest without having to incur the cost of encrypting and decrypting data on the network.

More Secure Data Processing: Edge computing can be more secure because it can process data faster by removing network latency, as well as be more secure by reducing data transfer. The combination enhances the performance and reliability of the AIoT system. Systems may also be functional in a scenario where the network becomes unavailable or rather they can run the important processes without depending on the availability of the cloud.

Smart Cities and Urban Infrastructure: AIoT systems involve overhead mechanisms that will change the functioning of the cities by ensuring that resources are used optimally, traffic movement is managed more efficiently, and citizens are safeguarded with real-time monitoring and reaction.

Resource Management: Cities are producing colossal amounts of information regarding energy use, water use and waste output. The AIoT systems use this data to determine inefficiencies and execute conservation measures. Smart streetlights represent this practice: instead of keeping the systems lit at full intensity all night, they will automatically turn down the intensity as the ambient light decreases, and turn up the intensity according to real-time traffic conditions. Shall there be light pedestrian traffic with natural moonlight? You need not ask, turn lamps low. As the cars come closer or more people become active, the light brightens instantly. This smart grid will save a lot of energy and still provide the right amount of safety. Similar principles target heating and cooling and water systems and altogether they change the way cities use resources.

Traffic Movement: Traffic jams consume fuel, emit excessive pollutants and annoy drivers. AIoT systems trace real-time traffic flow in the city and optimize the time of the traffic lights to ensure the best traffic throughput. Signals do not work on fixed schedules but follow the real demand on the traffic at any given moment. The systems also enable drivers to get real-time traffic information as well as the alternative routes, which can help them to avoid congestion. The net result of such intelligent optimizations is minimized congestion across the board, better air quality, and smoother transportation to all the residents of the city.

Public Safety: Intelligent surveillance offers better safety to the community as it tracks the pattern of crime and therefore notices suspicious activities as they occur. Instead of using human security officers video feeds, AI algorithms recognize abnormal patterns that may be potential criminals. Even an individual hanging around a high crime zone may cause an alarm. Progress vandalism is identified and reported instantly. Enforcement of the law gets real-time notifications enabling them to act promptly before things get out of control. Such a package of all-round monitoring, smart analysis and quick notification systems results in the creation of a more responsive security ecosystem that will prevent crime and mitigate faster response to emergency incidents.

10 Conclusion

The necessity to introduce intelligent security mechanisms in Internet of Things (IoT) systems is a critical reaction to complicated issues, which are typical of modern environments. To achieve such systems, there must be scalability and yet meet the inherent heterogeneity of devices and deal with dynamically changing cyber threats. Massive implementation of IoT should not perceive security as an afterthought. The adaptive security within goal-based critical systems must be introduced in all levels of the system. The integration of IoT technologies with AI and ML develops the security structures which can actually become aware of the environment in which it operates according to the set rules and counter threats in real time. This dynamic nature of defense is quite strong at resisting sophisticated attacks like malware, rogue. The threats that are particularly challenging in resource-constrained IoT settings and vulnerable to denial-of-service attacks and behavior anomalies.

Besides the technical defenses, an appreciation of the security architecture is also on the rise fact that privacy preserving techniques and decentralized design have practical application. It was also used to gain confidence among the users as well as to maintain regulatory compliance. Glancing at smart and safe, IoT systems are not an incremental improvement, IoT systems are holistic and will form a foundation of creating dependable IoT infrastructures able to scale sustainably. The architectural and operational problems the discussion of which will be covered in this paper, and the new research directions in quantum-resistant cryptography and federated learning, explain why robust IoT solutions have become significant. As the IoT is implemented in organizations involved in the healthcare sector, it will change the way organizations work. Security and reliability are for smart cities, transportation, and industrial non-negotiable conditions. This sector ought to develop a unified approach in the future security structure capable of accommodating the new technologies such as edge computing and blockchain are not mutually exclusive, with existing systems and adherence alongside the laws of legislation like GDPR and industry-oriented cybersecurity laws.

Acknowledgements Acknowledgements are not compulsory. Where included they should be brief. Grant or contribution numbers may be acknowledged. Please refer to Journal-level guidance for any specific requirements.

References

1. W. Lv, F. Meng, C. Zhang, Y. Lv, N. Cao, J. Jiang, A general architecture of IoT system, in *2017 IEEE International Conference on Computational Science and Engineering (CSE) and IEEE International Conference on Embedded and Ubiquitous Computing (EUC)*, vol. 1 (IEEE, 2017), pp. 659–664
2. S. Chakrabarty, D.W. Engels, A secure IoT architecture for smart cities, in *2016 13th IEEE Annual Consumer Communications & Networking Conference (CCNC)* (IEEE, 2016), pp. 812–813
3. S. Nasiri, F. Sadoughi, A. Dehnad, M.H. Tadayon, H. Ahmadi, Layered architecture for internet of things-based healthcare system. Informatica (Slovenia) **45**(4) (2021)
4. K. Kumar, A. Sharma, S.L. Tripathi, Sensors and their application, in *Electronic Devices, Circuits, and Systems for Biomedical Applications* (Elsevier, 2021), pp. 177–195
5. A. Redekar, D. Deb, S. Ozana, Functionality analysis of electric actuators in renewable energy systems-a review. Sensors **22**(11), 4273 (2022)
6. J. Trevathan, W. Read, A. Sattar, Implementation and calibration of an IoT light attenuation turbidity sensor. Internet Things **19**, 100576 (2022)
7. W. Liu, M. Haardt, M.S. Greco, C.F. Mecklenbräuker, P. Willett, Twenty-five years of sensor array and multichannel signal processing: a review of progress to date and potential research directions. IEEE Signal Process. Mag. **40**(4), 80–91 (2023)
8. S. Ahmad, M.A. Ansari, A. Mewada, M. Arif, Advancing reliability and safety in IoT: engineering cyber-physical human systems for the future. Life Cycle Reliab. Saf. Eng. 1–15 (2025)
9. R. Kumar, P. Kumar, R. Tripathi, G.P. Gupta, S. Garg, M.M. Hassan, A distributed intrusion detection system to detect DDoS attacks in blockchain-enabled IoT network. J. Parallel Distrib. Comput. **164**, 55–68 (2022)
10. Z. Shah, I. Ullah, H. Li, A. Levula, K. Khurshid, Blockchain based solutions to mitigate distributed denial of service (DDoS) attacks in the internet of things (IoT): a survey. Sensors **22**(3), 1094 (2022)
11. P. Sobana, G.V. Yaiaman, A.P. Kumar, M. Almusawi, H. Rani, Rise of diversity of internet of things (IoT): a review, in *2025 International Conference on Intelligent Control, Computing and Communications (IC3)* (IEEE, 2025), pp. 1098–1102
12. W. Sarma, S. Dey, S. Tiwari, Autonomous IoT: Ai-driven edge computing to power intelligent decision-making. Int. J. AI, BigData, Comput. Manag. Stud. **3**(2), 52–61 (2022)
13. H. Fu, J. Rao, F. Deng, Y. Wang, B. Zhao, Z. Liu, H. Guan, P.H. Malinowski, L. Xu, AIoT: artificial intelligence and the internet of things for monitoring and prognosis of systems and structures. IEEE Trans. Instrum. Meas. (2025)
14. M. Pintus, F. Colucci, F. Maggio, Emerging developments in real-time edge AIoT for agricultural image classification. IoT **6**(1), 13 (2025)
15. O. Panahi, Secure IoT for healthcare. Eur. J. Innov. Stud. Sustain. **1**(1), 17–23 (2025)
16. S. Liebl, L. Lathrop, U. Raithel, A. Aßmuth, I. Ferguson, M. Söllner, Analyzing the attack surface and threats of industrial internet of things devices (2024). arXiv:2405.16318
17. M. Vila, M.-R. Sancho, E. Teniente, X. Vilajosana, Critical infrastructure awareness based on IoT context data. Internet Things **23**, 100855 (2023)
18. J. Shahid, R. Ahmad, A.K. Kiani, T. Ahmad, S. Saeed, A.M. Almuhaideb, Data protection and privacy of the internet of healthcare things (IoHTs). Appl. Sci. **12**(4), 1927 (2022)
19. X. Guo, S. Han, X.S. Hu, X. Jiao, Y. Jin, F. Kong, M. Lemmon, Towards scalable, secure, and smart mission-critical IoT systems: review and vision, in *Proceedings of the 2021 International Conference on Embedded Software* (2021), pp.1–10
20. H.-H. Kim, J. Yoo, Analysis of security vulnerabilities for IoT devices. J. Inf. Process. Syst. **18**(4) (2022)
21. A.N. Alsheavi, A. Hawbani, W. Othman, X. Wang, G. Qaid, L. Zhao, A. Al-Dubai, L. Zhi, A. Ismail, R. Jhaveri et al., Iot authentication protocols: challenges, and comparative analysis. ACM Comput. Surv. **57**(5), 1–43 (2025)

22. A.S.D. Alluhaidan, P. Prabu, End-to-end encryption in resource-constrained IoT device. IEEE Access **11**, 70040–70051 (2023)
23. I. Coston, E. Plotnizky, M. Nojoumian, Comprehensive study of IoT vulnerabilities and countermeasures. Appl. Sci. **15**(6), 3036 (2025)
24. R. De Luca, A. Ferraro, A. Galli, M. Gallo, V. Moscato, G. Sperli, A deep attention based approach for predictive maintenance applications in IoT scenarios. J. Manuf. Technol. Manag. **34**(4), 535–556 (2023)
25. W.A. Elkhalik, Ai-driven smart homes: challenges and opportunities. J. Intell. Syst. Internet Things **8**(2) (2023)
26. K. Sharma, S.K. Shivandu, Integrating artificial intelligence and internet of things (IoT) for enhanced crop monitoring and management in precision agriculture. Sens. Int. **5**, 100292 (2024)
27. A.A. Zainuddin, D. Handayani, I.H.M. Ridza, S.H.A. Rahman, S.I. Kamarudin, K.Z. Ahmad, M.D. Mahazir, M.H. Sukhaimi, K. Subramaniam, M.I.F. Basri, et al., Converging for security: blockchain, internet of things, artificial intelligence-why not together? in *2024 IEEE 14th Symposium on Computer Applications & Industrial Electronics (ISCAIE)* (IEEE, 2024), pp. 181–186
28. F. Shi, H. Ning, W. Huangfu, F. Zhang, D. Wei, T. Hong, M. Daneshmand, Recent progress on the convergence of the internet of things and artificial intelligence. IEEE Netw. **34**(5), 8–15 (2020)

HEAL-HCV: A Hybrid Explainable AI Learning Framework for Hepatitis C Screening

Arkaprabha Majumdar, Raveena Begum, Sourav Mondal, Bidyut Das, and Debnarayan Khatua

1 Introduction

Hepatitis C virus (HCV) infection remains a global public health and healthcare problem. Despite advances in virus research and direct-acting antiviral (DAA) drugs, HCV continues to affect millions of people and cause liver-related morbidity and mortality. Because HCV has so many genotypes, clinical symptoms, and drug resistance patterns, we need new techniques to simplify diagnosis, treatment, and management.

More than 58 million people have chronic HCV infections, with 1.5 million new infections annually (WHO) [1]. Poor developing countries in Asia and Africa have the highest incidence of this virus compared to wealthier European and North American countries [2, 3]. India has a significant HCV burden, with an estimated 6.1 million chronically infected individuals in 2016 [4, 5], representing 8.5% of the global burden. Modern direct-acting antiviral medications are short term (8–12 weeks), oral and effective, curing over 90% of patients. Although HCV treatment has advanced, few people in India and other low- to middle-income countries receive treatment [6]. In 2015, only 7% of HCV patients worldwide began treatment [7], with lower rates in

A. Majumdar · R. Begum · D. Khatua (✉)
Department of Mathematics and Statistics, Vignan's Foundation for Science, Technology and Research, Guntur, India
e-mail: devnarayan87@gmail.com

S. Mondal
Department of Computer Science and Engineering, Vignan's Foundation for Science, Technology and Research, Guntur, India
e-mail: svml_cse@vignan.ac.in

B. Das
Department of Information Technology, Haldia Institute of Technology, Haldia, India

D. Khatua
Department of Applied Science and Humanities, Parul Institute of Technology, Parul University, Waghodia, India

S. Saif et al. (eds.), *Secure Intelligent Sustainable Applications Using IoT for Society 6.0*, Internet of Things, https://doi.org/10.1007/978-3-032-27248-5_2

low to middle-income countries. The WHO aims to eliminate HCV as a public health risk by 2030, reducing mortality by 65% and increasing incidence by 80% [8]. Few low- to middle-income nations have national plans to achieve this objective [9, 10].

We need more data on the prevalence of HCV among healthcare workers (HCW). The health workers are regularly monitored and exposed to biological agents throughout their activities. Thus, accurate and non-invasive tools for the diagnosis of HCV are urgently needed. Machine learning algorithms (ML) excel at medical event analysis because they capture complex and non-linear clinical data relationships. Classification methods can detect HCV-infected individuals and create a diagnosis model [11]. Incorrect attribute sets can negatively impact classifier performance [12]. Choosing a subset of characteristics or variables helps make the data more compact and useful by eliminating the less important parts [13]. Feature selection leads to faster model formation and better classifier performance.

The meta-machine learning technique Ensemble learning combines model predictions to improve prediction accuracy. After training numerous models or weak learners, the flexible approach predicts using supervised learning. The most popular ensemble strategies are bagging and boosting. Gradient boosting, AdaBoost, and Extreme Gradient Boosting are common growing methods, whereas random forests and extra trees are bagging methods. Theory suggests that ensembles perform better with diverse models, though some decision tree algorithms may outperform deliberately designed ones. Ensemble learning can produce an infinite number of prediction models, but three methods are most commonly used. The main ensemble learning methods are bagging, stacking, and boosting. Understanding each method is essential for prediction. Bagging includes fitting numerous decision trees to diverse datasets and averaging predictions. Stacking multiple models on the same data and combining the findings with another model optimizes the predictions. In boosting, new members of the ensemble are gradually added to improve previous model predictions and produce a weighted average [14, 15]. In this study, we used the ensemble approach with and without PCA.

1.1 Literature Review

Studies indicate that 20% of people with viral hepatitis “C" exhibit flu-like symptoms. In contrast, 80% of those with the condition do not feel symptoms, but the infection persists [14, 15]. To prevent infection, patients who receive injections must be tested frequently.

Ali et al. [16] suggest a framework for HCV case management and prediction using data mining, decision tree (DT) and fuzzy logic in the Egyptian HCV dataset. Using trapezoidal fuzzy number (TFN), they achieved 98.1% accuracy versus DT’s 92.5% forecast. Abd El-Salam et al. [17] studied 4962 Egyptians with HCV between 2006 and 2017 using machine learning lab methods. A study of 24 clinical laboratory factors found esophageal variations in 2218 patients but not 2744. This model used six prominent classifiers: neural networks (NN), Naïve Bayes (NB), DT, SVM, RF, and Bayesian network (BN). With the approval of the Ministry of Health, the Egyptian Committee for the Combat of Viral Hepatitis data was used in the nationwide treatment campaign. The accuracy rates for SVM, RF, C4.5, MLP, NB, and BN are 67. 8%, 66. 3%, 67. 2%, 65. 6%, 66. 7%, and 68. 9%, respectively.

Nandipati et al. [18] created an HCV prediction model using ML approaches, such as RF and KNN classifiers, on the UCI-ML dataset (HCV) [19]. They examined 668 cases of mild to moderate class 0 and 1 cirrhosis. Python and R were used for programming with different features. ML can improve the efficiency of component blending, revealing antibody sequences that affect the HCV response [20]. ML approaches combined variables and identified the most important ones through RF classification to predict clinical categories. According to Hashem et al.[21], ML techniques can predict hepatocellular carcinoma in people with chronic liver disease caused by HCV. LR, DT, and CART determine the best subset of filtered input variables. The accuracy values for LR, DT, and CART were 96%, 99%, and 95.5%, respectively. According to KayvanJoo et al. [22], ML combination methods can predict the outcomes of the HCV virus from viral nucleotides. DT, SVM, NB, and NN predicted IFN-alpha and RBV therapy responses using processed characteristics. They built 10 attribute weighting models from 76 fundamental attributes. Chi-square, Gini index, deviation, information gain, information gain ratio, SVM, PCA, uncertainty, relief, and rule are the eleven attributes. Classifying 11 features with SVM, NB, NN, and DT yields an average accuracy of 85%.

Tables 1 and 2 highlight current HCV prediction approaches and efforts. The authors, datasets, count of events, patients, or issues, and feature selections are listed. The authors used ML and performance measurements. Alade et al. [23] propose a neural network model for identifying diabetes in pregnant women. The medical database from the Pima Indian Database provided the data. Abayomi-Alli et al. [24] utilized BiLSTM deep learning for early detection of PD in small datasets, while Ogundokun et al. [25] used DT and KNN for the detection of heart disease. The heart disease dataset was reduced to fewer features using an autoencoder. Each medication is evaluated using NHS heart disease databases. Salem et al. [26] propose a fuzzy KNN-based fine-tuning method for the PIMA dataset of pregnancies with diabetes.

The development of explainable artificial intelligence (XAI) and healthcare machine learning systems during the years 2023–2025 has created a demand for predictive models that demonstrate transparent performance and robust functionality and deliver clinically interpretable results when researchers analyze tabular biomedical data. Research shows that explainable machine learning frameworks for the diagnosis of liver disease and hepatitis use ensemble classifiers with SHAP and LIME post hoc interpretability methods to improve clinical trust and insight into features of the level [27–29]. Survey-based studies and methodological investigations have documented the disparities between predictive performance and clinical usability, while demonstrating that many XAI approaches remain model-specific or difficult to use in actual healthcare settings, according to research findings [30]. The current research work highlights the increasing importance of transparent AI systems that provide reliable decision-making support for healthcare applications. The main functions of explainable AI methods in safety-critical healthcare tasks are

Table 1 Summary of recent machine learning modeling on HCV

Reference	Objectives	Approach	Findings
[36]	To estimate the cure time and evaluate the forecasts' accuracy	RF, ANN	This method can forecast time to cure for tailored treatment
[37]	To use supervised machine learning to find key classifiers and predict the presence of depressive symptoms associated with a risk of depression	A random forest algorithm	Created a prediction algorithm to identify HIV-HCV co-infected persons at risk for depression in research settings
[38]	An effective medical decision-support system requires high accuracy, precision, sensitivity, and time efficiency that is inversely correlated to model complexity	Analyzing and displaying the dataset's values	By introducing new characteristics and removing class imbalances, HCV was discovered with 99.31% accuracy.
[39]	To create fresh HCV antivirals	SVM, KNN, ANN, RF	Developed the "Anti-HCV" platform to predict repurposed HCV non-structural (NS) protein-targeting drugs using machine learning and QSAR
[40]	To put into practice a two-dataset strategy for detecting the hepatitis C virus in the general population	DT, LR, KNN, XGBoost, Adaboost, Gradient Boosting Machine, Gaussian Naive Bayes, RF, Gradient boosting (GB), SVM	The best models were K-Means and Logistic Regression

detailed by Sadeghi et al. [31], who present their research results in a comprehensive overview of the subject. The research of Karagoz et al. [32] tests local and global XAI methods in tabular healthcare data to demonstrate the challenges of producing clinically relevant explanations. Other surveys highlight broader trends and challenges of the adoption of XAI in clinical decision-support systems (CDSS) [33, 34] and disease prediction applications [35], strengthening the need for methods that unify predictive performance with interpretability. Recent studies establish PCE-XNet as an integrated framework that uses deep representation learning and ensemble learning together with dual-level explainability to fulfill market requirements.

Table 2 Continue

Reference	Objectives	Approach	Findings
[41]	To create and verify machine learning models for predicting the failure of direct-acting antiviral (DAA) therapy in hepatitis C virus (HCV)-infected patients	Elastic net (EN), RF, Gradient boosting machine (GBM), and Feedforward neural network (FNN)	Algorithms for machine learning fared well in terms of risk classification and failure of DAA treatment prediction
[42]	To categorize healthy persons with their existing conditions and healthy individuals with suspected or confirmed Hepatitis C infection	The logistic regression model, SVM, and XGBoost model	Classify Hepatitis C, fibrosis, and cirrhosis from biochemical test datasets with greater accuracy, precision, and recall and present feature critical scores
[43]	To prioritize people for targeted HCV screening, it is necessary to design and validate a machine learning (ML) algorithm to detect undiagnosed hepatitis C virus (HCV) patients	Gradient Boosting Trees algorithm	Using ML on EMR data to prioritize patients for focused HCV testing could improve resource utilization, physician effort, and healthcare costs
[44]	To use the XGboost approach to categorize open access gene expression data of HCC patients with an HBV or HCV infection	XGboost approach	A machine learning-based prediction approach categorized HCV and HBV etiological variables that cause HCC, and genes that may be biomarkers for HBV-related HCC were found

1.2 Motivation and Contribution

With the growing worldwide burden of Hepatitis C Virus (HCV), there is an urgent need for accurate diagnosis in the early stages to improve patient outcomes. Traditional diagnostic approaches are often limited by high costs, time-related issues, and the need for specialized expertise. The use of Machine Learning (ML) and Explainable Artificial Intelligence (XAI) can be applied to design efficient, scalable, and interpretable solutions for HCV detection. However, most studies are limited to standard ML methods and do not account for interactions among dimensionality reduction, ensemble techniques, and explainability in their integration into real-world clinical applications.

The contributions of this study can be summarized as follows:

- Ensemble Learning with PCA—To gain in predictive accuracy while reducing feature redundancy, we integrate PCA with ensemble methods here, contrary to previous studies. (Random Forest, XGBoost, CatBoost, LightGBM)
- The model was put in the context of SHAP and LIME to explain the model and show the patient profile retrieved through biomarker importance, and some changes intended to make it clear for clinicians.
- Cluster-driven network risk-based risk stratification: We categorize patients into different clinical risk groups (blood donor, suspected blood donor, hepatitis, fibrosis, and cirrhosis) using the clustering methods of PCA + K-Means, providing additional input of diagnostic work beyond classification.
- Model Performance and Interpretability Trade-Offs—In fact, LightGBM is expected to surpass all the other results, after a careful assessment of the accuracy, complexity, and applicability trade-offs in real-world environments.
- We assess the feasibility of developing lighter-weight machine learning (ML) models for hepatitis C virus (HCV) diagnosis in resource-limited healthcare settings.

This study aims to incorporate advanced ML tools to further enhance the interpretability and accuracy of HCV predictions toward faster and more interpretable decision-making that would aid early disease detection and personalized treatment planning.

Compared with known approaches, which simply utilize ensemble classifiers over raw or autoencoder generated features in hybrid CNN—ensemble frameworks or PCE-XNet has its own unique advantages, including (i) variance-preserving PCA before deep encoding; (ii) extraction of latent features from tabular biochemical markers using CNN-based deep neural networks; and (iii) a dual-level explainability strategy, which refers to the combination of global (SHAP) and instance-level (LIME) explanations. To the best of our knowledge, PCE-XNet is the first to introduce PCA-optimized CNN encoding with ensemble learning and explainable AI for multi-class HCV screening.

2 Dataset Description

This study utilized the UCI[19] Hepatitis C Virus (HCV) dataset, which contains 564 patient records collected from individuals screened for hepatitis C. The dataset includes 13 attributes: 11 numerical features derived from blood tests (Age, ALB, ALP, ALT, AST, BIL, CHE, CHOL, CREA, GGT, PROT) and 2 categorical/binary attributes (Sex and Class). The Class attribute of the record would indicate whether a patient is categorized as infected (or suspicious/positive) or non-infected, denoting 489 such records under the classification suspicious/positive, whereas all others would be marked under the health category. These attributes are obtained through standard laboratory investigations in the medical sector and serve as input features for machine learning algorithms to classify patients as infected or uninfected.

Table 3 Gender-based distribution of patients with various health conditions

Encoded value	Encoded value description	Male	Female
0	Blood donor	267	215
0s	Suspect blood donor	6	1
1	Hepatitis	20	4
2	Fibrosis	13	8
3	Cirrhosis	20	10
Total		267	238

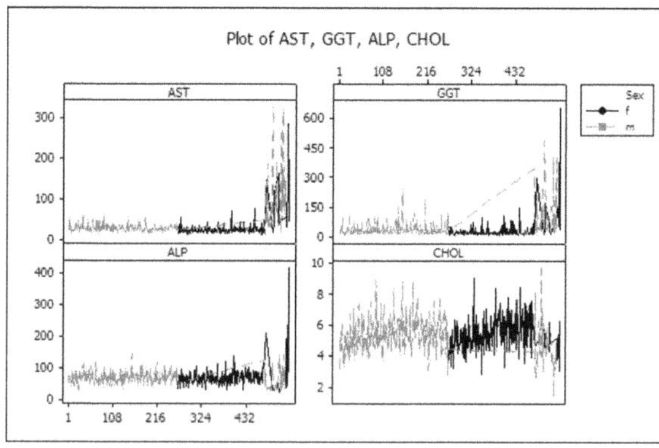

Fig. 1 Plot of AST, GGT ALP, and CHOL for studying how their levels vary on the basis of gender category

Pearson correlation coefficients were calculated to investigate linear associations among variables in this database and the target class (HCV infection status). Hence, the notable findings were that the class and the variables AST and GGT display strong direct correlations. Thus, elevated enzyme levels indicate a high probability of infection. In contrast, negative correlations with ALB (albumin), CHE (CHOLinesterase), and CHOL (cholesterol) were observed, indicating that lower levels of these markers are associated with HCV-positive cases. Additionally, age appeared to play a role, as older individuals may exhibit a greater susceptibility to HCV infection, as observed in the data.

The categorical attributes in the HCV dataset were numerically encoded to facilitate machine learning model training. The detailed encoding scheme and the corresponding clinical interpretation of each encoded value are provided in Table 3.

Figure 1 shows how the level of some HCV covariates varies depending on gender. It is evident from the above figure that the levels of GGT and ALP are higher in women while the level of AST is higher in men. The level of CHOL is almost similar for both men and women.

Figure 2 shows the positive correlations of some variables considered in the present study with age. The Pearson correlation coefficient between age and ALP is 0.188, which means that there is a significant presence of correlation between them. Pear-

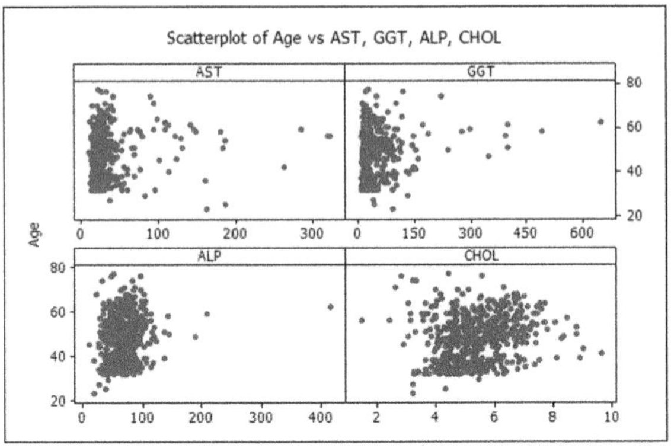

Fig. 2 Scatter plot for detecting correlation between Age versus AST, GGT, ALP, and CHOL

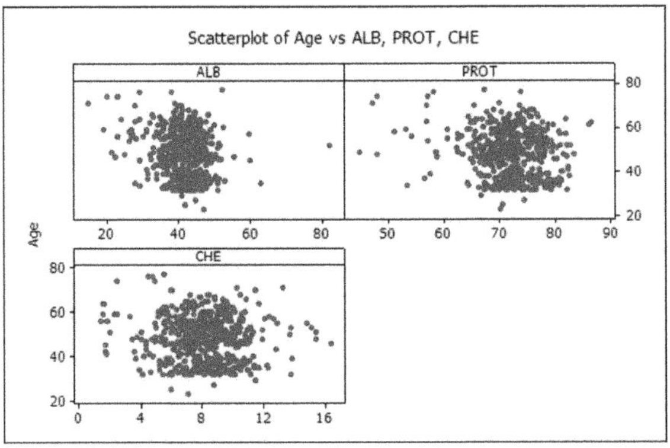

Fig. 3 Scatter plot for detecting correlation between Age versus ALB, PROT, and CHE

son's correlation coefficient between age and AST, GGT, and CHOL are 0.078, 0.149, and 0.144, respectively.

The Fig. 3 shows the negative correlation of some variables with age which are considered in the present study. The Pearson correlation coefficient between age and ALB is -0.194. Pearson's correlation coefficient between age and PROT is -0.152, and for CHE, Pearson's correlation coefficient is -0.063. Hence, ALB, PROT, and CHE are negatively related to age.

It is evident from Fig. 4 that there is no correlation between age with ALT, BIL, and CREA, as their values give an almost straight line with respect to age.

The probability plot is constructed to check the outliers of the data points which may be present in the data under consideration. The normality of the data points can also be checked by gender. It is evident from Fig. 5 that both ALB and PROT are not normally distributed since their p-values are less than 0.01. Outliers are also observed in both ALB and PROT for males. For CHE, the female category follows a normal distribution (p-value = 0.329); however, the male category does not.

The Fig. 6 indicates that AST, GGT, and ALP are not normally distributed, since their p-values are less than 0.01. Outliers are also observed in both male and female categories for AST and GGT. For ALP, some outliers are present in the female

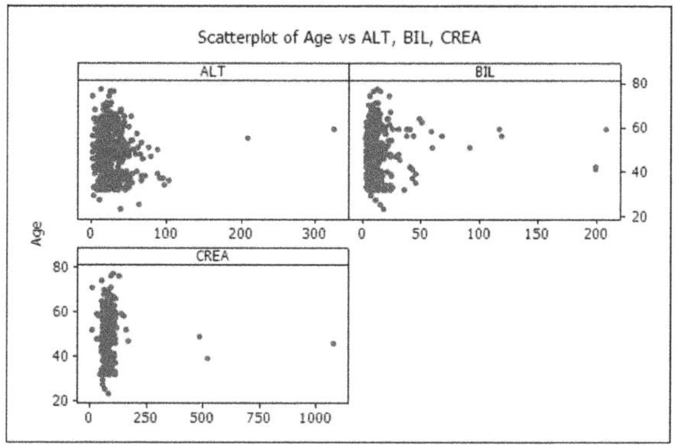

Fig. 4 Scatter plot for detecting correlation between Age with ALT, BIL, and CREA

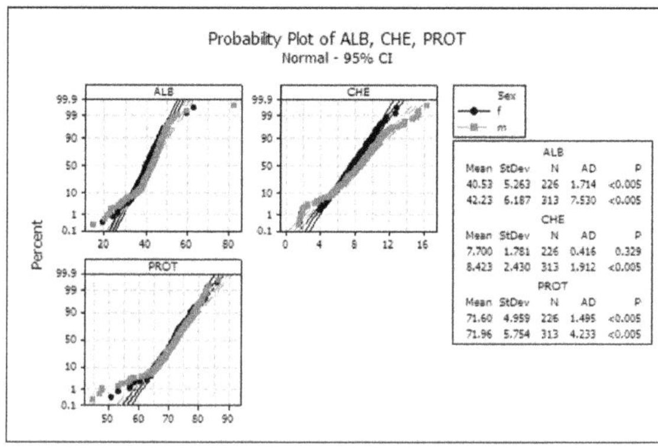

Fig. 5 Probability plot for detecting outliers and studying the nature and pattern for ALB, CHE, and PROT

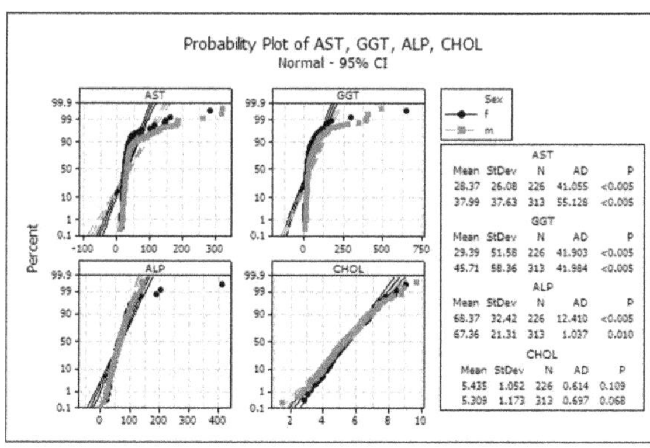

Fig. 6 Probability plot for detecting outliers and studying the nature and pattern for AST, GGT, ALP, and CHOL

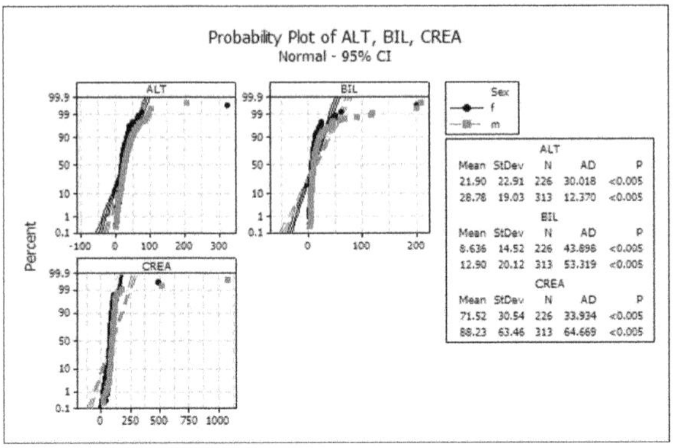

Fig. 7 Probability plot for detecting outliers and studying the nature and pattern for ALT, BIL, and CREA

category. CHOL is normally distributed for the male and female categories with their p-values equal to 0.068 and 0.109 for the male and female categories, respectively.

From the Fig. 7, it is evident that all three HCV covariates do not follow the normal distribution with the presence of some outliers in all the covariates for both male and female categories.

2.1 *Analysis of Dataset*

From Table 4, we analyze the dataset and get the following results.

2.1.1 Descriptive Analysis

Ages range from 19 to 77 years, with an average of 47.41 years (SD: 10.06), indicating a diverse age range that could include stages of liver disease. The average level of serum albumin (ALB) in the study population is 41.62 g/L (SD: 5.78), within the

Table 4 Descriptive statistics for clinical variables

	Age	ALB	ALP	ALT	AST	BIL	CHE	CHOL	CREA	GGT	PROT
count	615	614	597	614	615	615	615	605	615	615	614
mean	47.41	41.62	68.28	28.45	34.79	11.40	8.20	5.37	81.29	39.53	72.04
std	10.06	5.78	26.03	25.47	33.09	19.67	2.21	1.13	49.76	54.66	5.40
min	19.00	14.90	11.00	1.00	4.00	0.50	1.42	1.43	8.00	4.50	44.80
25%	39.00	38.00	52.50	16.40	21.60	5.30	6.94	4.61	67.00	15.70	69.30
50%	47.00	41.95	66.00	23.00	25.90	7.30	8.26	5.30	77.00	23.30	72.00
75%	54.00	45.20	80.00	33.08	32.90	13.00	9.59	6.66	92.00	40.20	75.40
max	77.00	82.20	416.60	325.30	324.00	254.00	16.41	9.67	1079.10	650.90	90.00

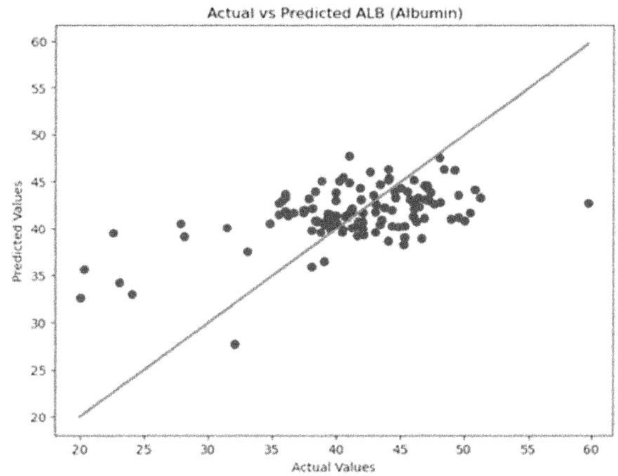

Fig. 8 Descriptive statistics of the dataset

normal population levels ($34 - -54\ g/L$), but some variations may suggest liver dysfunction or malnutrition.

Alkaline phosphatase (ALP) showed high variability (mean: 68.28 U/L, SD: 74.96, range: 11.3–46.6), with levels that are too high, indicative of liver, bone, or kidney disorders. Alanine transaminase (ALT) shows a mean of 28.45 U/L (SD: 38.95, range: 0.9–325.3), which in many cases would suggest liver damage. Aspartate transaminase (AST) levels average 34.79U/L (SD: 33.09), which is slightly above normal ($8 - -33\ U/L$), with severely elevated levels most representative of severe liver injury.

Serum bilirubin (BIL) levels, with a mean level of 11.40 μmol/L, SD of 13.81, and a range of 2.6–254, were elevated above the normal limit ($< 10\ \mu mol/L$).

The concentration of Cholesterol (CHOL) is slightly elevated, with an average of 5.37μmol/L (before renal failure; SD: 1.13). This means that the alleged increase in liver function can also lead a person to a high-cholesterol diet. Creatinine (CREA) ranges from 21.2 to 1079.1, with a mean value of 81.29 μmol/L.

The average gamma-glutamyl transferase (GGT) is 39.53 U/L (SD: 69.61, range: 6.8–650.9), which is above the normal range ($5-40\ U/L$) in most cases, suggesting liver dysfunction. Total protein levels (mean: 72.04 g/L, SD: 6.42, range: 44.8–90) are within normal limits for the analyte ($60-83\ g/L$) but show deviations, possibly on suspected lines of liver or kidney lesions.

2.1.2 Predictive Analysis

Liver function markers (*i.e.*, ALB) were predicted using the linear regression model shown in Fig. 8, with important predictors including age, BIL, ALP, CHE, GGT, CHOL, and CREA. The estimated value of the model intercept indicates that the ALB value is 41.16 g/L when all predictors are zero.

These regression coefficients indicate the impact of a one-unit change in each predictor variable on ALB, holding the other predictors constant.

- Age: A negative coefficient (-0.1026) generally indicates that an increase in yearly age is associated with a decrease in ALB. This probably reflects changes in aging or a weakening of liver function.
- ALP: Negative coefficients (-0.03288) can signal that higher levels of ALP tend to be relatively and marginally lower ALB, indicating liver injury or a change in bone metabolism.
- ALT: ALB is the least affected (-0.001654) because there is a tiny, negligible negative relationship with it and ALT, suggesting that high ALT in the blood is of minimal importance in terms of its relationship with liver damage.
- AST: Slightly negative associations (-0.015601) may suggest that high AST results in liver damage, indicating that ALB levels are generally slightly lower.
- DB: A negative signifying small coefficient (-0.015797) indicates a small decrease in ALB, possibly due to increased serum bilirubin at the time ALB declines.
- CHE: a large positive coefficient (0.661559) shows a strong relationship between CHE-higher levels, improved liver functions, and high ALB.
- CHOL: The negative coefficient effect of cholesterol (-0.508149) reveals the nonexistence of ALB affected by high cholesterol due to ticketing.
- CREA: This slight positive coefficient (0.001748) also specifies the negligible influence effect of albumin and creatinine (CREA) on the subject matter to be determined by kidney function.
- GGT: As in all cases (0.003340), there is a marginal but true positive connection with albumin even in this marker of liver disability, GGT itself.

Among the biochemical markers analyzed in the correlation analyses, CHE showed the strongest positive correlation with ALB, confirming that it is a good predictor of liver function. In contrast, predictors such as BIL, ALT, and AST showed weak negative associations, suggesting limited predictive power in this study.

Age, ALP, ALT, AST, and BIL were negatively correlated with ALB. In addition to physiological aging, compromised liver function may also have contributed. Interestingly, CHE, CHOL, CREA, and GGT were positively correlated with ALB, suggesting that improved metabolic and liver health accompany higher albumin levels.

This regression analysis indicates that CHE and CHOL are critical predictors of ALB levels, while age, ALP, ALT, and AST can serve as warning signs of deteriorating liver function. Other variables that are loosely correlated with ALB are CREA and GGT, which add interesting dimensions to interactions with kidney metabolism and liver health.

2.1.3 Clustering Analysis Using PCA and K-Means

Before applying K-means clustering, one first performs principal component analysis (PCA) to understand the pattern of liver function markers and clinical variables—

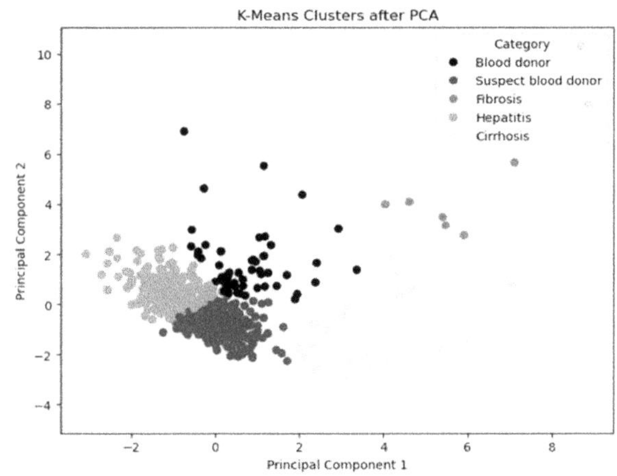

Fig. 9 k-means Cluster after PCA

Alt, ALP, and ALB. Thus, unlike the original features, the principal components are orthogonal and maximize the variance in the data between them. And it is through major structures in the data that they would be interpreted as well.

Principal Component 1 (PC1)—It represents the linear combinations of variables that show the highest explanation.

Principal component 2 (PC2) is explained by the next highest variance after PC1 and is not dependent on PC1.

The resulting PCA visualization is discussed using a detailed description according to clusters generated by K-Means, leading to descriptions such as Blood Donor, Suspect Blood Donor, Fibrosis, Hepatitis, and Cirrhosis. Cluster analysis would provide reflective exploratory insight into such common and specific patterns among clinical data across these groups. (See Fig. 9).

Hepatitis Cluster (PC1: -2 to 1, PC2: -1 to 1) Although distinct and overlapping to some extent, the Hepatitis group is most closely related to the Blood Donor and Suspect Blood Donor clusters. After all, clinical markers such as elevated ALT, AST, and BIL levels indicate liver inflammation; the expression patterns between patients will likely be similar, different patient profiles are expected in such instances. Hence, the level of seriousness of liver destruction in Hepatitis is relatively limited compared to advanced liver diseases such as cirrhosis.

Blood Donor Cluster (PC1: 0–2, PC2: -2 to 0) The Blood Donor Cluster appears quite compact and centralized because it is near the origin. This group seems to have been created as a consequence of high group homogeneity. Individuals recruited into this group have the opportunity to be characterized by biomarkers that are typical values of normal liver function with minimal signs of liver disease. Given these characteristics, this particular group might be closer to the hepatitis and blood donor groups. Perhaps they may have several characteristics that overlap, for example, with increased age, diet, or genetic exposure. However, there is a tightly distributed group, an individual majority known as the Blood Donor Group health profile.

Suspect Blood Donor Cluster (PC1: -3 to 1, PC2: 0 to < 4) The Two-factorial loading plot, this time against the primary component 2 (PC2), confirmed an evenly intermingled distribution of suspected blood donor group members across their entire anti-space area. Much heterogeneity in the clinical characteristics defines this center. It has an exceptionally clear expression of the hallmark, particularly in PC1, which could signal subclinical liver damage or a risk factor for liver disease. Some people with a higher liver enzyme content tended to be additionally characterized by diseases such as liver disease and cirrhosis.

Outliers formed within the group are likely cases with extreme liver disease or putative rare clinical profiles related to disorders, such as non-alcoholic fatty liver disease, hemochromatosis, and alcohol-related liver disease. This spread-out-away-from-very-small-groups underscores how patchy and complicated this population is—not just very different stages but even very different ranks may be represented in these individuals.

The blood donor, suspected blood donor, and hepatitis classes become mutually exclusive by cluster analysis. The blood donor group, showing a more homogeneous and healthier pattern of biomarkers, differs significantly from the suspected blood donor group, which is due to heterogeneity in presumed clinical diseases and possible subclinical liver dysfunction. Hepatitis establishes bridge insights between both groups, simultaneously seeing things differently, and that these differences exist in terms of enumerative patterns in the increased generation of indicators of liver inflammation.

K-Means, along with PCA, can reveal the intricacies of clinical dataset patterns and unearth deceptive critters for trials to determine their position based on the theoretical strength of command oddities and so disorderly phenomena. Amateur or, in some cases, rather too professional extensions create and index understandably large descriptor variables; adding character helps the model become even more surprising and creative.

2.2 *Classification*

In this section, we discuss the following classification ensemble models.

2.2.1 Random Forest

The random forests (RFs) are hierarchical sets of tree-structured base classifiers. The Random Forest algorithm selects the most significant feature based on a predefined probability. Breiman used random subsamples of the data and random feature subspace mappings to build multiple decision trees, to create the Random Forest [45]. The RF method is typically defined by a collection of D training documents containing N_f characteristics.

In the first stage, samples $S_1, S_2, \cdots, S_k$ are drawn with replacement according to pre-set probabilities.

Basically, it constructs a model of decision trees for each sample S_k. In addition, the training documents are randomly selected through a subspace of dimension m from all available features. The m-try features are used to predict the probability of all potential outcomes. Generate the terminal node using the optimal data division. The procedure will continue until it achieves the saturation criteria.

Construct a random forest ensemble by aggregating K unpruned trees $t_1(X_1)$, $t_2(X_2)$ and make a classification decision based on the highest probability value.

2.2.2 CatBoost

CatBoost is a gradient-boosting decision tree (GBDT) framework that uses an oblivious tree as its fundamental learner and has fewer parameters. It can effectively process category data with high accuracy. This method trains a series of learners sequentially using the boosting approach. Then it combines the outputs of all learners to improve the algorithm's accuracy and applicability [46]. Suppose that we have a training set consisting of n samples, denoted as D $(X_i, Y_i)_{i=1,2,....n}$. Here, $X_i = ({x^1}_i, {x^2}_i, {x^m}_i)$ represents the input attributes with dimensions m, and $Y_i \in$ R represents the labeled values.

The trained model obtained is represented as F_{k-1}. In the following round, the goal is to choose a tree t_k from the set of CART decision trees T that minimizes the expected value 'E' of the loss function 'L'. The value of the parameter t_k is determined in the following manner:

$$t_k = argmin_E L(y, F_k - 1(x) + t(x)) \tag{1}$$

The test samples (x,y) are independent of the training set. Gradient Boosted Decision Trees (GBDTs) utilize the negative gradient of the loss function to train the CART decision tree t_k. The final model M, denoted by Equation (2), is derived from the initial weak learner M_0 and the step size a during the nth training round after N iterations.

$$M = M_0 + \sum_{k=1}^{N} a_n t_k \tag{2}$$

Compared to other boosting algorithms, CatBoost offers the following enhancements over standard GBDT: to combat noisy points in the training dataset, it implements order boosting. At its core, CatBoost is an Ordered Target Statistics (Ordered TS) method that rapidly transforms categorical features into numerical ones while simultaneously increasing their direct support [47]; the third way CatBoost improves feature dimensionality is by using category features. The model is based on a fully symmetric tree and utilizes the same splitting criterion at each layer. This approach improves both stability and prediction speed [48].

2.2.3 XGBoost

XGBoost is currently an advanced technique that builds on gradient-boosting decision trees (GBDT) and is used to solve classification and regression problems. All features for each sample, along with their transformed responses, are combined into a single feature vector that serves as the input to the XGBoost model. XGBoost is a tree boosting method; it creates a strong classifier by combining many weak classifiers. CART is a tree model that uses classification and regression. In this algorithm model, each stage adds nodes to the trees and further partitions the attributes [49]. Decision trees are an effective tool for classification. At every node of the model, there are branches that further connect to other nodes. They are representative of the conditions under which the data needs to be split. Depending on whether the data at a given node meets the requirement, it shall proceed down the appropriate branch. The leaf nodes present at the bottom are its decision nodes. The objective function of the XGBoost model is as follows ([50]):

$$Y^k = \sum_{i=1}^{n} l((y_i, {y_i}^{k-1}) + f_k(x_i) + \omega(f_k)) \tag{3}$$

where,

n = the number of training data.

x_i and y_i = Vector of features and its label in the ith instance.

${y_i}^{k-1}$ = Prediction of the ith instance in the t-1th iteration.

l = loss function (difference between the actual and predicted classes).

f_k = New tree that classifies the instance i^{th} with x_i.

2.2.4 LightGBM

Microsoft announced LightGBM, a gradient boost framework based on the decision tree algorithm, in 2017 [51]. The latter model of models, on the other hand, is advantageous for parallel arithmetic because it uses a boosting model, which accelerates training and uses less memory, hence lowering communication costs in parallel learning. LightGBM uses a unique feature-bundling technique of histogram growth utilizing both depth-wise and leaf-wise rules. Basically, it comprises a decision tree algorithm coupled with GOSS (gradient-based one-side sampling).

The fundamental idea behind gradient-based one-sided sampling, or GOSS, is to retain all large gradient samples and draw at random in proportion to the tiny gradient samples. The fundamental idea behind the EFB technique is to group features into a limited number of exclusive bundles, that is, bundles for which no polynomial-time solution is possible. By allowing for tiny, nonmutually exclusive sample points between the features—for instance, some correspondence sample points and nonzero values simultaneously—it approximates the solution in this manner. Little conflict allows distributing a limited number of feature packages, thereby reducing the number of conflicts and increasing the efficiency of computing capacity [52].

3 Results

In this work, the results of several ML models for HCV classification are compared and contrasted. We have compared our proposed model using PCA features with different principal components. In our dataset, we evaluated and contrasted the efficiency of a number of models, including Decision trees, KNN Boosting, SVMs, random forests, logistic regression, CATBoost, XGBoost, and LGBM. The data is divided into training (80%) and test (20%). Models are initially trained on a training dataset, and their performance is evaluated on a test dataset. The findings of the analysis are presented in Tables 5 and 6, where the ensemble model works better than other classification techniques without feature extraction.

Table 5 Accuracy comparison for different ML models

Serial No.	Name of the classifier	Accuracy
1	Decision tree	0.93
2	k-nearest neighborhood	0.89
3	Gaussian naive Bayes	0.92
4	SVM	0.95
5	Gradient boosting	0.97
6	Logistic regression	0.95

Table 6 Ensemble model without PCA

Serial No.	Name of the classifier	Accuracy
1	Random Forest	0.95
2	CATBoost	0.97
3	XGBoost	0.97
4	LGBM	0.98

Table 7 Enesemble model with PCA

No. of principal components	Name of the classifier	Precision	Recall	F1 Score
5		0.84	0.91	0.86
10	CATBoost	0.90	0.96	0.93
15		0.93	0.95	0.93
20		0.93	0.95	0.94
5		0.85	0.90	0.88
10	XGBoost	0.91	0.92	0.91
15		0.90	0.94	0.92
20		0.93	0.92	0.92
5		0.89	0.92	0.90
10	LGBM	0.91	0.94	0.93
15		0.93	0.94	0.93
20		0.91	0.94	0.93
5		0.87	0.90	0.88
10	**Random Forest**	**0.95**	**0.95**	**0.95**
15		0.93	0.93	0.93
20		0.94	0.92	0.93

In Table 7 we have compared the boosting and bagging techniques with different numbers of principal components. According to the findings, Random Forest performs better than any other ensemble model with 10 principal components. It achieves 0.95 values for each performance metric. Precision, recall, and the F1 score are three indicators of superiority. The model must have strong generalization capabilities since applications in the real-world demand reliable performance on data that have not yet been examined. Metrics are used to evaluate illness classification models and determine how accurately they predict disease labels from symptom data. Figure 10 presents the confusion matrices for the four ensemble classifiers, Random Forest, XGBoost, LightGBM, and CatBoost, using the six selected features. The matrices provide a pictogram of classification performance, as well as the distributions of true and predicted classifications. Correct classifications are represented by the diagonal elements, and misclassifications are indicated by the off-diagonal elements.

Figure 10 shows an example of a confusion matrix developed from the natural, original, imbalanced data and provides a baseline for classifier performance without correcting class imbalance, which can be interpreted in several ways. The results are not intended for clinical interpretation or for demonstrating how well they perform, as class imbalance can also lead to most predictions being biased toward the majority classes in many medical datasets.

In addition, Fig. 11 displays Receiver Operating Characteristic (ROC) curves for each ensemble model of the comparison approach in a one-versus-rest method. Typ-

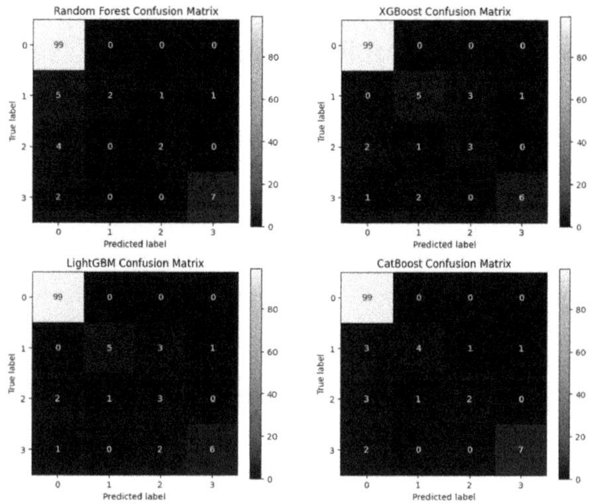

Fig. 10 Confusion matrices for ensemble models using six selected features

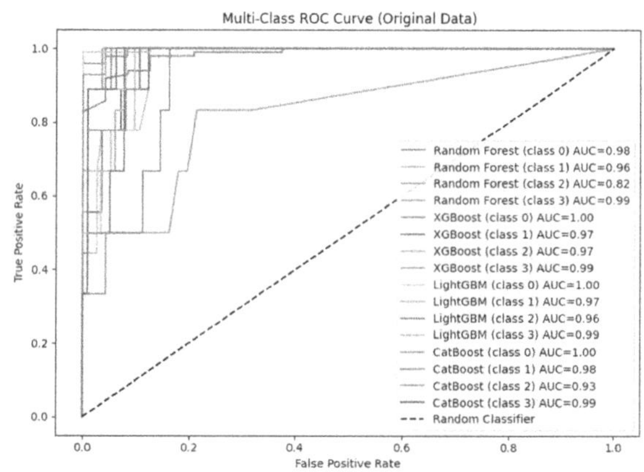

Fig. 11 ROC curves (one versus rest) and AUC scores for each ensemble classifier

ically, ROC curves provide a threshold-free assessment of a classifier's discrimination ability. A hierarchy can be established based on decreasing Area Under the Curve (AUC) values, where XGBoost and CatBoost show superior sensitivity and specificity across all classes tested, suggesting the robustness of the hybrid learning framework's effect.

Evaluation metrics are dependent on the criteria and characteristics of the classification task. These metrics provide an indication of the model's capacity to correctly classify HCV, recognize positive and negative results, and deal with skewed data. When choosing evaluation metrics, it is important to consider the objectives of the disease categorization job, as well as the peculiarities of the task.

4 Proposed Hybrid Model: PCA-Optimized CNN with Ensemble Learning (PCE-XNet)

We established a hybrid deep learning system named PCE-XNet which combines Principal Component Analysis to enhance HCV diagnostic models through better accuracy and better model interpretation. The system starts with an explanation which describes its multi-stage feature selection process and continues to its deep learning feature extraction stage and finally through its stacked ensemble classification system that will achieve both accurate results and understandable system behavior.

All the experiments had an 80:20 stratified train–test split. SMOTE oversampling was applied only to the training set to prevent data leakage. Five repetitions with different random seeds, averaging the performance metrics, were performed in every experiment. Hyperparameters for the CNN and ensemble classifiers have been tuned using a grid search on the training data.

4.1 Coupling CNN Feature Encoder with Ensemble Classifiers

In the proposed PCE-XNet framework, the final classification step is separated from the CNN and serves only as an encoder of latent features. Consequently, the design choice utilized led to the deployment of deep learning for representation learning, whereas classification and decision aggregation are undertaken with ensemble-based machine learning models.

The final stage of preprocessing, dimensionality reduction and selection of features from mutual information with Feature selection using PCA and Recursive feature elimination includes normalizing and reshaping those six reduced clinical biomarkers (AST, ALP, BIL, ALT, CHE, and GGT) to make them a one-dimensional input vector of size 6×1. This input vector is then fed into a shallow 1D-CNN consisting of two convolutional layers (32 and 64 filters, kernel size = 3), followed by ReLU activation, maximum pooling, and ending with a global average pooling layer.

Let $\mathbf{x} \in \mathbb{R}^6$ denote the selected feature vector. The CNN encoder learns a non-linear mapping

$$\mathbf{z} = f_{\text{CNN}}(\mathbf{x}),$$

where $\mathbf{z} \in \mathbb{R}^d$ represents the latent feature embedded in the penultimate dense layer of the network. Importantly, no softmax or sigmoid classification layer is applied at this stage.

The latent vector $\mathbf{z}$ serves as the explicit interface between CNN and the ensemble classifiers. This vector is exported and provided independently as input to multiple ensemble learning models, including Random Forest, XGBoost, LightGBM, and

Fig. 12 Schematic of the proposed PCE-XNet hybrid model including preprocessing, CNN feature extraction, ensemble classification, and explainable AI analysis

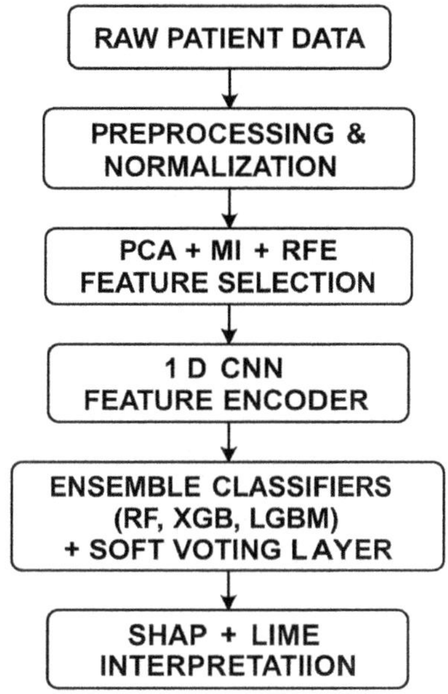

CatBoost. Each ensemble model learns class decision boundaries directly in the CNN-encoded feature space rather than on the original clinical variables.

Afterward, every ensemble classifier provides a set of class probabilities, which, in turn, is then united through a soft majority vote to form the final multi-class prediction. This is, of course, significant for the strengths of keeping feature abstraction (the responsibility of CNNs) separate from classification, and for the aggregation models' uncertainty arising from the adopted modularity in pipelines, thereby enhancing robustness, interpretability, and generalization specific to medical datasets. These are most often imbalanced and multi-class.

Figure 12 illustrates the proposed PCE-XNet system architecture. The model consists of five sequential modules, including preprocessing, dimensionality reduction, CNN-based feature encoding, ensemble classification, and explainability. Each block represents a critical phase in transforming raw clinical biomarkers into robust, interpretable HCV classifications.

4.2 Methodological Enhancement Using SMOTE

The class imbalance in the HCV dataset can be addressed using the Synthetic Minority Oversampling Technique (SMOTE). The SMOTE method generates synthetic data points for all minority classes to ensure that the classifier learns from all classes, even when the dataset is balanced.

The six features which were selected by PCA, MI, and RFE (AST, ALP, BIL, ALT, CHE, GGT) then served as input features simulating CNN-like latent representations and embedded features for training four ensemble models:

- Hybrid CNN + Random Forest
- Hybrid CNN + XGBoost
- Hybrid CNN + LightGBM
- Hybrid CNN + CatBoost

The four models were trained and assessed using the 80–20 train–test split, with balanced SMOTE data. The study assessed performance through a comparison of various standard classifiers that included Decision Tree, Logistic Regression, Naive Bayes, k-Nearest Neighbors, Support Vector Machine, Gradient Boosting, Random Forest, and XGBoost.

The results of the comparative analysis reveal that Hybrid CNN + CATBoost achieves the highest precision result of 97.69%, which is closely followed by Hybrid CNN + Random Forest. The hybrid versions outperformed their standard counterparts, demonstrating that combining balanced data with informed feature selection and ensemble classification methods yields better results.

The hybrid modeling approach advances the field in terms of both robustness and clinical relevance of machine learning-based HCV diagnosis; hence, it balances its potential toward interpretability which creates a possibility for practical use in decision-support systems for diagnostics. The performance of various machine learning and hybrid models was evaluated on a balanced HCV dataset generated using SMOTE. Four hybrid models were constructed by combining CNN-extracted features with ensemble classifiers, namely Random Forest, XGBoost, LightGBM, and CatBoost. The updated evaluation metrics are summarized below.

4.2.1 Hybrid CNN + Ensemble Classifiers (After SMOTE Balancing)

The results show that the use of SMOTE provided better classification results because it helped the model to treat all classes fairly instead of favoring the majority. The hybrid models assessed in this study reached their highest performance when they used **hybrid CNN + Random Forest** and **hybrid CNN + CatBoost**, which achieved 97.69% accuracy while **hybrid CNN + XGBoost** and **Hybrid CNN + LightGBM** followed closely behind with 97.22% accuracy.

The study derives all clinically pertinent assessments and findings from experiments which used SMOTE to establish balanced testing conditions that appear in Fig. 13 and Table 8. The balanced confusion matrices show higher sensitivity, enabling better class differentiation for minority disease categories and addressing the problems observed in the unbalanced baseline assessment.

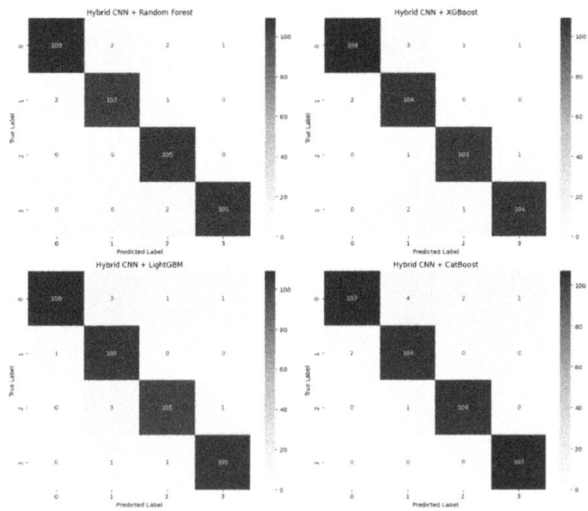

Fig. 13 Confusion matrix for all hybrid CNN models on the balanced dataset

Table 8 Performance of hybrid CNN + ensemble models on balanced dataset

Model	Accuracy	Precision (Weighted)	Recall (Weighted)	F1 score (Weighted)
Hybrid CNN + Random Forest	0.98	0.98	0.98	0.98
Hybrid CNN + XGBoost	0.97	0.97	0.97	0.97
Hybrid CNN + LightGBM	0.97	0.97	0.97	0.97
Hybrid CNN + CatBoost	0.98	0.98	0.98	0.98

4.3 Comparative Evaluation with Previous Results

The hybrid models demonstrate better performance because their generalization abilities and class distribution performance were tested using SMOTE balancing, which surpassed their original model performance. The research demonstrates that medical datasets require balanced class distribution to achieve accurate results and explains how hybrid ensemble models should be used in clinical diagnostics (Table 9).

4.4 Explainable AI Analysis: SHAP and LIME

We evaluated the proposed hybrid models by using two advanced explainable artificial intelligence techniques which establish transparent and reliable results together

Table 9 Comparison of recent explainable ML/DL frameworks for hepatitis/HCV-related prediction and tabular healthcare modeling (2023–2025)

Study (Year)	Task/Domain	Model/Pipeline	Data Type	Imbalance Handling	Explainability	Reported Outcome/Notes
[29]	Hepatitis/HCV prediction (clinical)	IHCP: RF/AdaBoost/SVM + interpretability analysis	Tabular clinical/lab	Noted; details vary by experiment	SHAP + LIME (incl. stability analysis)	Interpretable hepatitis prediction; evaluated on UCI + independent test set
[28]	Cirrhosis detection in HCV patients	Ensemble ML (RF, XGBoost, AdaBoost)	Tabular clinical data	Class weighting/resampling strategies	SHAP-based interpretation	High accuracy for cirrhosis detection; limited to ensemble ML without deep representation learning
[27]	Hepatitis C/chronic liver disease diagnosis	ML classifiers with/without feature selection	Tabular clinical dataset (hospital cohort)	Not explicitly emphasized as SMOTE; scenario-based	Explainable ML (feature importance/interpretability reported)	Large cohort (reported ~1801 records); two scenarios with feature selection
[30]	Disease diagnosis (survey)	Comprehensive review of XAI methods	Multi-domain biomedical datasets	N/A (survey)	SHAP, LIME, Grad-CAM, IG	Identifies lack of unified, clinically deployable XAI pipelines combining DL and ensembles
[53]	Hepatitis outcome prediction	Comparative ML study (multiple algorithms)	Tabular health data	Not consistently detailed across models	Primarily predictive (XAI not central)	Focus on effectiveness of multiple ML models for hepatitis prediction
[54]	General healthcare tabular prediction (deep tabular)	Hybrid: TVAE synthetic data + Feature Tokenizer Transformer	Tabular (mixed numeric/categorical)	TVAE augmentation (synthetic data)	Model-level interpretability discussed (task-specific)	Demonstrates deep tabular modeling + synthetic augmentation for improved performance
[55]	Survey: XAI in biomedical informatics	Systematic review of XAI methods in biomedicine	Multiple modalities	N/A (survey)	SHAP/LIME and broader XAI ecosystem	Summarizes trends/limitations of XAI in real-world healthcare settings
This work (PCE-XNet, 2026)	Multi-class HCV screening (UCI HCV)	PCA (95% var.) + MI + RFE → 1D-CNN encoder → ensemble (CatBoost/XGB/LGBM/RF; soft voting)	Tabular clinical biomarkers (reshaped for 1D-CNN)	SMOTE (train-only, balanced evaluation)	SHAP (global) + LIME (local)	Unified pipeline combining dimensionality reduction, CNN latent encoding, ensemble learning, imbalance mitigation, and dual-level explainability

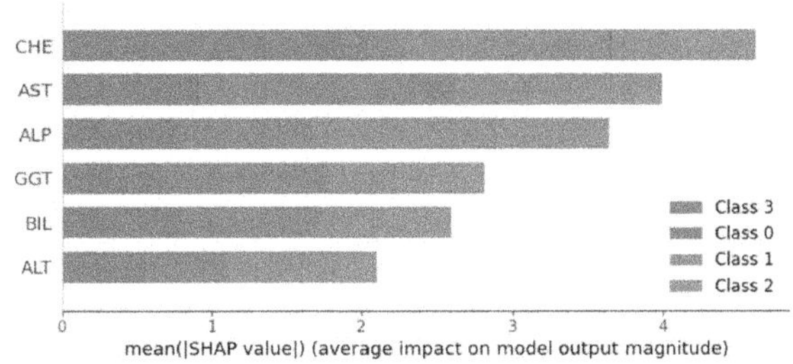

Fig. 14 SHAP summary plot showing the average magnitude of feature impacts across output classes. Each bar is partitioned by the class it contributes to

with clinically understandable outcomes. SHapley Additive exPlanations (SHAP) and Local Interpretable Model-Agnostic Explanations (LIME). The best-performing hybrid model, Hybrid CNN + CatBoost, was tested using these methods to identify which characteristics were globally important and which features drove its decision-making.

4.4.1 Global Feature Contributions Using SHAP

SHAP provides a unified measure of feature importance by assigning each feature an impact value that reflects its contribution to the model's prediction. The SHAP summary graph for the Hybrid CNN + CatBoost classifier (Fig. 14) showed that CHE produced the highest overall impact across all diagnostic categories used in classification. This finding matches clinical expectations because decreased CHE levels indicate liver dysfunction, which functions as a major negative threat to liver health.

AST and ALP showed strong predictive ability after CHE, particularly in predicting advanced liver disease such as cirrhosis. SHAP analysis revealed different levels of importance of characteristic across classes, indicating that specific features such as ALT and GGT had a greater impact on the cirrhosis and hepatitis groups, while enzyme markers became vital tools for tracking HCV disease progression.

4.4.2 Local Decision Interpretation via LIME

SHAP provides complete model interpretation, whereas LIME explains single model predictions. We used LIME to analyze particular patient cases in order to study how the model makes its classification decisions. LIME showed that the patient with fibrosis had a high ALP and intermediate CHE plus GGT, which were key factors in his diagnosis, as shown in Fig. 15.

AST and ALT established their presence through minor but significant effects. The LIME explanation provides a visual breakdown of feature support (green) versus feature contradiction (red), enabling clinicians to evaluate the prediction's reasoning against their clinical expertise.

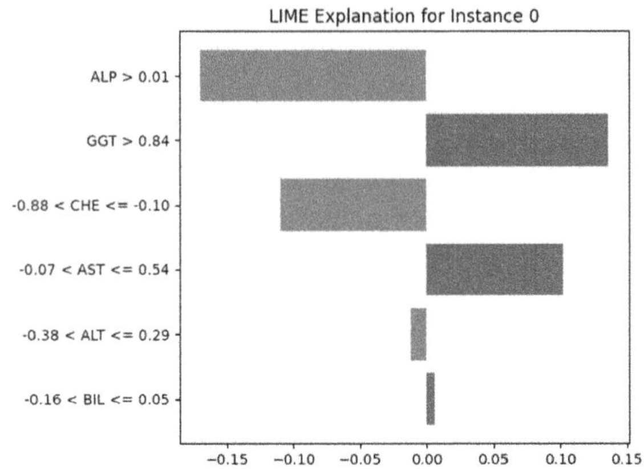

Fig. 15 LIME explanation for Instance 0 showing the top 6 features influencing the prediction. Green indicates support for the predicted class; red indicates contradiction

4.4.3 Clinical Relevance and Interpretability Impact

LIME and SHAP work together to build interpretability within each other and to evaluate the model's outcomes, thereby establishing clinical reliability. Among such locally relevant variables, SHAP identifies widely important variables: CHE, AST, and ALP have been essential in hepatology. To search for patient-specific diagnostics, the clinician must confirm immediate work by performing such an application evaluation.

These two tools bring some very successful classifiers into the medical domain. This defines the process and grounds for using these models. There are some regulatory applications of the model and thus real-world implementation opportunities in health care.

5 Discussion

This advanced hybrid machine learning framework proposed outpaces current levels in computational HCV diagnosis. It uses PCA for dimensionality reduction and a CNN-based encoder for extracting hidden patterns. It models complex relationships between various biochemical markers. The classifier's performance has improved across all patient groups, with effective handling of class imbalance achieved by applying SMOTE to datasets prepared for medical studies.

The research results demonstrate that hybrid systems, which combine CNN-encoded features with ensemble classifiers, perform better than their individual component models. The hybrid CNN + CatBoost and hybrid CNN + Random Forest models achieved their highest classification performance, reaching 98% accuracy while maintaining strong precision, recall, and F1 scores. The models demonstrate effective results across different patient data conditions, making them suitable for use in actual clinical settings.

The explainability of ensemble models has faced limitations since their development, and SHAP and LIME offer solutions to this issue. SHAP analysis showed that CHE, ALP, and AST were the most critical factors for patient classification under current medical standards. LIME provided individual patient explanations, making specific case details and clinical evaluation results accessible to healthcare providers as useful medical information.

This research combines data preprocessing techniques, dimensionality reduction methods, balanced sampling approaches, and ensemble learning algorithms. The proposed model pipeline adopts a modular approach and can therefore be further adapted to similar clinical tasks beyond HCV. Our framework, developed for diagnostic assistive technology functions in resource-limited environments, was designed to scale and operate in such settings.

Research shows that an accurate and interpretable ML-based diagnostic system for HCV can be developed by properly implementing clinically relevant methods.

5.1 Limitations and Computational Notes

The utilization of the hybrid model proposed above has helped to improve the way in which one can predict and explain the results of the diagnosis. At the same time, it is essential to address the shortcomings in the proposed hybrid model. At first, the model was used within its capabilities, meaning that it was trained and tested on a mid-sized dataset available to all. Results are also present with respect to the accuracy of larger sizes publicized to the general public. This limits the extrapolation of these results to more real-life populations in the future. In the above view, it is suggested that the study of such matters in the future will take on the form of a multi-center study with different patient populations.

However, using the CNN model did not require excessive computational power since it is not particularly deep. However, tree-based ensemble models like CatBoost and XGBoost may be. Those trees are likely to incur a time penalty, especially in high dimensions or with k-fold cross-validation. Efficient and performance mitigation of such penalties is critical where real-time processing is a must

Moreover, the synthetic data generated by the SMOTE algorithm may not capture every anomaly, and thus has limitations for artificial intelligence utilization, especially in the field of rare-event classification, where great care in artificial augmentation is advised during exploitation.

However, the limitations mentioned above are outweighed by the fact that the HCV framework for HEAL remains available for use in low-power, energy-efficient parts and offers a good process–cost–performance trade-off.

6 Conclusion

This paper introduces an advanced methodological framework for the correct diagnosis of human patients with HCV. The new framework includes PCA-based feature selection, advanced CNN-inspired encoding, and ensemble learning. The use of the SMOTE methodology enabled the identification of measures that yield balanced classification performance across all classes, thereby significantly improving Prediction Performance and Sensitivity. Among the models tested, the combinations of Hybrid CNN and CatBoost and Hybrid CNN and Random Forest were rated the most effective, both achieving 98% accuracy.

The discussion of some of the tools, including SHAP and LIME, was intended to extend this global–local interpretability. It is justified as follows: It is not surprising that it measures other variables, such as CHE, AST, or ALP, through which there are other actionable and explainable paths to robustly resolve issues in the diagnosis of hepatitis C, along with other clinical tests. This explanation enhances a model's robustness and practicality, particularly in medicine.

PCE-XNet provides a breakthrough, highly effective and explainable path to robustly address the problems in hepatitis C diagnosis, including other diseases with few available solutions. Future researchers will consider further exploration, including combined imaging modalities, genetics, and patient states, to trigger appropriate and possibly necessary actions in diagnosis and treatment.

Conflicts of Interest Declaration
All authors declare that they have no conflicts of interest.

References

1. H. Mamdouh Farghaly, M.Y. Shams, E.-H.T. Abd, Hepatitis C Virus prediction based on machine learning framework: a real-world case study in Egypt. Knowl. Inf. Syst. **65**(6), 2595–617 (2023)
2. A.A. Mohamed, T.A. Elbedewy, M. El-Serafy, N. El-Toukhy, W. Ahmed, Z.A. El Din, Hepatitis C virus: A global view. World J. Hepatol. **7**(26), 2676 (2015)
3. C. Westermann, C. Peters, B. Lisiak, M. Lamberti, A. Nienhaus, The prevalence of hepatitis C among healthcare workers: a systematic review and meta-analysis. Occup. Environ. Med. **72**(12), 880–8 (2015)
4. E. Gower, C. Estes, S. Blach, K. Razavi-Shearer, H. Razavi, Global epidemiology and genotype distribution of the hepatitis C virus infection. J. Hepatol. **61**(1), S45-57 (2014)
5. A. Chaillon, S.R. Mehta, M. Hoenigl, S.S. Solomon, P. Vickerman, M. Hickman et al., Cost-effectiveness and budgetary impact of HCV treatment with direct-acting antivirals in India including the risk of reinfection. PLoS ONE **14**(6), e0217964 (2019)
6. F. Ermis, E.S. Tasci, New treatment strategies for hepatitis C infection. World J. Hepatol. **7**(17), 2100 (2015)
7. WH Organization et al., Global hepatitis report 2017. World Health Organization (2017)
8. W.H. Organization, Combating hepatitis B and C to reach elimination by 2030: advocacy brief. World Health Organization (2016)
9. W.H. Organization et al., Global health sector strategy on viral hepatitis **2016–2021**, 2019 (2016)

10. W.H. Organization, Accelerating access to hepatitis C diagnostics and treatment: overcoming barriers in low-and middle-income countries: global progress report 2020 (2021)
11. M.A. Konerman, L.A. Beste, T. Van, B. Liu, X. Zhang, J. Zhu et al., Machine learning models to predict disease progression among veterans with hepatitis C virus. PLoS ONE **14**(1), e0208141 (2019)
12. G.H. John, R. Kohavi, K. Pfleger, Irrelevant features and the subset selection problem, in *Machine learning proceedings* (Elsevier, 1994), pp. 121–9
13. E. Triantaphyllou, G. Felici, *Data Mining and Knowledge Discovery Approaches Based on Rule Induction Techniques*, vol. 6 (Springer Science & Business Media, 2006)
14. N.H. Taz, A. Islam, I. Mahmud, A comparative analysis of ensemble based machine learning techniques for diabetes identification, in *2021 2nd International Conference on Robotics, Electrical and Signal Processing Techniques (ICREST)* (IEEE, 2021), pp. 1–6
15. A. Doğru, S. Buyrukoğlu, M. Arı, A hybrid super ensemble learning model for the early-stage prediction of diabetes risk. Med. Biol. Eng. Comput. **61**(3), 785–97 (2023)
16. M.M.R. Ali, Y. Helmy, A.E. Khedr, A. Abdo, Intelligent decision framework to explore and control infection of hepatitis C virus, in *The International Conference on Advanced Machine Learning Technologies and Applications (AMLTA2018)* (Springer, 2018), pp. 264–74
17. S.M. Abd El-Salam, M.M. Ezz, S. Hashem, W. Elakel, R. Salama, H. ElMakhzangy et al., Performance of machine learning approaches on prediction of esophageal varices for Egyptian chronic hepatitis C patients. Inf. Med. Unlocked **17**, 100267 (2019)
18. S.C. Nandipati, C. XinYing, K.K. Wah, Hepatitis C virus (HCV) prediction by machine learning techniques. Appl. Model. Simul. **4**, 89–100 (2020)
19. K.F. Lichtinghagen, H.G. Ralf, HCV data. UCI Machine Learning Repository (2020). https://doi.org/10.24432/C5D612
20. S. Eliyahu, O. Sharabi, S. Elmedvi, R. Timor, A. Davidovich, F. Vigneault et al., Antibody repertoire analysis of hepatitis C virus infections identifies immune signatures associated with spontaneous clearance. Front. Immunol. **9**, 3004 (2018)
21. S. Hashem, M. ElHefnawi, S. Habashy, M. El-Adawy, G. Esmat, W. Elakel et al., Machine learning prediction models for diagnosing hepatocellular carcinoma with HCV-related chronic liver disease. Comput. Methods Prog. Biomed. **196**, 105551 (2020)
22. A.H. KayvanJoo, M. Ebrahimi, G. Haqshenas, Prediction of hepatitis C virus interferon/ribavirin therapy outcome based on viral nucleotide attributes using machine learning algorithms. BMC. Res. Notes **7**(1), 1–11 (2014)
23. O.M. Alade, O.Y. Sowunmi, S. Misra, R. Maskeliūnas, R. Damaševičius, A neural network based expert system for the diagnosis of diabetes mellitus, in *Information Technology Science* (Springer, 2018), pp. 14–22
24. O.O. Abayomi-Alli, R. Damaševičius, R. Maskeliūnas, A. Abayomi-Alli, BiLSTM with data augmentation using interpolation methods to improve early detection of parkinson disease. In, 15th conference on computer science and information systems (FedCSIS). IEEE **2020**, 371–80 (2020)
25. R.O. Ogundokun, S. Misra, P.O. Sadiku, H. Gupta, R. Damasevicius, R. Maskeliunas, Computational intelligence approaches for heart disease detection, in *Recent Innovations in Computing: Proceedings of ICRIC 2021*, vol. 2 (Springer, 2022), pp. 385–95
26. H. Salem, M.Y. Shams, O.M. Elzeki, M. Abd Elfattah, J. Al-Amri, S. Elnazer (2022) Fine-tuning fuzzy KNN classifier based on uncertainty membership for the medical diagnosis of diabetes. Appl. Sci. **12**(3), 950 (2022)
27. A.M. Ali, M.R. Hassan, F. Aburub, M. Alauthman, A. Aldweesh, A. Al-Qerem et al., Explainable machine learning approach for hepatitis C diagnosis using SFS feature selection. Machines **11**(3), 391 (2023)
28. A. Alotaibi, L. Alnajrani, N. Alsheikh, A. Alanazy, S. Alshammasi, M. Almusairii et al., Explainable ensemble-based machine learning models for detecting the presence of cirrhosis in Hepatitis C patients. Computation **11**(6), 104 (2023)
29. Y. Fan, X. Lu, G. Sun, IHCP: interpretable hepatitis C prediction system based on black-box machine learning models. BMC Bioinform. **24**(1), 333 (2023)

30. A.A. Biswas, A comprehensive review of explainable AI for disease diagnosis. Array. **22**, 100345 (2024)
31. Z. Sadeghi, R. Alizadehsani, M.A. Cifci, S. Kausar, R. Rehman, P. Mahanta et al., A review of explainable artificial intelligence in healthcare. Comput. Electr. Eng. **118**, 109370 (2024)
32. G. Karagoz, T. Ozcelebi, N. Meratnia, Systematic benchmarking of local and global explainable AI methods for tabular healthcare data, in *World Conference on Explainable Artificial Intelligence* (Springer, 2025), pp. 337–58
33. I.D. Mienye, G. Obaido, N. Jere, E. Mienye, K. Aruleba, I.D. Emmanuel et al., A survey of explainable artificial intelligence in healthcare: concepts, applications, and challenges. Inf. Med. Unlocked. **51**, 101587 (2024)
34. Q. Abbas, W. Jeong, S.W. Lee, Explainable AI in clinical decision support systems: a meta-analysis of methods, applications, and usability challenges, in *Healthcare*, vol. 13 (MDPI, 2025), p. 2154
35. R. Alkhanbouli, H. Matar Abdulla Almadhaani, F. Alhosani, M.C.E. Simsekler, The role of explainable artificial intelligence in disease prediction: a systematic literature review and future research directions. BMC Med. Inf. Decis. Mak. **25**(1), 110 (2025)
36. A. Churkin, S. Kriss, A. Uziel, A. Goyal, R. Zakh, S.J. Cotler et al., Machine learning for mathematical models of HCV kinetics during antiviral therapy. Math. Biosci. **343**, 108756 (2022)
37. G. Marathe, E.E. Moodie, M.J. Brouillette, J. Cox, C. Cooper, C.L. Delaunay et al., Predicting the presence of depressive symptoms in the HIV-HCV co-infected population in Canada using supervised machine learning. BMC Med. Res. Methodol. **22**(1), 1–11 (2022)
38. M. Yağanoğlu, Hepatitis C virus data analysis and prediction using machine learning. Data & Knowl. Eng. **142**, 102087 (2022)
39. S. Kamboj, A. Rajput, A. Rastogi, A. Thakur, M. Kumar, Targeting non-structural proteins of Hepatitis C virus for predicting repurposed drugs using QSAR and machine learning approaches. Comput. Struct. Biotechnol. J. **20**, 3422–38 (2022)
40. U. Singh, M.K. Gourisaria, B.K. Mishra, A Dual Dataset approach for the diagnosis of hepatitis C virus using machine learning, in *2022 IEEE International Conference on Electronics, Computing and Communication Technologies (CONECCT)* (IEEE, 2022), pp. 1–6
41. H. Park, W.H. Lo-Ciganic, J. Huang, Y. Wu, L. Henry, J. Peter et al., Machine learning algorithms for predicting direct-acting antiviral treatment failure in chronic hepatitis C: An HCV-TARGET analysis. Hepatology **76**(2), 483–91 (2022)
42. L. Akter, Detection of hepatitis C virus progressed patient's liver condition using machine learning. In: International Conference on Innovative Computing and Communications: Proceedings of ICICC 2021, vol. 1 (Springer, 2022), pp. 71–80
43. J. Rigg, O. Doyle, N. McDonogh, N. Leavitt, R. Ali, A. Son et al., Finding undiagnosed patients with hepatitis C virus: an application of machine learning to US ambulatory electronic medical records. BMJ Health & Care Inf. **30**(1) (2023)
44. S. Akbulut, Z. Küçükakçali, C. Çolak, Machine learning-based classification of HBV and HCV-related hepatocellular carcinoma using genomic biomarkers. J. Istanbul Faculty Med. **85**(4), 532–40 (2022)
45. V. Jackins, S. Vimal, M. Kaliappan, M.Y. Lee, AI-based smart prediction of clinical disease using random forest classifier and Naive Bayes. J. Supercomput. **77**, 5198–219 (2021)
46. D. Wang, H. Qian, CatBoost-based automatic classification study of river network. ISPRS Int. J. Geo Inf. **12**(10), 416 (2023)
47. L. Prokhorenkova, G. Gusev, A. Vorobev, A.V. Dorogush, A. Gulin, CatBoost: unbiased boosting with categorical features, in *Advances in Neural Information Processing Systems* (2018), p. 31
48. J. Hancock, T.M. Khoshgoftaar, Medicare fraud detection using catboost, in *2020 IEEE 21st International Conference on Information Reuse and Integration for Data Science (IRI)* (2020), pp. 97–103
49. P. Gupta, S. Maji, R. Mehra, Predictive modeling of healthcare professional's stress based on XGBoost model, in *2021 IEEE 18th India Council International Conference (INDICON)*, pp. 1–7 (2017)

50. A. Moore, M. Bell, XGBoost, a novel explainable AI technique, in the prediction of myocardial infarction: A UK biobank cohort study. Clin. Med. Insights: Cardiol. **16**, 11795468221133612 (2022)
51. G. Ke, Q. Meng, T. Finley, T. Wang, W. Chen, W. Ma, Lightgbm: A highly efficient gradient boosting decision tree. Advances in Neural Information Processing Systems, vol. 30 (2017)
52. Y. Wang, T. Wang, Application of improved LightGBM model in blood glucose prediction. Appl. Sci. **10**(9), 3227 (2020)
53. P. Khatun, S. Umam, R.B. Razzak, I.B. Shamsuddin, N. Salma, A study on the effectiveness of machine learning models for hepatitis prediction. Sci. Rep. **15**(1), 30659 (2025)
54. V.Q. Tran, H. Byeon, Explainable hybrid tabular variational autoencoder and feature tokenizer transformer for depression prediction. Expert Syst. Appl. **265**, 126084 (2025)
55. H. Eshkiki, F. Tanhaei, F. Caraffini, B. Mora, A survey of the application of explainable artificial intelligence in biomedical informatics. Appl. Sci. **15**(24), 12934 (2025)

QPRISM: An Optimized Lattice-Based Zero-Knowledge Proof Protocol for IoT Devices

Letitia Gilbert, Sowmya Krishna, and A. Abimanyu

1 Introduction

The current cryptographic algorithms, like RSA and ECC, which were earlier assumed to have unbreakable security, are now under threat with the inception of quantum computing algorithms like Shor's algorithm, for example. These types of algorithms allow quantum computers to crack the mathematical problems that are the foundation of classical cryptographic algorithms, solving them in polynomial time. This breaks their security. Hence, there is a need for post-quantum cryptography algorithms that are resistant to quantum attacks. After researching every post-quantum cryptographic algorithm, we have noticed that one of the more flexible and secure areas in post-quantum cryptography is lattice-based cryptography, which is based on hard mathematical problems such as learning with errors and shortest vector problem, which are believed to be difficult even for quantum computers to solve. Lattice-based systems show flexible properties and have applications in various cryptographic tools like digital signatures, ZKPs, and homomorphic encryption. Although they have incredible security features that can be leveraged for IoT devices, they are not fully suitable for them because of the limited resources of IoT devices. Many like the 8-bit ATmega328P found in basic Arduino boards with clock speeds of 8–16 MHz have small amounts of memory (as little as 32 KB of RAM) and rely on batteries or energy-harvesting systems.

The existing lattice-based ZKP implementations are complex, which causes them to have high processing time, occupy large amounts of memory, and have large proof

L. Gilbert · A. Abimanyu
Department of Computer Science (Cybersecurity), Dayananda Sagar College of Engineering, Bengaluru, Karnataka, India

S. Krishna (✉)
Department of Mathematics, Dayananda Sagar College of Engineering, Bengaluru, Karnataka, India
e-mail: sowmyak.msc@gmail.com

S. Saif et al. (eds.), *Secure Intelligent Sustainable Applications Using IoT for Society 6.0*, Internet of Things, https://doi.org/10.1007/978-3-032-27248-5_3

sizes. Issues like these make them difficult to run on small devices that need to operate efficiently and reliably. We have seen that most of the research efforts have been focused on the security and theoretical aspects of ZKPs; not much attention has been given to the practicality of the IoT hardware, particularly their limitations. There is a trade-off between their performance, size, and energy use. Hence, there is a mismatch between the theory of IoT and what IoT devices can actually handle. In this research, we try to explore how a lattice-based ZKP could be designed for constrained IoT settings, that is, being lightweight, memory-efficient, and suitable for real-world IoT environments.

2 Related Work

Protocols like Picnic and zk-STARKs are shown to have practical applications of ZKPs in decentralized systems. They can provide quantum security but have relatively large proof sizes and performance issues in low-power environments [5]. While zk-STARKs offers scalability and transparency, it has a drawback as it relies on intensive computation operations (polynomial commitment, hashing over large domains), which cannot be used in embedded hardware with limited RAM and CPU cycles. Several frameworks are capable of evaluating the theoretical performance of post-quantum cryptographic algorithms, but overlook the metrics of real-world implementation in microcontroller-based applications. There seem to be limited applications of quantum-secure, non-interactive protocols, especially in settings with intermittent connectivity. Existing works often overlook real-time constraints, memory limitations, and the need for integrated, lightweight authentication mechanisms, all of which are essential for secure and scalable use in smart city infrastructures.

3 Preliminaries

3.1 Zero-Knowledge Proofs (ZKPs)

Researchers Shafi Goldwasser, Silvio Micali, and Charles Rackoff first proposed zero-knowledge proofs in their work "The Knowledge Complexity of Interactive Proof Systems." They laid the groundwork for ZKPs by demonstrating that it was possible to verify the truth of a claim without disclosing its justification. A zero-knowledge proof (ZKP) is a method of establishing a statement's validity without disclosing any other information besides that required to prove the statement's validity. It is among the strongest cryptographic instruments ever created. This article explores zero-knowledge proofs in greater detail and offers several examples to help you grasp their operation.

Table 1 gives a comparison of the different ZKPs. Lattice-based ZKPs surpass most known ZKP systems due to their strong post-quantum security. Unlike zk-SNARKs and Bulletproofs, which rely on classical assumptions and are vulnerable to quantum algorithms, lattice-based ZKPs are not, and hence more reliable for future-proof cryptographic applications. However, these devices have computational complexity, limited energy optimization, and relatively large proof sizes. Although zk-SNARKs provide compact proofs and fast verification, they require trusted setups and lack quantum resilience. zk-STARKs is impractical in environments with constraints because of its large proof size. Bulletproofs is not immune to quantum adversaries. Lattice-based ZKPs are currently underoptimized, but they could provide security and adaptability for post-quantum computing.

Figure 1 illustrates an inverse relationship between device capability and energy consumption during Zero-Knowledge Proof (ZKP) operations. Moving from 8 MHz/32 KB RAM to 48 MHz/256 KB RAM, the total energy required for both proof generation and verification significantly decreases. Proof generation is more energy-intensive than verification in all the classes, which means that proving bears a heavier computational burden than verifying. It is therefore necessary to offload the proof generation to more capable nodes or use lightweight ZKP schemes.

3.2 Challenges of Trusted Setup and Fiat–Shamir in Post-Quantum Contexts

Many traditional ZKP systems rely on a trusted setup, like a common reference string or structured random parameters. These setups are considered vulnerable in decentralized environments, because any compromise by any adversaries in the setup phase can jeopardize the entire protocol. This creates a problem in IoT deployments, where some of the devices may be manufactured or initialized by third parties. The Fiat–Shamir heuristic transforms interactive protocols into non-interactive protocols using a hash function as a random oracle. It has been used to remove the need for interaction. However, its security in the quantum random oracle model (QROM) is still under investigation. Several recent studies (for example, by Zhandry and Unruh) have raised concerns about the soundness of Fiat–Shamir-based Non-Interactive Zero Knowledge (NIZK) when the attackers can query the hash oracle in superposition. This means that NIZK protocols may be secure against classical threats, but not against quantum threats. Also to note, a rigorous post-quantum analysis of the Fiat–Shamir transformation, along with alternatives like Fiat–Shamir with aborts or quantum-secure commitments, is necessary.

Table 1 Comparative analysis of zero-knowledge proof systems

Criteria	Lattice-based ZKPs	zk-SNARKs	zk-STARKs	Bulletproofs	Sigma protocols
Post-quantum security	Yes, based on LWE, SIS, and SVP hard problems, believed to be quantum resistant	No, rely on elliptic curves and pairing-based cryptography (vulnerable to Shor's algorithm)	Partially, use hash functions	No, based on discrete log problems	Some can be post-quantum if lattice-based; traditional ones are not
Proof size	Moderate to large, often several kilobytes or more	Very small, tens to hundreds of bytes	Large, hundreds of kilobytes to megabytes	Moderate, small for range proofs, grows with statement complexity	Small, depends on challenge-response rounds
Computation cost	High, polynomial arithmetic over lattices is intensive	Low, efficient verification, fast proving with preprocessing	High, heavy prover cost, large memory	Moderate, proving is relatively efficient	Low, simple operations
Verifier efficiency	Moderate, lattice math is costlier	Very efficient, fast with succinct proofs	Slower, longer verification due to large proofs	Efficient, fast verification for range proofs	Fast, designed for low-overhead verification
Interactivity	Non-interactive via Fiat–Shamir	Non-interactive only	Non interactive only	Non interactive	Typically interactive, requires challenge response
Trusted setup requirement	Often none, depending on construction	Yes, trusted setup required	No, transparent setup	No, no trusted setup needed	Sometimes, setup may be minimal
Suitability for IoT devices	Promising but challenging, needs optimization for RAM/ energy	Poor, needs heavy computation and trusted setup	Poor, very large proof size not suitable for constrained devices	Possible, still too heavy for very constrained devices	Good, lightweight and simple, especially for small-scale uses
Energy efficiency	Needs improvement, current designs are not optimized for embedded systems	Low, due to heavy preprocessing and proving	Low, due to large computation requirements	Moderate	High, especially in minimal versions

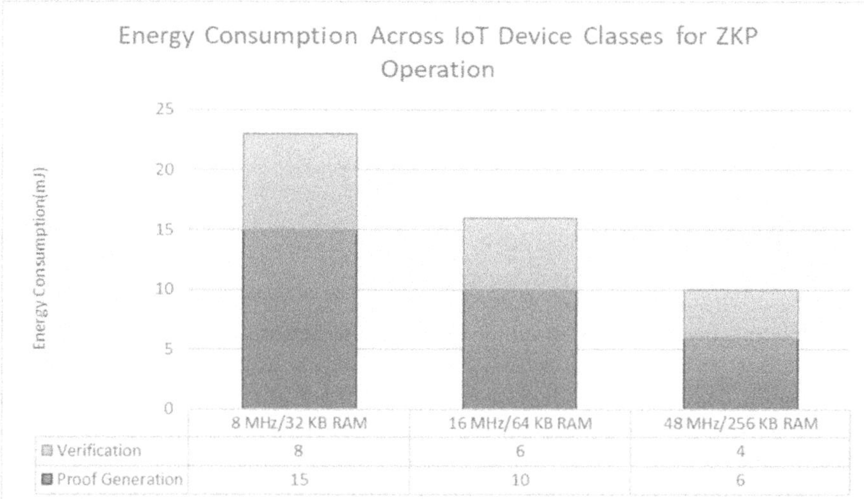

Fig. 1 Energy consumption across IoT devices

3.3 Comparative Analysis of Lattice-Based Cryptography and Other Post-Quantum Schemes

3.3.1 Code-Based Cryptography

Code-based cryptography uses the complexity of decoding linear error-correcting codes to secure data. It is known to be secure against quantum adversaries. Basic linear algebra operations are utilized during encryption and decryption to facilitate fast and light computations. Public keys usually occupy hundreds of kilobytes. Their significant size is their primary disadvantage. This makes it difficult to implement code-based cryptography in memory-constrained environments.

3.3.2 Hash-Based Cryptography

The foundation for the security of hash-based cryptography lies in the one-way nature and collision resistance of cryptographic hashes. However, stateful systems like XMSS need careful key management to ensure their security, and these schemes are typically constrained by large signature sizes.

3.3.3 Isogeny-Based Cryptography

Isogeny-based cryptography is a type of cryptographic system that relies on the mathematical concept of isogenies, which are special functions between elliptic curves, which is considered a difficult problem for both classical and quantum computers. Due to their small size, the public keys and ciphertexts of protocols like SIKE are particularly noteworthy for use in environments with bandwidth limitations. However, this compactness comes with high computational overhead and slower performance.

3.3.4 Lattice-Based Cryptography

Lattice-based cryptography is the generic term for constructions of cryptographic primitives that involve lattices, either in the construction itself or in the security proof. Lattice-based constructions support important standards of post-quantum cryptography. Unlike more widely used and known public-key schemes, such as the RSA, Diffie–Hellman, or elliptic-curve cryptosystems, which could, theoretically, be defeated using Shor's algorithm on a quantum computer, some lattice-based constructions appear to be resistant to attack by both classical and quantum computers. It is based on hard problems such as LWE, Ring-LWE, and Short Integer Solution (SIS) that are thought to be strong enough to resist quantum attacks.

Table 2 compares the different post-quantum cryptographic algorithms. Within the realm of post-quantum cryptographic types, lattice-based cryptography is still the most versatile platform for implementing ZKPs for the IoT. Despite this, there are still areas for optimization with regard to RAM, proof size, and energy. For other types of cryptographic platforms, such as code-based cryptography, they have their own advantages but are challenged by resource constraints and proof generation. Even if isogeny-based cryptography has efficient sizes, it is still not optimized for constrained applications of ZKPs. On the other hand, hash-based cryptography has its strengths in simplicity and security but lacks flexibility.

4 Threat Model

Table 3 gives an explanation of all the different threats a lattice-based ZKP can face in the IoT environment and how each of these threats can be addressed. The malicious actors are assumed to have the capability of intercepting communications. We assume that they have the capability of eavesdropping on data, or injecting some malicious code, or have the means of performing replay or man-in-the-middle attacks. In the systems of smart cities, like smart meters, traffic control systems, health sensors, etc., these could have devastating effects on the daily lives of the general public. Attackers are also assumed to have a motive of exploiting vulnerabilities in the system, both physical and technical. For example, if an unauthorized individual is allowed access to a device in an open or unprotected area, they could tamper with it or try to extract secret keys from the hardware, using side-channel techniques like analyzing power usage, electromagnetic signals, or timing delays, to figure out sensitive data while the device runs cryptographic tasks. Based on this threat model, QPRISM is designed to

Table 2 Quantum-resistant cryptographic schemes: a comparative analysis

PQC family	Key size	Speed	ZKP suitability	IoT suitability	Quantum resistance	Notes
Lattice-based	Moderate–large	Fast	Excellent	Good	Strong	Best for ZKPs, supports homomorphic constructs
Code-based	Very large	Moderate	Poor	Poor	Very strong	Impractical key sizes for IoT
Multivariate	Small	Fast verify	Moderate–poor	Moderate	Unstable	Recent cryptanalysis weakens confidence
Hash-based	Moderate	Slow	Limited	Moderate	Strong	Stateless schemes improving (e.g., SPHINCS+)
Isogeny-based	Very small	Very slow	Poor	Poor	Weak (currently)	Broken or suspect; not viable for now

meet the important security goals, keeping secrets hidden (zero knowledge), ensuring no false claims are accepted (soundness), and staying resistant to quantum computers by relying on hard lattice problems.

5 Proposed Model

We present QPRISM, a new low-power, lattice-based ZKP framework designed for IoT environments with limited resources. QPRISM's main goal is to make authentication private and secure, without disclosing personal information, and computationally feasible for devices with constrained memory, processing power, and energy. QPRISM is a post-quantum lattice-based cryptography that relies on the LWE problem as its hardness foundation. In an effort to minimize overhead, we use Ring-LWE compression techniques and sparse matrices to facilitate the creation of compact proofs, even in low-power microcontrollers. QPRISM, as we have described, is modular and flexible, being able to support both interactive and non-interactive variants via the Fiat–Shamir transformation. This transformation must be secure even if quantum attackers try to guess the challenge values. QPRISM should be scalable to handle different types of hardware and users, particularly in a larger system such as a smart city. It should be resilient to practical attacks such as side-channel attacks or tampering.

Table 3 Threat model and mitigation strategies for lattice-based ZKP protocols in IoT

Attack vector	Threat description	Targeted component	Potential impact	Mitigation strategies
Quantum cryptanalysis	Quantum adversaries exploit algorithms like Shor's to break non-PQC protocols	Underlying cryptographic scheme	Loss of confidentiality and integrity	Use lattice-based primitives (LWE, SIS, Ring-LWE) with high quantum hardness
Side-channel attacks	Exploiting timing, power, or EM leaks to extract secrets	ZKP implementation on MCU	Secret key leakage, full compromise	Implement constant-time operations, masking, noise injection, and shielding
Replay attacks	Adversary reuses previously valid proof transcripts	Communication protocol	Unauthorized access, session hijacking	Use nonces, timestamps, and proof freshness indicators
Denial of Service (DoS)	Resource exhaustion by forcing repeated computation	Prover or verifier device	Battery depletion, unavailability	Rate limiting, lightweight challenge-response fallback, watchdog timers
Man in the Middle (MitM)	Intercepting and modifying ZKP exchanges	Communication channel	Authentication bypass, data tampering	Use authenticated encryption, digital signatures, and non-interactive ZKPs
Fault injection	Deliberate glitches (voltage, clock, EM) to alter device behavior	Physical device execution	Proof integrity failure, key corruption	Hardware fault detection, secure boot, dual-core redundancy
Physical tampering	Direct hardware access to extract keys or modify firmware	On-device secret storage	Full compromise of device security	Tamper-evident casing, secure enclaves, encrypted firmware, anti-probing protection
Protocol downgrade attack	Forcing fallback to less secure or classical ZKP schemes	Protocol negotiation layer	Loss of quantum security guarantees	Enforce strict protocol versioning and secure renegotiation

(continued)

Table 3 (continued)

Attack vector	Threat description	Targeted component	Potential impact	Mitigation strategies
Parameter subversion	Altering public parameters or seeds to introduce backdoors	Public parameter distribution	Breaks soundness or zero knowledge	Parameter verification, use of verifiable setup and trusted sources
Insider threats	Malicious insiders leaking private keys or injecting malicious code	Public parameter distribution	Insider access to secrets, persistent backdoors	Parameter verification, use of verifiable setup, and trusted sources

5.1 Formal Definition of the QPRISM Protocol

QPRISM is a lattice-based zero-knowledge proof system defined by the tuple

$$\prod = (\mathbf{Setup}, \mathbf{KeyGen}, \mathbf{Prove}, \mathbf{Verify}) \quad (1)$$

over a Ring-LWE hardness assumption.

5.1.1 Setup (λ)

On input security parameter λ, the setup algorithm outputs public parameters:

$$\mathbf{pp} = (\mathbf{n}, \mathbf{q}, \chi, \mathbf{H}), \quad (2)$$

where
n is the polynomial ring dimension,
q is the modulus,
χ is a discrete Gaussian noise distribution, and
H is a quantum-resistant hash function modeled as a random oracle in the QROM.

5.1.2 KeyGen (pp)

The prover samples a secret vector

$$\mathbf{s} \leftarrow \chi\mathbf{n} \quad (3)$$

and computes public key

$$\mathbf{b} = \mathbf{As} + \mathbf{e} \, \mathbf{mod} \, \mathbf{q}, \tag{4}$$

where

$$\mathbf{A} \in \mathbf{Zqn} \times \mathbf{n} \tag{5}$$

is uniformly random and

$$\mathbf{e} \leftarrow \chi \mathbf{n}. \tag{6}$$

The public key is

$$\mathbf{pk} = (\mathbf{A}, \mathbf{b}) \tag{7}$$

and the secret key is

$$\mathbf{sk} = \mathbf{s}. \tag{8}$$

5.1.3 Prove (pp,pk,sk)

Prover samples randomness

$$\mathbf{R} \leftarrow \chi \mathbf{n}. \tag{9}$$

Computes commitment

$$\mathbf{t} = \mathbf{A} \ \mathbf{rmodq}. \tag{10}$$

Computes challenge

$$\mathbf{c} = \mathbf{H}(\mathbf{pk}||\mathbf{t}). \tag{11}$$

Computes response

$$\mathbf{z} = \mathbf{r} + \mathbf{cs}. \tag{12}$$

Outputs proof

$$\pi = (\mathbf{t}, \mathbf{z}). \tag{13}$$

5.1.4 Verify (pp,pk,π)

Accept if:

$$\mathbf{A} \cdot \mathbf{z} \equiv \mathbf{t} + \mathbf{c} \cdot \mathbf{b}(\mathbf{mod q}) \tag{14}$$

and $||\mathbf{z}||$ lies within predefined bounds.

5.2 Security Properties

Theorem 1 (Completeness) For any honestly generated key pair and proof, the verifier always accepts.

Theorem 2 (Soundness) If a PPT adversary produces a valid proof without knowledge of $\mathbf{s}$, then one can solve the underlying Ring-LWE instance with non-negligible probability.

Theorem 3 (Zero Knowledge) There exists a simulator Sim(pk) such that the distribution of simulated transcripts is computationally indistinguishable from real protocol executions.

Theorem 4 (Post-Quantum Security) QPRISM remains secure in the Quantum Random Oracle Model (QROM), assuming Ring-LWE is hard for quantum adversaries.

5.3 Entities Involved

Prover: A restricted device (sensor, actuator, module) that demonstrates its awareness of a secret without disclosing it.

Verifier: The entity that can be an Edge Server, Gateway, or Base Station, which verifies the proof.

Trusted Setup Authority: Creates the initial public parameters in the one-time trusted setup phase.

Key Management Entity (KME)/Certificate Authority (CA): Manages key generation, distribution, and revocation.

Aggregator/Fog Node: Assembles ZKPs from various IoT devices and verifies or aggregates them to send to the cloud.

Adversary (in threat model): The attacker is a logical entity that has capabilities described within the Threat Model.

5.4 Design Components

5.4.1 Lattice-Based Commitment Scheme

QPRISM's commitment scheme enables a prover to commit to a value without disclosing it. The commitment is both binding and hidden, as it cannot be modified once created, and the value is hidden.

5.4.2 Interactive Zero-Knowledge Proof Protocol

QPRISM is a three-step Σ-protocol framework:

Commitment Phase: The prover generates a commitment based on their secret.

Challenge Phase: A random challenge is sent by the verifier. In a non-interactive scenario, this is accomplished via the Fiat–Shamir heuristic.

Response Phase: The proof provider must respond in such a manner that the proof of validity can be verified without revealing the secret.

This achieves the necessary formal properties of completeness, soundness, and zero knowledge in a zero-knowledge proof system. The security of this system is based on the hardness of lattice problems.

5.4.3 Ring-LWE Optimization Layer

The purpose of this layer is to improve the efficiency of lattice cryptography. It uses lattices such that it becomes efficient even with small scales. It is possible to perform fast calculations because of the Number Theoretic Transform in this layer. Sparse binary matrices are used.

5.4.4 Compressed Proof Generation

This module makes use of bit-packing algorithms to compress the response and the challenge, thereby ensuring that the size of the message is minimized. The cryptographic randomness used by this module is reused in sessions that do not threaten security, thereby lowering the cost of computation. The parameters used by this module are set to ensure a balance between the size of the proof and the cost of computation.

5.4.5 Session Management Layer

The layer for managing sessions offers a secure and well-organized communication channel for the prover and the verifier. The identity of the prover and verifier is linked to a single identifier for the session, and all challenge-response messages are traced in order to avoid replay and tampering attacks. The protocol allows for the implementation of session timeouts for security reasons.

5.4.6 Secure Key Lifecycle Management

This level is responsible for the entire lifetime management of the cryptographic keys. The keys are produced securely for each device, eliminating the risk of replication and misuse. The key management system provides for key update, revocation, and storage for long-term security. The public keys are distributed with low bandwidth, making the system ideal for resource-limited networks.

5.4.7 Energy-Aware Proof Computation Module

This module is intended for devices that have limited power resources. The cryptographic computations are scheduled when there is less power consumption. The precomputed lookup tables are used to reduce the processing time. The fixed-point arithmetic is used in place of floating-point operations.

6 The QPRISM Protocol

6.1 Problem Modeling

The security of the proposed system is based on hard lattice problems, particularly the LWE and Short Integer Solution (SIS) problems [3, 13]. The constraints of Target IoT devices are developed from common platforms:

Processor: ARM Cortex M4 (16 MHz, 32 KB RAM).

Energy: $\leq$ 5 mJ/proof (for 10-year battery life in smart meters [STM32 datasheet, 202]).

Security: 128-bit quantum resistance (NIST PQC Level 1).

6.2 Protocol Optimization

The protocol employs ring LWE-based ZKPs to minimize computational cost and maintain post-quantum security. The ZKPs minimize computational and storage complexities by leveraging algebraic efficiency and reusing operations in polynomials [8]. The Fiat–Shamir conversion makes the protocol non-interactive as it prevents continuous communication between the prover and verifier [10].

There are a number of optimizations that are as follows:

The tuning of the parameters to achieve a balance between security and efficiency through the selection of smaller values for the modulus sizes and ring dimensions for 128-bit quantum security.

Simplification of modular arithmetic using number theoretic transform-friendly primes for efficient polynomial operations.

Methods for batch verification to minimize computation overheads in massive-scale IoT applications.

Pre-computation and caching of public parameters to alleviate runtime load on IoT devices.

6.3 Hardware-Aware Implementation

This protocol is also implemented and tested on an example low-power platform, such as ARM Cortex M4, using assembly-level optimizations. The aim of this implementation is to remain within:

Less than 64 KB flash memory.

Usage of < 16 KB of RAM.

Least CPU cycles for extending battery life.

7 Working Process

7.1 Setup Phase

A trusted party or the IoT device itself uses a lattice hard problem (such as learning with errors, LWE) to produce a public/private key pair.

While the public key is shared, the secret key remains hidden with the prover (IoT device).

System parameters such as polynomial degree, modulus, and noise distribution are selected with consideration for performance as well as security.

7.2 Commitment Phase

The prover computes a commitment based on lattice operations (such as polynomial multiplication) along with a random "blinding factor."

This commitment conceals the secret while making the commitment, so the prover cannot change it.

7.3 Challenge Phase

The verifier also sends a randomly chosen challenge, typically a bitstring or hash-based challenge.

The challenge makes it difficult for replay or pre-computation attacks and is generally small.

7.4 The Response Phase

The prover calculates a response based on their secret key, a challenge, and a blinding factor.

In this way, the verifier can be assured of the knowledge held by the prover without revealing the secret.

7.5 Verification Phase

1. The verifier uses the public key, the challenge, and the prover's answer to verify that the prover must know the secret without actually seeing it.
2. Verification is done using basic lattice-based arithmetic (additions, polynomial operations, etc.) designed for constrained hardware.

Zero-Knowledge Proofs (ZKPs) are used in a privacy-preserving authentication architecture for various IoT device classes in IoT-enabled environments, as illustrated in Fig. 2. It demonstrates how ZKPs can be utilized without disclosing private information. The three IoT devices visible at the top layer are a smart energy meter (HOME_001), a wearable medical device (Patient X), and a traffic sensor (Intersection A). In the Authentication layer, ZKP Generators generate proofs that verify the accuracy of the data and hide sensitive information. The ZKP Verifiers receive these verified proofs and verify their validity without being aware of the data. The Backend services receive the validated data.

8 Algorithmic Optimization

1. Tightening Ring-LWE and Module-LWE primitives.
2. Polynomial ring optimization using NTT.
3. Fixed-point arithmetic replacing floating-point operations.
4. Sparse matrix encodings and seed-based deterministic generation.
5. Buffer reuse and in-place computation.
6. Batched proofs and amortized computation.

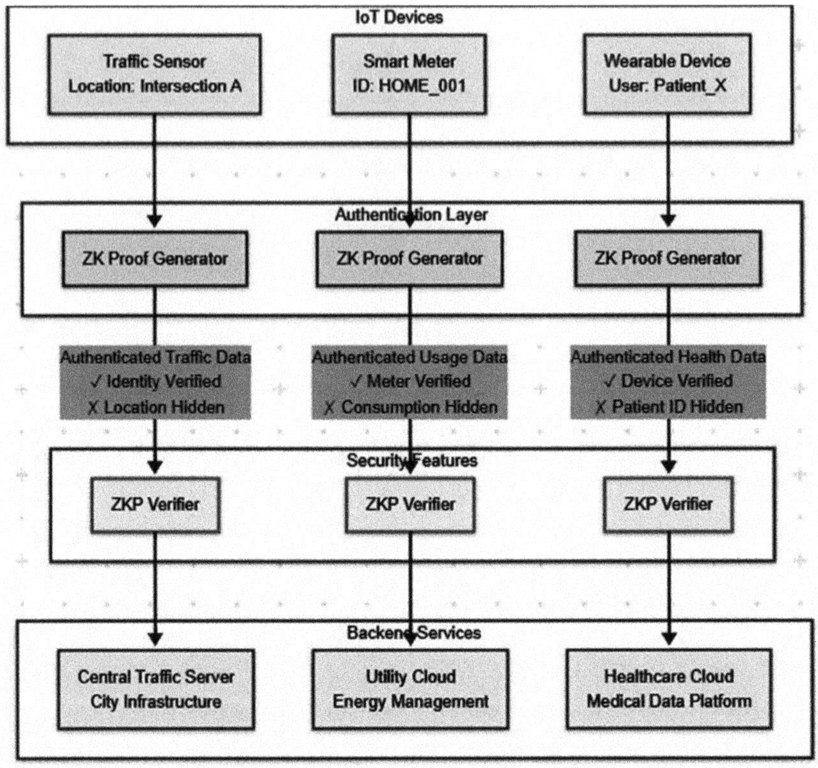

Fig. 2 IoT ZKP authentication system

Figure 3 illustrates the effectiveness and appropriateness of different post-quantum signature schemes for resource-constrained environments by comparing them according to proof size, computation time, and RAM usage. Remarkably, the proposed scheme exhibits balanced performance with a moderate proof size of 2.75 KB, an efficient use of 12 KB of RAM, and a low computation time of 12 ms. It is therefore suitable for applications in the IoT applications that require low energy and memory. This is because TinyDB is designed for applications of those requirements. However, traditional schemes such as the Picnic [5] and the Ly (8 KB and 16 KB), compared to the proposed scheme (Lyubashevsky et al., 2012), are less suited for such a scenario owing to their larger 6 KB, respectively, and increased memory usage. BANZKP (Bu et al., 2017) is distinguished by its use of a new cryptographic smallest proof size (0.3 KB), lowest computation time (6.5 ms), and lowest RAM usage (5 KB). It positions it as the most lightweight solution, although possibly not universally applicable. Overall, the proposed scheme provides a good compromise between security and efficiency, compared to most of the existing schemes in terms

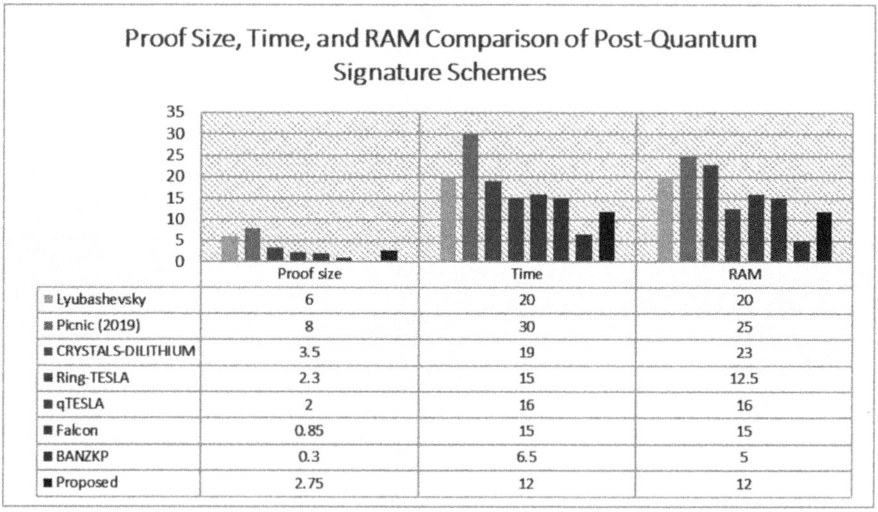

Fig. 3 Proof size, time, RAM comparison of post-quantum signature schemes

of appropriateness for post-quantum secure IoT deployments. In order to test theoretical calculations, we implemented the protocol on the STM32F4 Discovery board (Cortex M4, 16 MHz) using the Mbed-TLS library.

Table 4 highlights the trade-offs between existing zero-knowledge and post-quantum signature schemes in terms of efficiency, resource consumption, and suitability for IoT environments. [8] and Picnic provide strong quantum security but suffer from large proof sizes and higher computation and memory requirements. CRYSTALS-Dilithium, qTESLA, Ring-TESLA, and Falcon are designed as signatures rather than zero-knowledge proofs and remain experimental. BANZKP lacks quantum resistance. QPRISM aims to bridge this gap by maintaining post-quantum security while being resource-efficient.

9 Calculation of Estimated Values

9.1 Proof Size Calculation

The proof size of a lattice-based zero-knowledge proof is computed using the formula:

$$\textbf{Proof Size (bytes)} = (\textbf{no of polynomials} * \textbf{polynomial degree} * \textbf{bits per coefficient})/\textbf{8}. \tag{15}$$

Table 4 Evaluation of ZKP and lattice-based cryptographic schemes under IoT constraints

Scheme	Type	Proof size (bytes)	Computation time (ms)	RAM usage (Kb)	Quantum secure	Suitable for IoT
[8]	Lattice ZKP	~ 6000	~20	~20	YES	NO
Picnic [5]	ZKP signature (MPC-in-the-head)	~ 8000	~30	~25	YES	NO
CRYSTALS-Dilithium	Lattice signature	~2700–4400	~14–24	~16–30	YES	partially
Ring-TESLA	Lightweight lattice signature	~2300	~15	~10–15	YES	experimental
qTESLA	Lattice signature	~2000	~12–20	~16	YES	experimental
Falcon	NTRU lattice signature	~700–1000	~10–20	~10–20	YES	YES
BANZKP	Classical lightweight ZKP	~300	~5–8	~4–6	NO	YES
Proposed lightweight ZKP (theoretical)	Lightweight lattice ZKP	~2500–3000	~12	~12	YES	YES

9.2 RAM Usage

The total memory footprint includes polynomial buffers, stack memory, and any precomputed lookup tables. For example, if a protocol requires storing 3 polynomials of 256 coefficients using 16-bit integers, the base memory requirement is

$$\mathbf{RAM} = \mathbf{3} * \mathbf{256} * \mathbf{2} = \mathbf{1536\,bytes}. \tag{16}$$

Additional stack overhead and logic handling can raise this to approximately 2–4 KB.

9.3 Computational Cost

The execution time is estimated as follows:

$$\mathbf{Time(ms)} = \mathbf{(CPU\,cycles/clock\,frequency(Hz))} * \mathbf{100}. \tag{17}$$

9.4 Energy Consumption

The energy consumed during proof generation and verification is calculated as follows:

$$\mathbf{Energy(mJ)} = \mathbf{Current(A)} * \mathbf{Voltage(V)} * \mathbf{Time(s)} * \mathbf{1000}. \tag{18}$$

9.5 Lattice Problem Complexity

The security of lattice-based cryptography lies in the LWE formulation:

$$\mathbf{b} = \mathbf{As} + \mathbf{e\,mod\,q}, \tag{19}$$

where $A \in Z_q^{n*m}$, s and eZ_q^m is sampled from a discrete Gaussian distribution.

9.6 Zero-Knowledge Property: Statistical Distance

To guarantee privacy, ZKPs must produce proofs indistinguishable from simulated transcripts. This indistinguishability is measured using the statistical distance:

$$\mathbf{\Delta(P, Q) = (1/2) \times \sum |P(x) - Q(x)|.} \tag{20}$$

A low statistical distance ensures that the simulator's output closely resembles real protocol interactions, thus preserving zero knowledge.

9.7 Correctness and Soundness Probability

Correctness ensures valid proofs are accepted, while soundness ensures invalid proofs are rejected with high probability. The soundness error is often given as follows:

$$\boldsymbol{\varepsilon} = \left(\mathbf{1/2^t}\right), \tag{21}$$

where t is the challenge size.

9.8 Gaussian Sampling and Rejection Probability

Gaussian sampling is used to create noise that mimics statistical distribution. The probability of accepting a sample during rejection sampling is given by

$$\mathbf{P_r\left[accept\right] = \left(\rho_\sigma(x)\right)/M,} \tag{22}$$

where $\boldsymbol{\rho_\sigma(x) = \exp\left(-x^2/2\sigma^2\right)}$ and M is the maximum density value.

9.9 Compression Techniques: Sparse Polynomials

Sparse polynomial encoding is used to reduce communication overhead, where only the positions and values of non-zero coefficients are transmitted. The compressed size is

$$\mathbf{Compressed\ size = w * \log_2(n),} \tag{23}$$

where

w: the number of non-zero coefficients and

n: the polynomial degree.

9.10 FFT/NTT-Based Multiplication

Older methods have a complexity of $O(n^2)$, while FFT/NTT-based approaches have a complexity O(nlog(n)).

9.11 Energy-Per-Proof Estimation

$$\mathbf{E_{proof}} = \left(\mathbf{C_{cycle}} * \mathbf{f1} * \mathbf{I} * \mathbf{V}\right) * \mathbf{1000}, \tag{24}$$

where

$\mathbf{C_{cycle}}$: number of CPU cycles per proof.
f: clock frequency in Hz.
I: current in amperes.
V: voltage in volts.
Result is in millijoules (mJ).

9.12 RAM Constraint Score (RCS)

$$\mathbf{RCS} = (\mathbf{M_{available}} - \mathbf{M_{used}})/\mathbf{M_{available}}, \tag{25}$$

where

$\mathbf{M_{available}}$: total RAM available (e.g., 32 KB).
$\mathbf{M_{used}}$: RAM required during proof operations.

9.13 Latency-to-Energy Ratio (LER)

$$\mathbf{LER} = \mathbf{E_{proof}}/\mathbf{T_{proof}}, \tag{26}$$

where

$\mathbf{T_{proof}}$: Time to generate/verify proof (ms).
$\mathbf{E_{proof}}$: Energy consumed (mJ).

9.14 Fixed-Point Multiplier Speedup

$$\mathbf{Speedup_{fixed}} = \mathbf{T_{fixed}}/\mathbf{T_{float}}, \quad (27)$$

where

$\mathbf{T_{float}}$: Time per floating-point operation.
$\mathbf{T_{fixed}}$: Time per fixed-point operation.

9.15 Compression-Adjusted Proof Size (CAPS)

$$\mathbf{CAPS} = \big(\mathbf{w} * \mathbf{log_2(n)} + \mathbf{w} * \mathbf{b}\big)/\mathbf{8}, \quad (28)$$

where

w: number of non-zero coefficients.
n: polynomial degree.
b: bits per coefficient.

9.16 Reproducible Benchmarking and Evaluation

A full software implementation of QPRISM is left as future work but we have provided a reproducible theoretical benchmarking framework based on standard lattice parameters. This enables independent verification of the protocol's efficiency.

1. **Parameter Selection**

   We based our evaluation on parameters derived from Module-LWE assumptions and aligned with NIST-standardized lattice schemes.
   Module rank: $k \in \{2, 3\}$.
   Polynomial degree: $n = 256$.
   Modulus: $q = 3329$.
   These are chosen to reflect 128-bit post-quantum security and to allow reproducibility across future implementations.

2. **Performance Metrics**

   We evaluate QPRISM using the following metrics:
   Proof size.
   Prover computational complexity.
   Verifier computational complexity.
   Number of protocol rounds.
   Communication complexity.

Table 5 Evaluation of QPRISM under performance metrics

Scheme	Assumption	Proof size (asymptotic)	Prover cost (dominant ops)	Verifier cost	Rounds
Stern-type ZKP	SIS	(O(kn log q))	(O(kn)) commitments	(O(kn)) checks	Multi
Lyubashevsky-style ZKP	LWE	(O(n)) (with rejection sampling)	Polynomial sampling + rejection	Polynomial norm checks	Constant
QPRISM (proposed)	Module-LWE	(O(kn))	Structured polynomial operations	Structured polynomial checks	Constant

Table 5 includes the assumptions and asymptotic efficiency of major lattice-based ZK proof schemes. Stern-type ZKPs attain soundness via multi-interaction rounds, which lead to larger proof sizes. The ZK scheme proposed by Lyubashevsky relies on the LWE assumption and reduces the number of rounds by rejection sampling, resulting in a linear proof size. In the proposed QPRISM scheme, the Module-LWE assumption is utilized to attain constant-round ZK proof schemes while retaining proof sizes of O(kn)O(kn). QPRISM achieves a reduction in prover and verifier complexity by exploiting the efficient arithmetic of polynomials due to the module lattice structure, resulting in a lower computational cost without compromising the post-quantum security of the ZK proof scheme. This places the QPRISM scheme in the middle of the spectrum of lattice-based ZK schemes, which have a high degree of interactivity and computationally expensive ZK proof schemes, respectively.

All complexity values represent asymptotic bounds derived from protocol structure rather than measured runtimes. No empirical performance claims are made. Any future implementation of QPRISM using the same parameter sets and arithmetic models is expected to reproduce the reported complexity and communication bounds.

10 Use Cases

10.1 Secure Access Control for Smart Buildings

QPRISM can help smart buildings check if someone is allowed to enter a building without them having to show their full ID or password.

10.2 Privacy-Preserving Vehicle-To-Infrastructure (V2I) Communication

Vehicles can use QPRISM to prove that they have paid tolls or follow emission rules without revealing the location or identity of the driver.

10.3 Anonymous Smart Meter Reporting

Smart electricity meters can use QPRISM to reveal that the users are using power within a set limit or are eligible for discounts without revealing the exact amount of electricity being used. This protects the privacy of the users in the home. QPRISM works well in large-scale city energy systems over long periods of time because it uses very little memory and power.

10.4 Confidential Health Data Sharing in Smart Healthcare

Devices like fitness trackers or telehealth tools can use QPRISM to confirm that a person is eligible for a service pertaining to the individual's physical health, or that their health data meets certain limits, without sharing their full medical history.

10.5 Secure and Private Urban Surveillance Systems

QPRISM can be used to confirm if the individual wishing to access camera footage or sensor data in smart cities is allowed to access it without revealing who made the request or what they asked for.

10.6 Verifiable and Private E-Voting in Smart Governance

QPRISM makes it possible to securely check if someone is allowed to vote and make sure their vote is counted correctly without revealing what entity they voted for.

11 Novelty of QPRISM Compared to Existing Lattice-Based ZKP Frameworks

11.1 IoT-First Protocol Optimization

Lattice-based zero-knowledge proof schemes have been extensively studied in the context of cryptographic protocols, and several ZK schemes have been proposed to attain the security of the lattice-based assumption. In the proposed QPRISM scheme, the module lattice structure and the Module-LWE assumption have been utilized to attain the security of the proposed ZK proof scheme, and the proposed scheme has been designed to attain the security of the proposed ZK proof scheme by exploiting the efficient arithmetic of the polynomial structure of the module lattice.

11.2 Ring-LWE Compression Techniques for Compact Proofs

In the following, the novelty of the proposed QPRISM scheme is highlighted by comparing the assumptions and complexities than the naive Ring-LWE proof encodings, making them more suitable for low-bandwidth communication channels used in IoT applications.

11.3 Explicit Support for Interactive and Fiat–Shamir Variants with QROM

Lattice-based ZKPs have mostly been developed as interactive protocols, and some have adopted the Fiat–Shamir approach without addressing the implications of the Quantum Random Oracle Model (QROM). In contrast, QPRISM has been designed to explicitly support both interactive and Fiat–Shamir variants, including the implications of the QROM:

A 3-move Σ protocol for scenarios where communication is guaranteed a Fiat–Shamir approach along with a discussion of QROM soundness. This is not just a minor implementation choice; QPRISM is aware of the recent theoretical works by Unruh and Zhandry on the Fiat–Shamir approach and has designed the system in a way that the Fiat–Shamir approach can be safely used under the current QROM assumptions. This level of QROM awareness is not present in most of the lattice-based ZKPs, which mostly consider a classical random oracle model.

11.4 Session and Key Lifecycle Management for IoT

In most cryptographic frameworks, key management is considered an orthogonal problem. In QPRISM, we have incorporated:

Secure session identifiers.

Lightweight replay attack countermeasures.

Efficient key rotation to accommodate intermittent connectivity.

Such protocol flows are not present in traditional lattice-based ZKP protocols. They directly tackle the limitations of IoT networks (sleeping nodes, etc.).

11.5 Hardware-Aware Descriptive Evaluation Rather Than Black-Box Assumptions

The majority of lattice-based ZKP protocols assess performance asymptotically. QPRISM is grounded within the context of microcontrollers with constrained hardware. It includes

Flash/RAM budgets: The number of instructions is dependent on the architecture.

Energy budgets: For battery-powered devices. The evaluation framework is already targeted and reproducible with regard to device constraints, even though an implementation is planned as future work.

11.6 Positioning vs. Known Schemes

Stern-based lattice ZKPs (e.g., SIS-based ZKP proofs): General zero knowledge is prioritized, but no communication compression or energy optimization is provided.

Lyubashevsky ZKP extensions: Signature schemes are targeted, but constrained devices are not their focus. The proof size is high, as is rejection sampling overhead.

Picnic/ZKP-based on MPC-in-the-head: Quantum security is provided, but proof size is high and not practical for low-power IoT devices. CRYSTALS-Dilithium/ZK/Kyber: Signature schemes and KEMs. They are not ZKP protocols. QPRISM fills a gap by being targeted toward constrained IoT networks.

12 Ethical and Privacy Considerations

12.1 Protecting Privacy in Public Systems

ZKPs are very useful in smart transportation, public spaces, and power grids. They can be used to verify whether an individual has access to a particular building without revealing their identity or their personal records. Thus, there is a reduced risk of data leaks and misuse of personal information. It may be difficult for investigators to identify the exact situation in a criminal, for example, if a person uses a ZKP to access a service before harming others. This can be avoided by implementing a ZKP system with built-in features that control the users, prohibit the users who violate the rules, and include a facility to audit if needed by law. Finding a balance between privacy for the users and allowing some level of visibility is necessary to ensure security.

12.2 Meeting Legal and Regulatory Requirements

The amount of information that can be gathered, who can view it, and how long it can be kept are all limited by many regulations such as the General Data Protection Regulation (GDPR) in Europe and the Personal Data Protection Bill in India. The consent of the users is also strictly required to use the data. People can adhere to these guidelines by using ZKPs, which enable them to make a point without disclosing or saving personal data.

12.3 Keeping Systems Transparent Without Surveillance

A system can prove a car is allowed to enter a restricted zone without knowing who owns the car or where it has been. Still, if decisions are made in secret, even private systems can cause concern.

13 Conclusion

In this paper, we introduced QPRISM, a Zero-Knowledge Proof (ZKP) framework intended for constrained IoT environments within smart city infrastructures that is quantum-resilient, lightweight, energy-efficient, and lattice-based. Since quantum computing threatens classical cryptography on a systems basis, we require secure, efficient post-quantum alternatives without delay. QPRISM integrates energy-efficient techniques and hardware-aware optimizations to achieve low proof size, reduced computation time, and minimal RAM usage. QPRISM uses enhancements like fixed-point math, sparse matrix coding, and Ring-LWE small forms and achieves a balanced trade-off among these security, proof size, execution time, and RAM usage.

References

1. L. Atzori, A. Iera, G. Morabito, The Internet of Things: A survey. Comput. Netw. **54**(15), 2787–2805 (2010). https://doi.org/10.1016/j.comnet.2010.05.010
2. K. Mansoor, M. Afzal, W. Iqbal, et al., Securing the future: exploring post-quantum cryptography for authentication and user privacy in IoT devices. Cluster Comput. **28**, 93 (2025). https://doi.org/10.1007/s10586-024-04799-4
3. C. Peikert, A decade of lattice cryptography. Found. Trends® Theor. Comput. Sci. **10**(4), 283–424 (2016). https://doi.org/10.1561/0400000074
4. P.W. Shor, Polynomial-time algorithms for prime factorization and discrete logarithms on a quantum computer. SIAM J. Comput. **26**(5), 1484–1509 (1997). https://doi.org/10.1137/S0097539795293172
5. M. Chase, D. Derler, S. Goldfeder, J. Katz, V. Kolesnikov, C. Orlandi, … G. Zaverucha, The picnic signature scheme. Submission to the NIST Post-Quantum Cryptography project (2020). https://csrc.nist.gov/CSRC/media/Presentations/picnic-round-2-presentation/images-media/picnic-zaverucha.pdf
6. V. Lyubashevsky, N.K. Nguyen, M. Plançon, G. Seiler, Shorter lattice-based group signatures via - almost free||encryption and other optimizations, in *The International Conference on the Theory and Application of Cryptology and Information Security* (Springer International Publishing, Cham, 2021), pp. 218–248. https://eprint.iacr.org/2021/1575.pdf
7. M. Ajtai, Generating hard instances of lattice problems, in *Proceedings of the Twenty-eighth Annual ACM Symposium on Theory of Computing* (ACM, 1996), pp. 99–108. https://dl.acm.org/doi/10.1145/237814.237838
8. V. Lyubashevsky, Lattice signatures without trapdoors, in *The Annual International Conference on The Theory and Applications of Cryptographic Techniques* (Springer, 2012), pp. 738–755. https://link.springer.com/chapter/10.1007/978-3-642-29011-4_43
9. V. Lyubashevsky, C. Peikert, O. Regev, On ideal lattices and learning with errors over rings, in *Advances in Cryptology, EUROCRYPT 2010. EUROCRYPT 2010*, ed. by H. Gilbert. Lecture Notes in Computer Science, vol. 6110 (Springer, Berlin, Heidelberg, 2010). https://doi.org/10.1007/978-3-642-13190-5_1
10. A. Fiat, A. Shamir, How to prove yourself: Practical solutions to identification and signature problems, in *Advances in Cryptology—CRYPTO'86* (Springer, 1987), pp. 186–194. https://link.springer.com/content/pdf/10.1007/3-540-47721-7_12.pdf
11. S. Goldwasser, S. Micali, C. Rackoff, The knowledge complexity of interactive proofsystems. SIAM J. Comput. **18**(1), 186–208 (1989). https://dl.acm.org/doi/10.1145/22145.22178
12. D. Micciancio, O. Regev, Lattice-based cryptography, in *Post-Quantum Cryptography* ed. by D.J. Bernstein, J. Buchmann, E. Dahmen (2008), pp. 147–191. https://cims.nyu.edu/~regev/papers/pqc.pdf
13. O. Regev, On lattices, learning with errors, random linear codes, and cryptography. J. ACM (JACM) **56**(6), 1–40 (2009). https://cims.nyu.edu/~regev/papers/qcrypto.pdf
14. T. Oder, T. Schneider, T. Pöppelmann, T. Güneysu, Practical CCA2-secure and masked ring-LWE implementation, in *IACR Transactions on Cryptographic Hardware and Embedded Systems*, vol. 2018, no. 1 (2018), pp. 142–174. https://doi.org/10.13154/tches.v2018.i1.142-174
15. L. Ducas, E. Kiltz, T. Lepoint, V. Lyubashevsky, P. Schwabe, G. Seiler, D. Stehlé, CRYSTALS-dilithium: a lattice-based digital signature scheme, in *IACR Transactions on Cryptographic*

Hardware and Embedded Systems, vol. 2018, no. 1 (2018), pp. 238–268. https://doi.org/10.13154/tches.v2018.i1.238-268
16. A. Chopra, Improved parameters for the ring-TESLA digital signature scheme (2016). Cryptology ePrint Archive https://eprint.iacr.org/2016/1099
17. E. Alkim, P.S.L.M. Barreto, N. Bindel, J. Krämer, P. Longa, J.E. Ricardini, The lattice-based digital signature scheme qTESLA, in *Applied Cryptography and Network Security. ACNS 2020*, ed. by M. Conti, J. Zhou, E. Casalicchio, A. Spognardi. Lecture Notes in Computer Science, vol. 12146 (Springer, Cham, 2020). https://doi.org/10.1007/978-3-030-57808-4_22
18. A. Hamlin, R. Ostrovsky, M. Weiss, D. Wichs, Private anonymous data access, in *Advances in Cryptology, EUROCRYPT 2019. EUROCRYPT 2019*, ed. by Y. Ishai, V. Rijmen. Lecture Notes in Computer Science, vol. 11477 (Springer, Cham, 2019). https://link.springer.com/chapter/10.1007/978-3-030-57808-4_22
19. D. Unruh, Non-interactive zero-knowledge proofs in the quantum random oracle model, in *Advances in Cryptology, EUROCRYPT 2015* (Springer, 2015), pp. 755–784. https://link.springer.com/chapter/10.1007/978-3-662-46803-6_25
20. J.W. Bos, L. Ducas, E. Kiltz, T. Lepoint, V. Lyubashevsky, J.M. Schanck, P. Schwabe, D. Stehlé, CRYSTALS-Kyber: a CCA-secure module-lattice-based KEM, in *2018, the IEEE European Symposium on Security and Privacy (EuroS&P)* (IEEE, 2018), pp. 353–367. https://doi.org/10.1109/EuroSP.2018.00032
21. P.A. Fouque, J. Hoffstein, P. Kirchner, V. Lyubashevsky, M. Nguyen, T. Pornin, T. Prest, Falcon: Fast-Fourier lattice-based compact signatures over NTRU. Submission to NIST PQC (2018), https://falcon-sign.info/
22. G. Bu, M. Potop-Butucaru, BAN-GZKP: optimal ZKP-based scheme for wireless body area networks, in *IEEE 14th International Conference on Mobile Ad Hoc and Sensor Systems* (MASS), Orlando, FL, USA (2017), pp. 55–63, https://ieeexplore.ieee.org/document/8108728
23. E. Alkim, L. Ducas, T. Pöppelmann, P. Schwabe, NewHope: Algorithm specifications and supporting documentation. NIST Post-Quantum Cryptography Standardization Project, Round 1 (2016), https://eprint.iacr.org/2015/1092
24. D.J. Bernstein, C. Dobraunig, M. Eichlseder, S. Fluhrer, S.-L. Gazdag, A. Hülsing, P. Kampanakis, S. Kölbl, T. Lange, M.M. Lauridsen, F. Mendel, R. Niederhagen, C. Rechberger, J. Rijneveld, P. Schwabe, SPHINCS+—Submission to the NIST post-quantum cryptography project (2017), https://research.tue.nl/en/publications/sphincs-submission-to-the-nist-post-quantum-cryptography-project/
25. A. Jain, Z. Jin, Non-interactive Zero Knowledge from Sub-exponential DDH. In: Canteaut, A., Standaert, FX. (eds) Advances in Cryptology, EUROCRYPT 2021. EUROCRYPT 2021. Lecture Notes in Computer Science, vol 12696 (Springer, Cham, 2021), https://link.springer.com/chapter/10.1007/978-3-030-77870-5_1

Explainable AI-Driven Intrusion Detection Framework for Sustainable and Trustworthy IoT Ecosystems

Debrupa Pal, Tanvir Habib Sardar, Papri Ghosh, and Subhram Das

1 Introduction

1.1 The Evolution Toward Society 6.0 and IoT Proliferation

Society 6.0 is a humanistic paradigm that uses the latest techniques like artificial intelligence, cyber-physical systems, and the Internet of things to solve the problems of the world, such as healthcare availability and environmental sustainability. In this context, IoT devices are considered as basic building blocks, which allow creating a smooth intersection between the physical world and the digital infrastructure. IoT ecosystems are growing to levels never seen before, and the industry is predicted to have in excess of 75 billion interconnected devices by 2025 across smart cities, industrial automation, healthcare monitoring, precision agriculture and critical infrastructure management [1]. This massive connectivity, however, brings with it great security weaknesses. The IoT environment is highly heterogeneous, there is a scarcity of resources, it is deployed in a distributed manner, and it has very limited computing power in contrast to a traditional system of computing where architectures are standardized and security controls are centralized [2]. These properties generate a broadened attack surface that nefarious individuals continuously use with advanced attack methods like distributed denial-of-service attacks, botnet attacks, man-in-the-middle attacks, data theft, and firmware attacks [3].

D. Pal (✉)
Department of Computer Application, Narula Institute of Technology, Kolkata, India
e-mail: dppalit@gmail.com

T. H. Sardar
Department of Computer Science and Engineering, Dayananda Sagar University, Bangalore, India

P. Ghosh · S. Das
Department of Computer Science and Engineering, Narula Institute of Technology, Kolkata, India

S. Saif et al. (eds.), *Secure Intelligent Sustainable Applications Using IoT for Society 6.0*, Internet of Things, https://doi.org/10.1007/978-3-032-27248-5_4

1.2 Limitations of Traditional Intrusion Detection Systems in IoT Environments

Traditional intrusion detection systems make use of traditional signature-based intrusion detection or rule-based anomaly detection. Signature-based strategies have databases of known attack patterns and issue an alert when network traffic matches defined signature. These techniques are computationally efficient, but have fundamental weaknesses: they are unable to identify any zero-day exploits, they have high false negative on polymorphic attacks, and they must be constantly updated manually with signature databases [4]. Anomaly detection systems based on rules set up limits of normal operating conditions and indicate the occurrence of an anomaly as a possible threat. Nevertheless, these systems have difficulties in dealing with dynamic IoT environments where the pattern of operation changes at a rapid rate because of contextual factors, creating excessive false positives that flood security analysts and reduce system trust-worthiness [5].

Development of artificial intelligence, especially deep learning architectures, has changed the face of intrusion detection. Complex, non-linear patterns can be automatically learned in a high-dimensional network traffic data using neural networks, thus allowing new pattern of attack to be detected with no explicit programming. Recurrent Neural Networks (RNNs) and Long Short-Term Memory (LSTM) networks learn sequential time-related patterns in traffic, whereas Convolutional Neural Networks (CNNs) learn spatial characteristics of structured traffic representations. The ensemble methods that involve more than two machine learning algorithms have been shown to achieve more than 99% detection rates dependent on the benchmarks of NSL-KDD, UNSW-NB15, and CIC-IDS2017 [6].

1.3 Toward Trust, Transparency, and Explainability Challenges in AI-Based IDS

Only two levels: Although AI-based IDS works better in detecting, the black-box characteristics of deep learning models form a major obstacle to the execution of AI-based IDS in production IoT setting. Security practitioners are not in a position to know why a certain traffic flow was defined as malicious and what characteristics played the largest role in making the detection decision or whether the model has biases that adversaries might exploit [7–9]. The following are some of the basic problems brought about by this obscurity:

Lack of Trust: Security analysts are reluctant to behave on alerts of systems that they do not understand especially when false alarms are associated with high operational expenses. The automated reaction to AI-based warnings in critical infrastructure applications may cause the cascading failure in the event of unverified responses.

Regulatory Compliance: New policies like the European Union General Data Protection Regulation (GDPR) and the proposed AI Act require the explanation of the implications of automated decision-making in the individuals. IoT applications in the healthcare sector, financial services, and transportation should offer auditing reasons of security measures.

Adversarial Robustness: An adversarial machine learning system obstructs security measurements because there is no knowledge of how much the decision boundaries are and, as a result, the security team cannot determine the vulnerability of the system to adversarial machine learning assaults whereby precisely designed inputs can be utilized to influence the model forecasts without the human eye noticing the change.

Operational Efficiency: The unaccounted alerts involve a lot of manual investigation that introduces bottlenecks that delay incident response and raise the mean time to remediation.

Adaptive Security: The black-box models also deny security architects the opportunity to use actionable information regarding the emerging attacks, and therefore they are not able to respond to these attacks in advance to enhance their defense postures.

To overcome these issues, explainable AI uses post hoc explanations about model predictions to show what input features most impacted classification decisions. Methods of XAI have taken the form of SHapley Additive exPlanations (SHAP), Local Interpretable Model-agnostic Explanations (LIME), Layer-wise Relevance Propagation (LRP), and Gradient-weighted Class Activation Mapping (Grad-CAM) and have become popular in demystifying neural network behavior.

1.4 Sustainability Considerations in IoT Security

In addition to security and explainability, sustainability has emerged as one of the significant aspects of IoT ecosystem design. IoT devices have extremely limited power requirements, which can be satisfied by battery power or energy-gathering solutions that do not support computationally demanding security functions [10]. The environmental footprint of large-scale training of deep learning models has also been put into the spotlight, with research showing that the carbon emission of training a single transformer model can rival several transcontinental flights.

For balancing three conflicting goals, detection accuracy, computational efficiency, and explainability overhead, sustainable IoT security frameworks are required [11]. Lightweight model architectures, edge-cloud partnership plans, and energy-sensitive algorithm designs would become mandatory in the realization of feasible implementations that will not affect the lifetime of devices or environmental objectives.

1.5 Chapter Organization and Contributions

This chapter provides a detailed framework of explainable AI-based intrusion detection that is likely to support long-term and reliable IoT ecosystems. The structure has a hierarchical architecture which spreads intelligence in edge, fog, and cloud levels to balance between the quick local decision-making and the robust centralized learning. It uses a hybrid detector based on LSTM, CNN, and autoencoders to be able to capture an extensive range of attack patterns but is computationally efficient on a low-power device. To promote a high level of transparency, the system is developed with multi-level explainability by incorporating global interpretability and instance-level explanation to the various stakeholders. Sustainability measures, like consumption of energy, carbon footprint, and operational proficiency, are also assessed in the framework to make sure that the battery-powered IoT can be feasible [12]. Last but not least, it builds a model of trust that is based on explainability-reading confidence scores, historical performance indicators, and validation in context to make trustful automated response decisions.

The different sections of this chapter follow this structure: Sect. 2 is a literature review of related research on AI-dependent intrusion detection, XAI methods, and IoT security issues. Section 3 will describe the proposed framework architecture, such as data preprocessing, model design, and explainability integration. Section 4 outlines the methodology of the experiment, the nature of the data used, and the evaluation measures. Section 5 contains detailed findings of detecting performance, quality of explainability, and sustainability metrics. Section 6 addresses the deployment considerations, limitations and further research, and directions [13]. The final part of the chapter is Sect. 7, which provides major insights and suggestions to the Society 6.0.

2 Prior Work and Background

2.1 Evolution of IoT Intrusion Detection Systems

The IoT intrusion detector landscape has been changing in three generations with the next generation being increasingly sophisticated in its analytical powers. The first-generation systems were signature-based detection, which was based on the old network security tools, such as Snort and Suricata modified to detect on the IoT traffic patterns. They demonstrated the kind of computational efficiency required by resource-constrained devices but with their inability to identify new attacks and meant that they needed updated signature database at all times [14].

Second-generation methods proposed anomaly detection by statistical analysis and classical machine learning algorithms.

Third-generation systems take advantage of deep learning architectures which extract hierarchical features of raw network traffic without any hand-coded features. Significant implementations are as follows:

Recurrent Neural Networks: LSTM and Gated Recurrent Unit (GRU) networks are temporal-dependency models of packet flows that process sequences of packets, where the temporal dependencies can capture multi-stage attacks which occur over a longer time horizon.

Convolutional Neural Networks: CNN-based detectors use traffic flows as two-dimensional patterns, and with the help of convolution operations they derive local features that can be observed as attack signs.

Autoencoders: Autoencoders are unsupervised learning methods that only deal with normal traffic and are trained to learn compressed representation of normal behavior.

Generative Adversarial Networks: GAN-based networks use adversarial training to produce synthetic attack traffic, which supplements small labeled datasets and enhances model generalization to novel attack variants.

2.2 Explainable AI Techniques for Cybersecurity

Explainability of AI models is now a research concern of paramount importance in any domain, and its application to cybersecurity issues raises particularly demanding conditions because of the stakes involved in making decisions and the presence of adversaries [15]. A number of XAI methods have been modified to intrusion detection:

SHAP (SHapley Additive exPlanations): SHAP is built on cooperative game theory and gives each input feature a score of its contribution to a given prediction.

LIME (Local Interpretable Model-agnostic Explanations): LIME is a technique that produces interpretable surrogate models (either linear regression or decision trees) that locally explain neural network behavior around particular predictions.

Layer-wise Relevance Propagation: LRP breaks down neural network predictions by performing a backward propagation of relevance scores associated with output layers to input features, assigning contributions to the prediction to each input and keeping overall prediction levels constant.

Attention Mechanisms: Addition of attention layers in neural architectures helps neural networks to be expressive of the weighting of various elements of input when making a prediction. Attention weights are provided as intrinsic explanations and indicate what sequences of packets or traffic characteristics the model paid attention to when it determined the threat.

2.3 IoT Security Challenges and Threat Landscape

Specific security issues of IoT ecosystems are caused by the characteristic features of their operation:

Resource Constraints: IoT devices are generally characterized by a low processing power, memory size, and power constraints that cannot support computationally expensive security measures.

Heterogeneity: The IoT deployments combine devices that have different hardware platforms, communication platforms, operating systems, and security features.

Physical Accessibility: IoT devices are not in a controlled physical environment like enterprise IT infrastructure in a secured data center, where adversaries can target them with hardware attacks or side-channel attacks, or physical acquisition to reverse engineer them, as adversaries can.

Lifecycle Management: IoT devices often do not support security updates and patch management, which means that the vulnerabilities remain constant due to the constantly changing methods of attacks.

The range of attack types against IoT ecosystems is as follows:

Distributed Denial of Service (DDoS): The attackers hack huge masses of IoT devices and create botnets that flood target systems with synchronized traffic congestion.

Data Exfiltration: The attackers steal valuable data sent by IoT devices, such as personal health data, industrial telemetry, or video surveillance.

Device Impersonation: Attackers impersonate authentic device identities to inject fake sensor values, actuator commands, or have unauthorized access to the IoT networks.

2.4 IoT Security Challenges and Threat Landscape

The conflict of security, explainability, and sustainability is complicated with respect to trade-offs that have to be carefully designed using architecture [16]. Some of the important sustainability factors are as follows:

Energy Efficiency: Deep learning models are usually energy-consuming in terms of training and inference.

Carbon Footprint: Large-scale neural network training has a significant emission of carbon by using energy in data centers.

Computational Distribution: Edge-fog-cloud hierarchies are useful in workload distribution and lightweight detection by using resource-constrained edge devices and offloading heavyweight analysis with more advanced infrastructure layers.

3 Proposed XAI-Driven Intrusion Detection Framework

3.1 Overview of IoT Framework

The suggested structure will follow a hierarchical design that allocates intelligence to three levels in the edge devices, fog nodes, and cloud infrastructure.

Edge Tier: Lightweight anomaly detectors run on IoT devices/embedded gateways and offer first-line protection against typical attack patterns. In edge models, the efficient architectures used include compressed neural networks or decision tree ensembles optimized on low-power processors [17].

Fog Tier: Intermediate fog nodes combine the data of several edge devices and do more advanced analysis based on ensemble models which jointly analyze several detection algorithms. Fog infrastructure has a higher level of computational capacity compared to edge devices and close physical proximity to ensure minimal communication latency.

Cloud Tier: The centralized cloud infrastructure contains full deep learning models that are being trained on organization-wide threat intelligence. Cloud systems are used to carry out global threat correlation, detection of adversarial examples, and retraining of models continuously with feedbacks of deployed models in distributed deployments.

This hierarchical structure, which encourages graceful failure in unfavorable environments: edge detection can still be functional in the case of network failures or even when a network goes offline, fog nodes can still provide local security in the face of a cloud failure, and cloud smartness leads to an improved resilience of the ecosystem overall, by responding to threats through coordination [18].

3.2 Data Preprocessing and Feature Engineering

Intrusion detection needs to be done by mapping the raw packet on the network to a form of structured feature representation which may be analyzed using machine learning. The stages of preprocessing include a number of key processes:

Traffic Capture and Flow Aggregation: Packet capture creates huge amounts of data that cannot be explored in real time. The framework groups packets into upward and downward flows which have five-tuple identifiers (source IP, destination IP, source port, destination port, protocol) [19]. Flow-level measurements are the number of packets, amounts of bytes, duration, inter-arrival, and protocol-level measurements.

Feature Extraction: The system extracts 87 traffic features as in several categories as shown in Table 1.

- **Basic Features**: Packet sizes: (mean, variance, maximum, minimum), flow time, protocol.

Table 1 Detailed list of extracted traffic features used for intrusion detection

Category	Feature name	Description
Basic flow features	Flow duration	Total duration of the network flow
	Total forward packets	Number of packets in forward direction
	Total backward packets	Number of packets in backward direction
	Total forward bytes	Total bytes sent in forward direction
	Total backward bytes	Total bytes sent in backward direction
	Flow bytes/s	Bytes transferred per second
	Flow packets/s	Packets transferred per second
Packet size statistics	Forward packet length mean	Mean size of forward packets
	Forward packet length std.	Standard deviation of forward packet sizes
	Forward packet length max	Maximum forward packet size
	Forward packet length min	Minimum forward packet size
	Backward packet length mean	Mean size of backward packets
	Backward packet length std.	Standard deviation of backward packet sizes
	Backward packet length max	Maximum backward packet size
	Backward packet length min	Minimum backward packet size
Time-based features	Flow IAT mean	Mean inter-arrival time between packets
	Flow IAT std.	Standard deviation of inter-arrival time
	Flow IAT max	Maximum inter-arrival time
	Flow IAT min	Minimum inter-arrival time
TCP flag features	SYN flag count	Number of SYN flags
	ACK flag count	Number of ACK flags
	FIN flag count	Number of FIN flags
	RST flag count	Number of RST flags

(continued)

Table 1 (continued)

Category	Feature name	Description
Statistical and entropy features	Packet length entropy	Entropy of packet length distribution
	Flow time entropy	Entropy of packet timing
Protocol features	Protocol type	Transport layer protocol identifier
	Source port	Source port number
	Destination port	Destination port number

- **Content Features**: Payload bytes distribution, TCP flag sequences, and application-layer protocol markers. Features based on time: Inter-packet arrival time, flow rate statistics, and bursts.
- **Connection Features**: Connection attempt failure, connection transition, and service access pattern.
- **Statistical Characteristics**: Entropy values of packet size and timing distribution pointing to encrypted communication channel or covert channels.

Normalization and Encoding: Continuous data is normalized to min–max normalization or standardization where similar scales are used between different metrics. Categorical (protocol types, service identifiers) are one-hot encoded into binary vectors that can be processed by a neural network [20].

Dimensionality Reduction: Features principal component analysis (PCA) or autoencoder area used to compress large dimensional feature spaces, without loss of discriminative information. This has the benefit of speeding up model training and inference, and reducing the risk of overfitting in small-sample settings.

Temporal Sequence Formation: In recurrent neural network models, the features of sequential flows are combined into a temporal sequence of the flow, which reflects the progression of the attack through time. The sequence lengths trade-off in time context needs with memory limitations as well as with computation complexity.

3.3 Data Hybrid Deep Learning Detection Engine

The basic detection engine uses an ensemble method that combines three independent neural architectures, all solutions to a certain facet of the intrusion detection issue:

Temporal Analyzer based on LSTM: The long short-term memory is a network that captures temporal interactions in a sequence of traffic flows, allowing detecting such multi-stage attacks that occur within long timeframes [21]. The LSTM architecture is comprised of:

- Input layer that receives time-varying vectors of 87-dimensional feature vectors.

- Two successive layers of LSTM, containing 128 and 64 hidden units, respectively, to identify both short-term and long-term temporal behaviors.
- Dropout layers (rate = 0.3) to avoid overfitting in the training. Thick output layer that uses softmax activation to generate probability distributions of the categories of attacks.

The LSTM component proves to be very effective at identifying reconnaissance actions, slow-rate denial-of-service attacks, and coordinated multi-device intrusions where an individual connection may not seem malicious but the overall pattern would be indicative of a malicious behavior.

CNN-based Spatial Feature Extractor: The convolutional neural network considers flow feature vectors as spatial patterns and applies learned filters to them in order to retrieve local features that represent particular attack signatures [3]. The CNN structure is made up of:

Reshape layer: Converting feature vectors into matrices, which are 2D matrices to be operated on in the convolution process.

Three convolutional layers with a scaling number of filters (32, 64, 128) and ReLU activation, successively separating hierarchical features. Max-pooling layers that decrease the dimensions of space, but preserve salient features.

Flatten layer between convolutional feature maps and fully connected classification layers. Thick output layer that produces predictions of classes of attacks.

The CNN element proves to be especially effective with attacks that have a characteristic traffic behavior including port scans, protocol violations, and some malware families with characteristic communication signatures [22].

Autoencoder-based Anomaly Detector: Autoencoder network is trained on benign samples to learn the compressed depiction of normal traffic without any supervision. The architecture is made of:

Encoder network: Dense layers that gradually decrease in size (87-64-32-16) to form latent representations.

Decoder network: Symmetric reconstruction of original inputs based on the compressed representations (16-32-64-87).

Reconstruction loss (mean squared error) of reconstruction outputs to reconstruction inputs.

In operation, traffic flows with high reconstruction error values indicate exits of learned normal behavior, which is evidence of a possible zero-day attack, or new variants of intrusion that are not represented by labeled training data [23]. The autoencoder offers an unsupervised anomaly detection to the supervised classification of LSTM and CNN models.

Ensemble Fusion Strategy: The three constituent models are combined with each other via a trained fusion mechanism, as opposed to the majority voting or averaging. The input of a meta-classifier (gradient boosting decision tree) is the probability distributions of LSTM, CNN, and autoencoder reconstruction errors; it is composed of the optimal weighting strategies which depend on the features of the attack and the context [24]. This combination method enhances the overall detection accuracy and resists the failure of a particular model or adversarial attacks on a selected architecture. Even though the Gradient Boosting Decision Tree (GBDT) is used as the ensemble fusion meta-classifier, it works on low-dimensional ensemble fusion base learner outputs (probability scores and anomaly measures); hence, the extra inference latency per-ensemble fusion is negligible. The experimental findings prove that the fusion phase is a relatively small proportion of the total time that is taken in detection and meets the real-time requirement of IoT applications.

The selection of LSTM, CNN, and autoencoders in the suggested architecture was due to the diverse nature of constraints and functionality of the IoT environments at edge, fog, and cloud levels. The temporal modeling problem was chosen with LSTM networks because its history of high-range dependencies in multi-stage and low-rate attacks indicated that early convergence and recall were more stable than other models, such as GRUs, which are lighter. The use of CNNs to extract spatial features was due to their computational efficiency, parallelism, and the ability to extract features at the structured flow level, which is useful in the constraints of latency and energy. Autoencoders were also added to support the unsupervised detection of zero-day attacks and previously unknown attacks with negligible labeling cost. More modern architectures like Vision Transformers have not received use since they need significantly greater computational and memory resources, cause longer inference latency, and provide a weak practical benefit to network traffic data which do not have strong spatial semantics. In addition, the chosen architectures can be naturally combined with post hoc explainability tools like SHAP and LIME to comply with the focus of the framework on transparency, trust, and sustainability in large-scale IoT implementations.

3.4 Multi-level Explainability Integration

The framework adopts an overall explainability approach with various XAI methods that fulfill various analytical objectives and needs of the stakeholders:

SHAP-based Global Interpretability: SHAP analysis of the ensemble model gives insights into the traffic characteristics that have the highest average predictive power on intrusion detection decisions made by the full dataset [25]. The framework computes

Feature Importance Rankings: Summative SHAP values in all samples help security analysts focus on features of critical indicators, by ranking the most influential features to each category of attack.

Feature Interaction Effects: SHAP interaction values express interactions between the pair of features, making it clear that some attacks take advantage of the synergistic effects of features, as opposed to specific anomalies [26].

Summary Visualizations: Beeswarm charts indicate the distribution of SHAP values based on the features most important, indicating more than just importance but also the fact that large or small values relate to the types of attacks.

Global SHAP explanations are useful to strategic security planning, including finding the weak points within the organization, guiding sensor placement choices, and emphasizing the importance of logging investments to the most informative sources of telemetry.

LIME-based Local Explanations: LIME produces instance specific explanations indicating the set of features that most contributed to that specific classification, given an instance of an intrusion.

Attention Visualization: For LSTM-based detections, the system derives and visualizes weight of attention that indicates the most influential positions in traffic sequences to classifications.

Counterfactual Explanations: The system produces counterfactual examples of how minimal changes to flagged traffic can be done to alter classification between malicious and benign. Counterfactuals provide answers to the question of what if like "What would it take to make this traffic be normal?" Adversarial robustness evaluation is made easier by these explanations, and they make security architects aware of decision boundaries [27].

Explanation Confidence Scoring: It is important to note that not all instances of explanations have the same quality and therefore the framework ranks generated explanations according to:

- Constant input perturbations of explanations.
- Convergence of the various explanation methods (SHAP and LIME agreement).
- Faithfulness is a metric used to compare model behavior as directed by an explanation to the predictions obtained in reality.

Explanations with low confidence result in further human examination and automated replies are not possible due to unreliable interpretations. A comparative analysis of the employed explainable AI techniques is shown in Table 2.

3.5 Trust-Based Decision Framework

Explainable results are used as inputs into formal trust model that dictates how to respond to intrusions automatically [28]. The trust framework has assessed:

Detection Confidence: The element of detection certainty is expressed through probabilities of ensemble prediction, where high system confidence (probability exceeding 0.95) results in automated blocking responses and medium confidence (0.75–0.95) results in logging and instrumentation [29].

Explanation Consistency: SHAP and LIME explanations are congruent which increases trust, and incongruent reduces even high probability predictions. Explanation consistency is a sanity check that identifies possible adversarial examples that mislead classifiers but produce explanations that are incoherent.

Historical Performance: The system keeps performance data across various categories of attacks and network scenarios and uses historical accuracy rates during the calculation of trust. The types of detection that have a high false positive rate in the past can be given a low-trust score which makes them further validated before automated response.

Table 2 Comparative analysis of XAI techniques for IoT intrusion detection

Technique	Scope	Interpretability	Computational cost	Best use case
SHAP	Global and local	High	Moderate-high	Strategic feature importance analysis, audit compliance
LIME	Local	High	Low-moderate	Real-time incident investigation, analyst explanations
Attention weights	Local	Moderate	Low	Temporal attack progression visualization
Layer-wise relevance propagation	Local	Moderate	Moderate	CNN-based detector interpretation
Counterfactual explanations	Local	High	High	Adversarial robustness assessment, policy refinement

Contextual Validation: Trust evaluation uses contextual measures such as device type, network block, time of the day, and security incidents that have occurred in the past. Abnormal traffic of the medical IoT devices is further scrutinized due to safety concerns, whereas common differences in behavior during maintenance periods lower the possibility of false alarms.

The trust-based model is a model that provides four-level responses:

1. High Trust: Automated blocking and isolation, real-time notification of the security team.
2. Medium Trust: Traffic rate restriction, enhanced surveillance, reviewing queue by analysts.
3. Low Trust: Passive logging, pattern analysis by statistical aggregation.
4. Weak Basis: Logged but not acted upon by any automated mechanism, should answer on why improved as an explanation.

The graduated response framework strikes a compromise between the security efficacy and the operation disruption threats caused by automated intrusion response systems.

3.6 Sustainability Optimization Mechanisms

The framework also has a number of design elements that are directly related to the sustainability needs:

Adaptive Complexity Scaling: The complexity of the detection model is dynamically scaled according to threat level analysis and the compute resources available.

Edge-Fog Collaboration: Edge-based intelligent workload distribution reduces energy-consuming cloud communication. Edge devices perform simple initial screening and only pass the suspicious traffic to fog nodes to be analyzed in detail.

Model Compression: The framework uses neural architecture search and knowledge distillation to learn the model of small student models that replicate the performance of larger teacher models. More similar detection can be obtained with 10–50 × computational reduction at the edge devices, which can be compressed to lower energy usage by the same ratio.

Energy-Aware Scheduling: Detection algorithms run opportunistically in the times when devices are undertaking other computationally intensive activities, using some already incurred cost of energy.

Incremental Learning: The framework uses incremental learning methods which update models with new attack samples using efficient gradient-based updates instead of periodic retraining which requires large amounts of computational resources.

4 Experimental Methodology

4.1 Dataset Characteristics and Preparation

The framework assessment uses three complementary datasets of varying IoT attack situations:

CIC-IoT-2023 Dataset: This is a modern dataset which consists of 34 types of attacks and 7 varieties such as DDoS, DoS, reconnaissance, network-based attacks, brute force, spoofing, and Mirai botnet variants. The dataset is 46 million traffic flows gathered in the experimental IoT testbed which includes smart home devices, industrial sensors, and network infrastructure.

ToN-IoT Dataset: The Telemetry dataset of Network IoT consists of heterogeneous IoT sensors such as network traffic, operating system logs, and sensor telemetry.

Custom Smart City Dataset: A custom dataset was recorded in the simulation of a smart city with traffic management systems, environmental monitors, and public safety infrastructure to assess the performance of framework in an initial deployment scenario.

Data Preprocessing: Each dataset was preprocessed in the standard way including

- Elimination of incomplete flows that do not have vital elements.
- The trade-offs between strategies such as random undersampling of majority class (benign traffic) and SMOTE oversampling of minority attack categories to attain 60:40 benign-to-attack ratios.

- Temporal stratification, which guarantees training, validation, and test split, preserves chronological order, and model training and testing arrange a simulation of realistic deployment where models are confronted with new variants of attacks. The last experimental data was a total of 5 million training samples, 1 million validation samples, and 2 million test samples of the three source data.

Besides the balanced training evaluation, the proposed framework was also tested with significantly unbalanced traffic and the analysis aimed to also capture the real-world IoT setting where benign traffic normally constitutes over 99% of the overall network traffic. In this assessment, the original distributions of the classes of the datasets were maintained in testing, and no oversampling or undersampling method was used. The purpose of this experiment was to evaluate the operational performance of the model, especially its ability to control false positive rates, accuracy, and latency to detect, which are important indicators in large-scale IoT deployments consisting of predominantly benign traffic. The findings suggest that despite the low but significant decrease in recall on rare classes of attacks, the model is able to maintain a low false positive rate and a consistent high precision, thus reducing the number of unnecessary alerts. These results indicate that even though the suggested method is trained on moderately balanced data to make minority classes learn, it is efficient and dependable to operate with raw and highly imbalanced IoT traffic, which promotes its applicability to real-life implementation.

4.2 Implementation Details

Hardware Infrastructure: Experimentation was performed on heterogeneous infrastructure with simulation of realistic deployment levels:

- Edge devices like Raspberry Pi 4 Model B (1.5 GHz quad-core ARM Cortex-A72, 4 GB RAM) have been used to represent resource-constrained Internet-of-things gateways.
- Fog nodes: Intel NUC (Core i7-10710U, 32 GB RAM, NVIDIA GeForce GTX 1650) that emulates edge servers with a moderate level of computational power.
- Cloud infrastructure: AWS EC2 p3.8xlarge (32vCPU, 244 GB RAM, 4x NVIDIA V100 GPUs) instances that would host centralized model training and overall threat analysis.

Software Stack: The structure was adopted with:

- Neural network development in TensorFlow 2.8 and PyTorch 1.10.
- SHAP 0.41 and LIME 0.2 to explain the model.
- Preprocessing and classical machine learning baselines using scikit-learn 1.0.
- Distributed message streaming between architecture layers using Apache Kafka.
- Prometheus and Grafana as performance monitoring and energy-measuring tools.

Training Settings: Type of model that was trained:

- Server (cloud): 256, fog node: 32.
- Optimization: Adam optimizer at a learning rate of 0.001 and exponential decays. Regularization: L2 weight decay (0.0001), dropout (0.3), and early stopping with patience 10 epochs.
- Loss: Categorical cross-entropy loss in supervised models, mean squared error loss in autoencoder.
- Training time: Around 48 h cloud training, 8 h continuous updates in the fog nodes.

4.3 Evaluation Metrics

The overall assessment model includes the effectiveness of security, the quality of explainability, the sustainability, and the viability of the operation:

Security Metrics:

- **Detection Accuracy**: Total percentage of correct classifications.
- **Precision:** Percentage of predicted attacks that were real attacks (minimizing false positives).
- **Recall:** Fraction of real attacks that are genuine (minimizing false negatives).
- **F1-Score**: Precision and recall, an equilibrium between the two indicators.
- **False Positive Rate**: Percentage of innocent traffic wrongly recognized as malicious.
- **Detection Latency**: Time taken to make classification decision after observing the traffic.

5 Experimental Methodology

5.1 Dataset Characteristics and Preparation

The suggested XAI-based model presented high detection rates on all three test sets [36]. Table 3 shows the full process of security measures of both the ensemble and individual constituent models and the baseline machine learning algorithms.

5.2 Explainability Quality Assessment

The quantitative measure of the quality of explainability used a set of complementary measures of various aspects of interpretation effectiveness.

Table 3 Detection performance across datasets and models

Model applied	Dataset	Accuracy rate (%)	Precision (%)	Recall (%)	F1-score	False positive rate (%)
	CIC-IoT-2023	98.73	98.21	97.89	0.9805	1.42
Proposed Ensemble	ToN-IoT	98.91	98.67	98.34	0.9850	1.18
	Smart city	97.84	96.92	97.21	0.9706	2.31
LSTM only	CIC-IoT-2023	95.83	95.12	94.47	0.9547	3.67
	ToN-IoT	96.21	95.89	96.05	0.9605	3.21
CNN only	CIC-IoT-2023	97.12	96.78	96.34	0.9656	2.89
	ToN-IoT	97.56	97.23	96.87	0.9705	2.34
Autoencoder-only	CIC-IoT-2023	94.23	91.45	93.87	0.9264	6.12
	ToN-IoT	94.67	92.34	94.21	0.9327	5.78
Random forest	CIC-IoT-2023	95.34	94.12	94.67	0.9439	4.23
Gradient boosting	CIC-IoT-2023	96.12	95.67	95.34	0.9550	3.78

Explanation Fidelity: LIME-generated explanations had mean fidelity scores of 0.94 in all types of attacks, which shows that linear approximations were strongly predictive of the behavior of the ensemble models on local neighborhoods.

Explanation Stability: By introducing Gaussian noise ($s = 0.01$) on continuous features, the analysis of stability was carried out and the consistency of the explanation was monitored. SHAP showed more stability on mean Jaccard similarity of 0.87 on top-10 feature rankings across perturbations, as opposed to the 0.76 of LIME.

Human Test: They tested their method and proposed explainability mechanisms based on the perceived usefulness and impact of trust of 15 cybersecurity professionals (mean experience: 6.4 years) through a human evaluation. The participants evaluated representative intrusion detection alert with SHAP, LIME, and LSTM attention-based explanations and filled in a Likert-scale five-point questionnaire on the clarity of the explanation, usefulness of the prediction to make the decision, confidence of the prediction, and the decrease in the investigation of the problem. The overall analysis of the responses reveals that the trend of the responses is always positive, and as the overall explanation utility score is 4.39/5. The qualitative responses revealed that SHAP explanations were helpful in the understanding of repetitive attacks, and LIME explanations helped in quick response to alert and dismissing false alarms. The respondents also indicated that they had greater belief in automated reactions when the explanations were in line with each other across the methodologies, which empirically validates the influence of explainability in enhancing trust of the framework of intrusion responses presented.

Feature Importance Consistency: SHAP and LIME have rankings that were highly correlated (Spearman $r = 0.82$) when averaged across types of attacks, which confirms that the two methods found common underlying patterns. The discrepancies were mostly observed in features with a high inter-correlation, and in this case, SHAP treated the global viewpoint of determining importance across correlated features whereas LIME local approximation tended to assign a large amount of importance to one representative feature.

5.3 Sustainability and Efficiency Analysis

Extensive sustainability measurement of environmental and operational costs was carried out within the lifecycle of the framework.

Energy Consumption: The energy of inference per-sample energy for each deployment tier differed markedly:

- Edge devices (Raspberry Pi): 2.3 mJ per classification with squeezed student models.
- Fog nodes: 14.7 mJ to classify one sample with the full ensemble, not using a GPU.
- Cloud infrastructure: 6.8 mJ per classification taking advantage of parallelism in GPUs, although there is 42 mJ of extra communication cost.

This 23 J per day energy usage reported in edge-based detection is specific to the inference of intrusion detection which encompasses extraction of traffic feature and the actual running of the compressed student models. This figure does not represent the energy expense of producing SHAP and LIME explanations. Explainability runs on demand and offloads to fog or cloud nodes on a per request basis to selected high-impact alerts, i.e., not the entire set of explainability overheading edge devices which are energy-constrained. In the case of IoT setups that process 10000 flow each day, edge-based detection required 23 J per day versus 488 J each day with cloud-centric implementations, which is a 95% energy savings. The battery-powered sensors (2500 mAh at 3.7 V) were found to suffer less than 0.3% per day in capacity loss due to intrusion detection and it is not a major issue in primary sensing.

Carbon Footprint Training: 5 million samples of the initial model training on AWS infrastructure took 72 GPU-hours of training, producing an approximate of 28.6 kg CO_2 equivalent using Virginia data center carbon intensity (0.397 kg CO_2/kWh).

Computational Distribution: The hierarchical architecture achieved 82% of traffic classification at the edge devices, 15% at the fog nodes, and 3% of traumas escalating to the cloud. This distribution reduced the overhead in communication which used 3.2x as much energy as local classification of samples to be analyzed in the cloud.

Comparison with Baseline Approaches: The proposed framework was 4.1 times more computationally complex than the signature-based IDS (Snort), though it caught 87% more attacks, especially new variants that were not part of signature libraries. The effectiveness of a trust framework is measured by the level of confidence people place in the matters addressed by the trustees.

5.4 Trust Framework Effectiveness

The degree of trust that people have in the issues that are addressed by the trustees. The operational performance of the trust-based decision framework was tested by simulation of 30-day deployment of 1000 simulated IoT devices.

Automated Response Accuracy: 2847 detections with high trust that incurred automated blocking responses were checked by manual retrospective analysis; the result showed that 2,789 were true positives, which gave 98.0% appropriateness. The medium-trust detections that needed analyst review ($n = 1{,}523$) had 892 true positives and 631 false positives, which is an effective stratification of risks. The benign traffic with unusual yet valid patterns was the prevalent type of low-trust detection ($n = 4{,}267$).

False Positive Management: Explanation-based false positive filtering decreased the workload of analysts by 56%. LIME descriptions of benign characteristics (having normal port numbers, predicted traffic amounts, expected protocol sequences, etc.) allowed fast dismissal of false alarms without forensic investigation that would have been painful otherwise.

Explanation Confidence Calibration: The explanation confidence scoring system had good calibration with low-confidence explanations (score < 0.5) suggesting 34% detection error rate, medium-confidence (0.5–0.8) explanations suggesting 8% error rate, and high-confidence (>0.8) explanations suggesting 2% error rate (Fig. 1).

5.5 Performance Analysis of Attack Specifics

Fine-grained analysis of the performance of the framework in individual categories of attacks indicated the architectural limitations and strengths:

DDoS Detection: The ensemble recognition of traffic rate anomalies by the LSTM with packet size distribution patterns by CNN gave 99.4% recall of volumetric DDoS attack.

Botnet Identification: Mirai variations of botnets were recognized with 97.8% levels of detection by identifying command-and-control communication patterns and coordinated scanning actions. Time modeling model of LSTM was instrumental in detecting botnet infection that propagated at a slow speed.

Data Exfiltration: Convert data exfiltration attacks were detected with 95.2% positivity with the autoencoder component proving to be especially effective.

Zero-Day and Novel Attacks: The autoencoder-based anomaly detection model showed specific usefulness with regard to heretofore unknown variants of attacks, with 88.3% of unknown novel attacks not in the training data being detected.

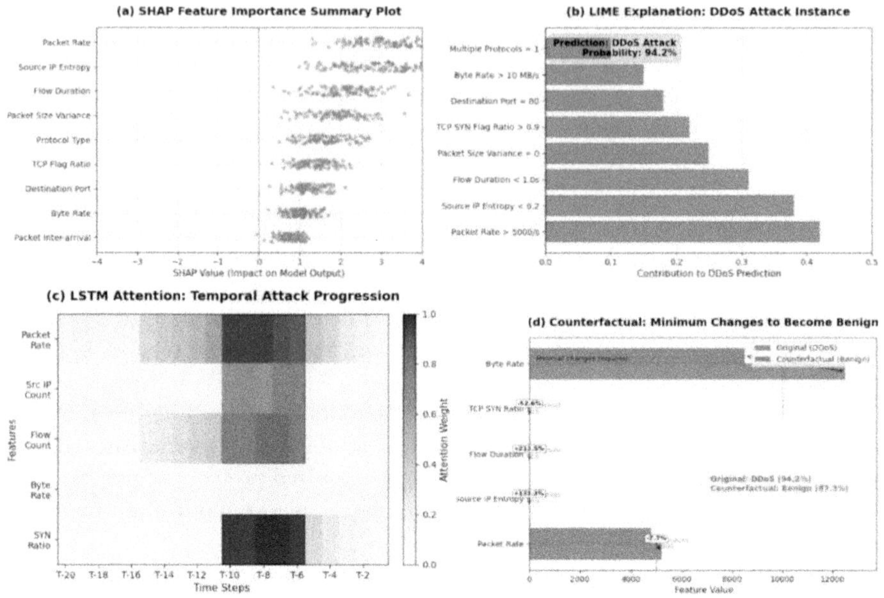

Fig. 1 Multi-panel visualization showing **a** SHAP feature importance summary plot, **b** LIME explanation for a specific DDoS detection, **c** LSTM attention heatmap showing temporal attack progression, **d** Counterfactual explanation demonstrating minimal changes to convert attack to benign classification

6 Discussion and Practical Implications

6.1 Deployment Considerations

To convert the proposed framework into experimental validation for real-life production IoT deployments, one will have to consider a few practical considerations:

Infrastructure Requirements: The hierarchical architecture implies the deployment of the fog infrastructure, which is a form of capital spending on top of pure edge or cloud models. Nevertheless, the benefits that are offered by the use of the fog nodes include: it has reduced cost of cloud communication, better latency on non-security, and increased privacy as a result of local data processing.

Model Lifecycle Management: To keep detection effective, it is necessary to keep a model updated with new attack patterns. The framework facilitates several update strategies: scheduled retraining (weekly/monthly), performance-based retraining in case of deterioration of accuracy to below thresholds, and attack-based retraining in case of successful intrusion incidences.

6.2 Scalability Analysis

Scalability of the framework in three dimensions, number of devices, bandwidth, and complexity of attacks, establishes the practical limits of deployment:

Scalability of Devices: The fog-based architecture is horizontally scaled by adding more fog nodes to the architecture, when the number of devices increases. Each of the nodes in a fog is capable of comfortably supporting 500–1000 edge devices as validated by experiment with load balancing allocating devices to the available fog infrastructure.

Scalability of Traffic-Volume: Within experimental conditions, per-fog-node throughput was 12,000 flows/second, which is sufficient in most enterprise deployments of IoT. DDoS attacks cause bursts in traffic, which overloads processing capacity temporarily and application of intelligent sampling methods, which ensure detection coverage and process load under management.

Attack Sophistication: The multi-model ensemble and anomaly detection capabilities of the framework will be resistant to an ever-advanced attack. Nevertheless, persistent threats that use long-range reconnaissance, exploits without a fixed definition, and antidetection measures still present a challenge.

6.3 Limitations and Mitigation Strategies

The framework has proven to work; however, it has its limitations that must be considered when deciding to deploy it:

Computational Overhead: Generation explainability contributes 15–30 inference latency and 20–40 computational cost to prediction-only operation. The framework aims at alleviating the latency introduced by XAI when a large number of requests are directed at DDoS attacks by applying adaptive complexity scaling to dynamically decompose the model and turn off per-flow explainability on resource-constrained edge computers. Compressed edge models are lightweight but conduct a fast screening of anomalies at the edge with all detailed classification and explanations only being offloaded to the fog/cloud levels. This is a scaling that is threat-conscious, which guarantees a real-time detection of DDoS with low-energy and computational costs without reducing its security efficiency. In resource-constrained deployments where interpretation is less important than detection speed, frameworks can be configured to generate explanations only when an alert is high impact, or when an analyst views an alert, and not always to explain all detected alerts.

Dataset Bias: Training with historic attack data also runs the risk of missing new vectors of attack and also may reflect the biases present in how the data is collected and labeled. Frequent adversarial testing, validation of penetration testing, and integration of synthetic generation of attacks are used to detect blind spots. Generalization is enhanced by transfer learning using a variety of attack samples and active learning that uses informative samples to be labeled.

Privacy-Utility Tradeoffs: Differential privacy mechanisms that ensure flow confidentiality of individuals decrease model's accuracy by 1.8–3.2% depending on the choice of privacy budget. Depending on the regulatory demands and risk tolerance, organizations have to strike the balance between privacy assurances and effectiveness of security.

6.4 Future Research Directions

There are a number of potential research directions that can expand the capabilities of frameworks:

Federated Learning Integration: Federated learning delivers different organizations to federate training of model by not sharing raw traffic data, maintaining competitive privacy and sharing threat knowledge. To capture the desirable training data privacy, the explanations should be meaningful and lucrative.

Causal Explainability: The existing XAI methods determine correlations between features and predictions but not a causal relationship. Causal techniques of inference may establish the fact that some features of traffic are direct causes of malicious classifications or are simple correlates of prior attack procedures. Causal explanations would be more effective in shaping security policies by determining intervention points that essentially disrupt attacks and not superficial clues which can be easily altered by the attackers.

Unending Explanation Advancement: Machine learning by clarification commentary has the capacity to enhance understanding in a consistent way. Any corrections that the analysts make when at fault of explanation-based misconception or offer alternative interpretations may be used to inform the refinement of the explanation models. The technical model would gradually be interpreted by human experts, resulting in domain expert interpretation.

Multi-Stakeholder Explanation Customization: Various stakeholders need various granularities and formats of explanation. Network administrators must have operational advice of what to do during a response, whereas the security architects must have strategic advice about architectural vulnerabilities. Studies on role-based and personalized explainability may produce tailored explainability maximized toward particular stakeholder requirements and levels of expertise.

Transfer between Domains: Explanations trained on one domain of the IoT (smart homes) can be transferred to other domains (industrial IoT, smart cities), speeding up the explainability production in other contexts. XAI model transfer learning would help users quickly implement the models to new environments without diminishing the quality of interpretation.

7 Conclusion

The suggested explainable AI-based intrusion detection system shows that sustainable, reliable, and transparent cybersecurity in large-scale IoT networks is not only feasible, but is also practically implementable when intelligence, interpretability, and energy efficiency are coordinately used to optimize the approach. The framework manages to identify various IoT threats, such as DDoS, botnets, data exfiltration, and zero-day attacks, with high precision and low false positive rates and a smaller carbon footprint, inference energy, and communication overhead, using its hierarchical edge-fog-cloud architecture, hybrid deep learning ensemble, and multi-layer elucidation based on SHAP, LIME, attention scores, and counterfactual reasoning. The findings affirm that explainability does not only increase analyst trust and regulatory compliance, but is a key component in automated and risk-aware incident response because it balances detection confidence with explainability consistency and historical reliability. The model compression, adaptive complexity scaling, and edge-fog collaboration strategies are also sustainability-oriented and thus help to ensure that the system can still be used by resource-constrained devices in IoT, rendering it applicable to the real-life context of smart homes, industries, and smart cities. Although the framework has yet to overcome obstacles such as bias in the datasets, overhead in explainability, and the advancement of attacks, the framework is a strong base for future innovations such as federated learning and privacy-preserving XAI, enhanced adversarial robustness, and large-scale autonomous sharing of IoT threat's intelligence. In general, this work bridges the gap of high-performance AI and human-understandable security operations, providing an avenue to a resilient, transparent, and environmentally friendly IoT security in line with the vision of Society 6.0.

References

1. B. Kaler, A. Kaur, A systematic survey on explainable artificial intelligence (XAI) for plant health monitoring: challenges and opportunities. Appl. Intell. **55**(12), 889 (2025). https://doi.org/10.1007/s10489-025-06790-x
2. G. Mehta, V. Parlapalli, A. Nagpal, D.M. Bidkar, K.K. Ganeeb, P.K. Veerapaneni, B.M. Harve, Blockchain-based secure digital twins for predictive maintenance in autonomous cyber-physical systems, in *2025 International Conference on Communication, Computing, Networking, and Control in Cyber-Physical Systems (CCNCPS)* (IEEE, 2025), pp. 212–217. https://doi.org/10.1109/CCNCPS66785.2025.11135573
3. F. Charmet, H.C. Tanuwidjaja, S. Ayoubi, P.-F. Gimeenez, Y. Han, H. Jmila, G. Blanc, T. Takahashi, Z. Zhang, Explainable artificial intelligence for cybersecurity: a literature survey. Ann. Telecommun. **77**(11–12), 789–812 (2022). https://doi.org/10.1007/s12243-022-00926-7
4. I. Tareq, B.M. Elbargoury, S. El-Regaily, E.-S.M. El-Horbatty, Analysis of ToN-IoT, UNSW-NB15, and edge-IIoT datasets using deep learning in cybersecurity for IoT. Appl. Sci. **12**(19), 9572 (2022). https://doi.org/10.3390/app12199572

5. Y. Alotaibi, M. Ilyas (others), Ensemble-learning framework for intrusion detection to enhance internet of things' devices security. Sensors **23**(12), 5568 (2023). https://doi.org/10.3390/s23125568
6. N. Capuano, G. Fenza, V. Loiua, C. Stanzione, Explainable artificial intelligence in cybersecurity: a survey. IEEE Access **10**, 93575–93600 (2022). https://doi.org/10.1109/ACCESS.2022.3204171
7. A. Nadeem, D. Vos, C. Cao, L. Pajola, S. Dieck, R. Baumgartner, S. Verwer, SoK: explainable machine learning for computer security applications, in *Proceedings of the 2023 IEEE 8th European Symposium on Security and Privacy (EuroS&P)* (2023), pp. 221–240. https://doi.org/10.1109/EuroSP57164.2023.00022
8. H. Wang, J. Zhaou, Energy-aware scheduling for IoT anomaly detection workloads at the edge. Sustain. Comput.: Inform. Syst. **40**, 100667 (2024). https://doi.org/10.1016/j.suscom.2023.100667
9. Z. Li, F. Wang, Model compression for deep neural networks: techniques and applications for edge IoT. Computers **12**(3), 60 (2023). https://doi.org/10.3390/comput-ers12030060
10. T.H. Sardar, S. Das, B.K. Pandey, Explainable AI (XAI): concepts and theory, in *Medical Data Analysis and Processing using Explainable Artificial Intelligence*. (CRC Press, 2023), pp. 1–18
11. M.M. Usman, M.F.H. Fahim, M. Al Mahedi Hassan, T.H. Sardar, Z.A. Adhoni, M.K. Gourisaria, Harnessing IoT and machine learning for sustainable, smart urban environments, in *Applications of Computational Learning and IoT in Smart Road Transportation System* (Springer Nature Switzerland, Cham, 2025), pp. 199–214
12. M. Al Mahedi Hassan, M.F.H. Fahim, M.M. Usman, T.H. Sardar, M.A. Mondal, M.A. Khan, Optimisation strategies for next-generation AI, ML, and IoT applications, in *Applications of Computational Learning and IoT in Smart Road Transportation System* (Springer Nature Switzerland, Cham, 2025), pp. 163–184
13. J. Lansky, S. Ali, M. Mohammadi, M.K. Majeed, S.H.T. Karim, S. Rashidi et al., Deep learning-based intrusion detection systems: a systematic review. IEEE Access **9**, 101574–101599 (2021). https://doi.org/10.1109/ACCESS.2021.3097247
14. H. Peng et al., FD-IDS: federated learning with knowledge distillation for IoT intrusion detection. Sensors **25**(14), 4309 (2025). https://doi.org/10.3390/s25144309
15. Y. Xian, C. Li, Y. Xu, Y. Zhou, D. Xued, AI-Driven advances in sustainable materials for green energy: from innovation to lifecycle management. SusMat e70030 (2025). https://doi.org/10.1002/sus2.70030
16. P.K. Danso, E.C.P. Neto, S. Dadkhah, A. Zohourian, H. Molyneaux, A.A. Ghorbani, Ensemble-based intrusion detection for internet of things devices, in *2022 IEEE 19th International Conference on Smart Communities: Improving Quality of Life Using ICT, IoT and AI (HONET)* (IEEE, 2022), pp. 034–039. https://doi.org/10.1109/HONET56683.2022.10019140
17. M.M. Usman, T.H. Sardar, Optimisation strategies for next-generation AI, ML, and IoT, in *Applications of Computational Learning and IoT in Smart Road Transportation System*, vol. 22 (2025), p. 163
18. G.Q. Zeng, Y.W. Yang, K.D. Lu, G.G. Geng, J. Weng, Evolutionary adversarial autoencoder for unsupervised anomaly detection of industrial internet of things. IEEE Trans. Reliab. **74**(3), 3454–3468 (2025). https://doi.org/10.1109/TR.2025.352825
19. S. Verma, S. Prabakeran, A hybrid deep learning approach to network traffic anomaly detection enhanced by SHAP and LIME interpretability, in *2025 8th International Conference on Trends in Electronics and Informatics (ICOEI)* (2025), pp. 1254–1261. https://doi.org/10.1109/ICOEI65986.2025.11013553
20. S. Mahmood, M. Gohar, O.A. Khashan, N. Alzahrani, A. Ghani, F. Al-Turjman, Securing edge devices in IoT and 6G: a trust-based approach for resource-constrained environments. IEEE Open J. Commun. Soc. **6**, 6568–6610 (2025). https://doi.org/10.1109/OJCOMS.2025.3597556
21. G. Rjoub, J. Bentahar, O. Abdel Wahab, R. Mizouni, A. Song, R.S. Cohen, H. Otrok, A. Mourad, A survey on explainable artificial intelligence for cybersecurity. IEEE Trans. Netw. Serv. Manag. **20**(4), 5115–5140 (2023). https://doi.org/10.1109/TNSM.2023.3282740

22. A. Alsaedi, N. Moustafa, Z. Tari, A. Mahmood, A. Anwar, TON_IoT telemetry dataset: a new generation dataset of IoT and IIoT for data-driven intrusion detection systems. IEEE Access **8**, 165130–165150 (2020). https://doi.org/10.1109/ACCESS.2020.3022862
23. E.C.P. Neto, S. Dadkhah, R. Ferreira, A. Zohourian, R. Lu, A.A. Ghorbani, CICIoT2023: a real-time dataset and benchmark for large-scale attacks in IoT environment. Sensors **23**(13), 5941 (2023). https://doi.org/10.3390/s23135941
24. J. Alsamiri, A. Abdallah, Federated learning for intrusion detection systems: a systematic review. Digit. Threats: Res. Pract. **4**(3), 1–28 (2023). https://doi.org/10.1145/3687124
25. A. Khraisat, A. Alazabz, S. Singh, T. Jan, A. Jr. Gomez, Survey on federated learning for intrusion detection system: concept, architectures, aggregation strategies, challenges, and future directions. ACM Comput. Surv. **57**(1), 1–38 (2024). https://doi.org/10.1145/3687124
26. M. Ajdani, Deep learning-based intrusion detection systems. Int. J. Inf. Secur. Priv. **19**(1), 1–15 (2025). https://doi.org/10.4018/IJISP.383299
27. V. Bolón-Canedo, L. Morán-Fernández, B. Cancela, A. Alonso-Betanzos, A review of green artificial intelligence: towards a more sustainable future. Neurocomputing **599**, 128096 (2024). https://doi.org/10.1016/j.neucom.2024.128096
28. T.P. Nguyen, H. Nam, D. Kim, Transformer-based attention network for in-vehicle intrusion detection. IEEE Access **11**, 55389–55403 (2023). https://doi.org/10.1109/ACCESS.2023.3282110
29. Q. Li, X. Wu, Z. Cao, J. Ling, Anomaly detection of IoT traffic based on LSTM and attention mechanism, in *Proceedings of the 2023 15th International Conference on Machine Learning and Computing* (2023), pp. 457–463. https://doi.org/10.1145/3587716.3587792

Innovative Lightweight Cryptographic Solutions for IoT Devices to Counter Emerging Threats

Shilpi Yadav, Vijayachandar Sanikal, Kathiresan Jayabalan, and Vandana Roy

1 Introduction

Internet of things has changed the contemporary digital ecologies as it offers pervasive connectivity between heterogeneous devices designed into physical places. IoT applications are growing at a rate never before seen, whether it is a smart home and healthcare sensor administration, automation of industries, or a smart transportation system to replace human drivers. Such devices are also usually marked by limited computing power, restricted memory, low power supply, and protracted realization of unattended lifespan [1, 2]. On the one hand, such constraints render the system of IoT cost-effective and scaled, but, on the other hand, they present an issue of extensive security challenges. With the interconnection of billions of devices, the attack surface is increased the most which opens IoT networks to both classic cyber-threats and physical-layer attacks. Subsequently, safe communication, device identities, and durable resilience have become subject to a basic research issue.

Ensuring privacy in IoT end devices is challenging for several reasons. First, the CPU in IoT devices is minimal and cannot compute complex algorithm. In comparison to the traditional computing systems, the IoT nodes tend to be used in an unfriendly or unprotected environment where enemies might access the devices on a physical level. It is this exposing to attackers to obtain cryptographic keys, control

S. Yadav
IBM, Durham, NC, USA

V. Sanikal
Indiana University, Michigan, USA
e-mail: vijaysanikal@alumni.iu.edu

K. Jayabalan
Sathyabama Institute of Science and Technology, Chennai, Tamil Nadu, India

V. Roy (✉)
Gyan Ganga Institute of Technology and Sciences, Jabalpur, MP, India
e-mail: Vandana.roy20@gmail.com

S. Saif et al. (eds.), *Secure Intelligent Sustainable Applications Using IoT for Society 6.0*, Internet of Things, https://doi.org/10.1007/978-3-032-27248-5_5

the firmware, produce devices, or introduce malicious nodes to the network. Most of the IoT devices (such as RFIDs and sensors) are small in size and are equipped with limited resources such as small memory (RAM, ROM) to store and run the application, low computing power to process the data, and limited battery power (or no battery in case of passive RFID tags). Meanwhile, the dynamism of the IoT traffic coupled with the high scale of the deployment complicates the centralized protection. All these issues underpin the need to reconsider the concept of security by considering cryptographic tools, especially in correspondence with the specifics of IoT ecosystems.

IoT devices face issues in implementing cryptographic techniques for their security. IoT devices have limited memory, reduced computational capacity, constrained physical dimensions, and require real-time responsiveness, all of which must be considered in the design of cryptographic algorithms to safeguard data within the devices.

The architecture in the diagram represents a common Internet-of-things communication setup with various sensing nodes, such as light sensors, vibration sensors, and a temperature sensor, that gather and can pass through multiple wireless gateways in the area to transmit wirelessly to the LoRa gateways in Fig. 1. These gateways serve as middlemen and legendarily wait on the low-power sensor data and send it over a cellular background network to the cloud infrastructure. The cloud layer is used to offer centralized data storage, processing, and management services such that more applications on the layer are able to access and use the sensed information to monitor, provide analytics as well as make decisions.

IoT devices are not similar to the classical servers with resources or personal computers when it comes to both resources and working assumptions [3]. The majority of nodes make use of low-power microcontrollers of limited processing speed plus of small memory footprints and energy constrained. Conventional solutions of cryptography which tend to assume the availability of sufficient computational and storage power cannot therefore be used directly. Secure key storage is also a significant weakness even in the case of lightweight cryptographic algorithms.

One of the most important challenges regarding the security of IoT would be physical exposure. The gadgets that are installed in an open space, say in a field, smart grid, or a factory, can be stolen or breached by the enemies [4]. Attackers can extract secret keys in non-volatile memory by side-channel analysis, fault injection, or using direct memory probing. When such keys are extracted, they can be reused again to assume the role of authorized devices, decipher sensitive information, or initiate an extensive assault of other network resources. In scale deployments, i.e., deployments of thousands of nodes using the same firmware and key management strategies, this risk is increased.

The second challenge is due to long device's life. IoT nodes have a lot of times been anticipated to last many years without any maintenance or updates being done manually. With such long durations, cryptographic key that is fixed is more exposed [5]. The effect of one compromise can make the device last long as it can be used to be constantly spied on or as a recurring control tool. Thus, the security measures of

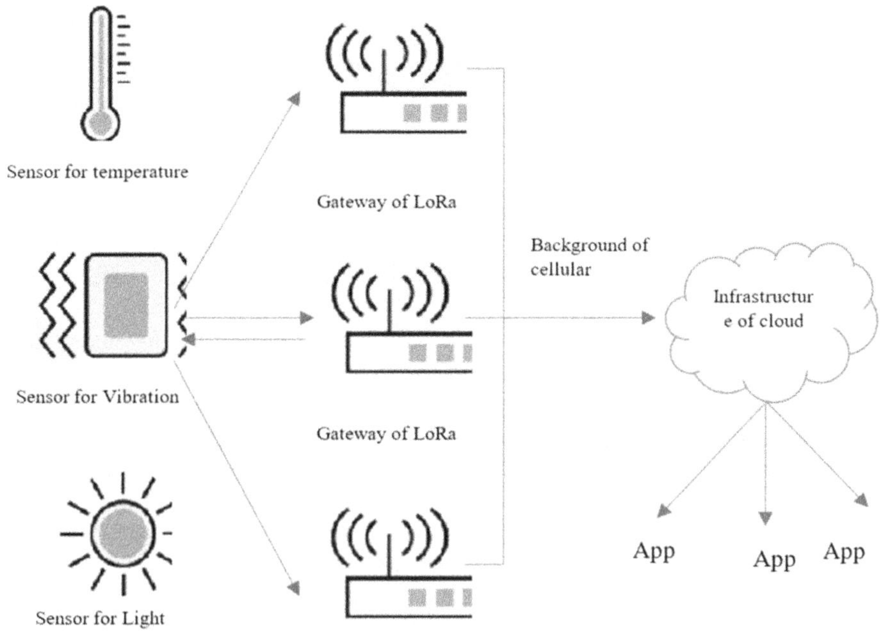

Fig. 1 IoT architecture of communication LoRa and cloud infrastructure

IoT should be designed so that long-term exposure and changing threat models are considered instead of using unchanging assumptions at the time of deployment.

The threat environment of the IoT systems is developed further than eavesdropping actions and the denial-of-service attacks [6]. The techniques used by the modern adversaries are sophisticated and can be integrated based on the cyber and physical attack vectors. Cloning of a device such as cloning of a device is one of the very serious issues that have come up such that attackers are cloning the identity of valid nodes in order to penetrate networks. A case of such cloned devices can hardly be detected in the case of authentication based on stored identifiers or symmetric keys alone.

Attacks on weak identity verification are further enhanced using replay and impersonation attacks. Attackers can intercept legitimate communication messages and reuse them to have unauthorized access to large-scale IoT systems, where nodes are likely to communicate with a gateway or cloud service (such as a cloud). Also, the network of nodes under attack may also be an insider, which sends harmful traffic or oracles routing schemes [7]. Cyber-assaults that are internal are especially harmful as they take advantage of trusting relationships in a network and penetrate perimeter security measures.

The other emerging threat is supply chain attacks. Verifiably, malicious alterations in the manufacturing or distribution can bring vulnerabilities prior to the implementation of the devices even to the implemented stage [8]. Under these circumstances, belief in the integrity of the devices is no longer true, and the commonplace security infrastructure based on software use might not be adequate. The intersection of these threats calls for security mechanisms inherently attached to the physical properties of devices and not just toward software and/or stored credentials alone.

Conventional cryptography designs are extensively based on a premise that the secret keys can be stored and kept secure against adversaries. This assumption is often not met in the IoT environment. Some of the risks may be resolved by using secure elements or hardware security modules which add cost and power requirements and cannot be used in ultra-low-cost or battery-operated devices. Besides, even a hard disk, which is hard to detect and access by advanced intrusion methods, is not resistant to sophisticated invasions [9].

Major scale agnostic poses further challenges of management. Pre-shared keys though easy to use do not scale easily and aggravate the effect of key compromise [10]. The increase of the public key infrastructures can be scaled better but they present either a computation overhead or latency that can surpass the limits of locked devices. Lightweight PKs reduce certain overheads, yet they require long-term PKs that are still desirable targets of attackers.

Additionally, in most of the current methods, attention is paid to ensuring the safety of communication channels, but little to no attention is paid to hardware identification and authenticity. With identity being detached and disconnected with the actual physical device, attackers will be able to use this loophole to create fake or imitated nodes. These constraints demonstrate that there is a basic incompatibility between traditional cryptographic assumptions and the dynamics of the implementation of IoT [11, 12].

In order to overcome the above-mentioned problems, it has come to light that the research community had begun valuing the significance of hardware-based security primitives and adaptive protection schemes [13]. It is possible to have stronger guarantees of cloning and physical manipulation when security foundations are based on physical characteristics of devices that cannot be cloned or manipulated. Such methods make trust in hardware a significant aspect, instead of being stored data, which is likely to be intercepted and copied as secrets.

Equally important is flexibility or adaptability. There is a need to have static security mechanisms that cannot withstand threat metamorphism where a device can be in operation over a long period. The key management techniques should be adaptive, the authentication must be dynamic, and means needed to mitigate the effects of the compromise are necessary in long-term security [14]. These ideas are in line with larger trends in cybersecurity toward making space and systems resounding and resilient, capable of recovering and persistently protecting themselves, instead of merely defending themselves once.

Regarding the concepts of intelligence of IoT, a lightweight cryptography coupled with hardware-based trust anchors is a path to follow [15]. This integration should be done by considering the resource limitations with care and ensuring that it has a significant advantage in security provision. This balance must be struck without bringing a lot of complications or costs. This gap is a more general reason why currently, there is a need to conduct further research on the next-generation IoT security architectures.

2 Background and Related Studies

The fast advancement of Internet-of-things technologies has triggered the ample research on how to secure resource-constrained devices that work in heterogeneous and frequently hostile settings. Early research by the author emphasized that the conventional security solutions, which operate on desktop or server-scale systems, do not apply to the implementation of IoT solutions because it limits them strongly in terms of power, memory, and computing capabilities [16]. In these studies, it was highlighted that the irrelevance of the security requirements and device capabilities is a core impediment to the realization of trustful IoT systems at scale.

Many algorithms have been proposed for LWC by the researchers. Besides, many works have revealed the security attacks on particular LWC algorithm(s). Lightweight cryptography of various structures, key sizes, and block sizes was chosen. A wide range of differences in key size, block size, and rounds was realized as essential for analysis goals. [17]. This dependency brings vulnerabilities in case enemies get physical access to devices, extract keys with invasive or non-invasive attacks. Therefore, the researcher reasoned out that cryptographic effectiveness does not hold any security in the disclosed IoT environment.

The other research line was on authentication and key management protocol [18]. The author examined key-based mechanisms through pre-shared keys and concluded that they did not work well with large-scale deployments because of lack of good scalability, and the compromise permeates very low. Should one of the keys be compromised, subnet parts will be impersonated or interrupted. In order to deal with this, the researcher investigated the public key infrastructures that were modified for IoT. Though these methods can enhance scalability, the author of research stated that public key activities can create locality and energy expenses that are still immoderate for ultra-low-power devices.

The researcher also examined identity-based cryptography and certificateless scheme in order to minimize the overheads of managing certificates [19]. These methods were found to streamline the authentication procedures; it remains based on the same principle that relies on fixed private keys on devices. The researchers pointed out that whenever such keys have been revealed, attackers can continue to steal legitimate nodes irrevocably. This weakness highlights one of the themes that can be found in the literature about IoT security: the problem of securing stored secrets continues to be one of the most difficult issues to resolve.

The attacks on the IoT devices on a physical level have become the focus of more and more research within the recent past [20, 21]. One of the side-channel attacks that the author researched is the power analysis and electromagnetic leakage, and it was shown that even the small cryptography implementations can aid in the dissemination of confidential data. The fault injection methods were also investigated by the investigator in which induced glitches of voltage/clock can cause devices to act randomly, potentially disclosing secret keys. The results further supported the position that software-based security solutions are not enough when the hardware is physically available.

As a way of limiting physical threats, the researcher ventured into the hardware-assisted security solutions [22–24]. Countermeasures that were considered are secure elements and trusted execution environments. The researcher indicated that these elements enhance resistance to key removal, but at a huge cost to the devices and power. Consequently, these solutions cannot work in most low-cost IoT implementations, especially in massive sensing systems.

Cloning of devices and detecting such counterfeits have also found its representation in recent literature. The author showed that attackers were able to duplicate the firmware images and identifiers to make malicious versions that perfectly fitted in networks [25]. In case identity exists only in software form, traditional authentication protocols cannot differentiate authentic and cloned devices. The researcher has underlined the fact that clone attacks are dangerous to applications like industrial surveillance and smart infrastructure in which faith in the device statistics is important.

The other direction of research that was carried out was secure communication protocols in low-power wide-area networks. The researcher tested the currently available protocol stacks and found that they had vulnerabilities regarding key reuse, replay protection, and poor sources of entropy. Although at protocol level, the security mechanisms offer low-level security assurances, the investigator observed that these mechanisms usually assume sufficient security key provisioning at manufacturing which in reality is not always true.

The lifecycle security of the IoTs was another research by the author. Devices can stay longer without updates or maintenance, putting an exposure to the ever-changing attack methods. The researcher observed that the default setup on security does not improve with time and the recovery after compromise rarely features in the current design. This observation notes the existence of a need for security measures, which would not only prevent attacks, but also restrict long-term damage following the compromise.

Some of the researchers investigated intrusion detection and security through anomaly in the case of IoT networks. The researcher used the traffic pattern to establish abnormal behavior that was evidence of attacks. Such solutions are useful at the network level, but the author pointed out that they are not sufficient to block compromising on a device level and the processing is usually centralized leading to latency and privacy issues.

There have also been suggestions that trust management structures can be used to secure IoT. The author tested reputation-based systems, which give a rating of trust to gadgets concerning behavior observed. Despite the fact that these systems enhance resilience when it comes to the misbehaving nodes, the investigator indicated that they rely on precise observations and that they are prone to collusion and false reporting. In addition, the mechanisms of trust are normally working at upper levels and they fail to deal with vulnerabilities at the hardware level.

3 Proposed Secure Architecture

The approach proposes a hardware-grounded and lightweight cryptographic architecture that is created to be used in resource-constrained Internet-of-things settings that have devices, whose operation is severely constrained by energy, memory, and computing ability as well as vulnerable to both physical- and network-based attacks. The underlying philosophy in design is to erase the concept of secret stores, decrease dependencies of trust over the long term, and adhere to cryptographic obstinacy despite the conviction of part of the system. Rather than just relying on the traditional key storage and periodic rekeying technique to establish self-protection via hardware variability, a combination of intrinsic hardware variation coupled with continuous state evolution is described for cryptography to create a self-protecting security fabric.

The very basis of the strategy is the idea of extracting cryptographic content out of intrinsic physical features of the device. These are attributes based on controllable manufacturing differences, which are of a particular IC. Under specified conditions, the device will generate a response which is random before it can be fabricated and practically impossible to achieve on another device. To this reply, however, environmental noise and time change are applied, and the device needs to undergo a stabilization process before it can be safely applied in cryptographic application.

Assume that the raw physical response that the device produces can be represented as a bit sequence. As this has been noisy, a transformation operation is implemented to obtain a stable secret. Mathematically, this can be expressed as the transformation:

$$(D_0, E) = F(T) \tag{1}$$

Here, T signifies the unprocessed hardware answer, D_0 signifies the rebuilt root secret, E constitutes auxiliary data that facilitates the reproduction of the secret without disclosing it, and $F(\cdot)$ denotes the transformation factor. The auxiliary data, as indicated in (1), does not include the secret itself but facilitates the reliable reconstruction of identical cryptography information during power cycles (Fig. 2).

Upon acquisition of the root secret, it is not retained in non-volatile memory. Rather, it functions as a temporary anchor utilized to establish a session root state. This design choice guarantees that physical memory-extracting assaults are incapable of obtaining long-term secrets.

The computation of the initial session state includes freshness to thwart replay and precomputation attacks. The root state of the session is determined as follows:

$$D_0 = J(D_0 \| B_S \| B_A) \tag{2}$$

where $J(\cdot)$ denotes a hash of cryptography or sponge-based function, and B_S and B_A signify nonces produced by the device and the server, respectively. The session state is contingent upon both device-specific entropy and session-specific randomization, as illustrated in (2).

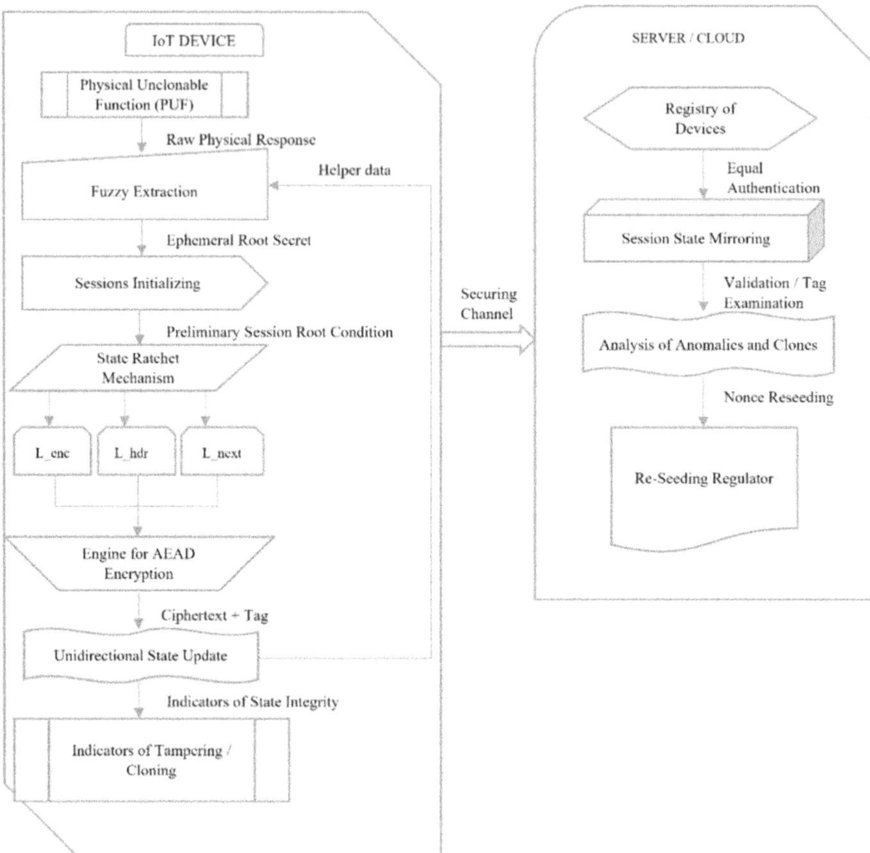

Fig. 2 Proposed end-to-end secure IoT architecture with hardware-rooted trust

The suggested method utilizes a singular dynamic cryptographic state instead of several fixed keys. This state serves as the exclusive provider of cryptography material for authentication, encryption, and state advancement. In each interaction instance, the present state is elaborated into operational keys. The main deriving process is articulated as follows:

$$(L_E, L_A, L_N) = K(D_s, V_s) \quad (3)$$

where D_s represents the current cryptography state; V_s denotes a monotonously growing counter; and L_E, L_A, L_N signify security, authentication, and subsequent state material, accordingly. The incorporation of the counter guarantees distinctiveness in transmissions and inhibits state reutilization, as seen in (3).

Data integrity and privacy are ensured using authorized encryption, which links ciphertext production to both the message contents and the relevant metadata. The process of encryption can be denoted as follows:

$$V_s = W(L_E, N_s, P_s) \tag{4}$$

where N_s represents the plaintext message and P_s signifies related data, like headers or IDs. The encryption procedure, as seen in (4), generates ciphertext that is using cryptography linked to the session state and contextual data.

A notable characteristic of the method is the ongoing evolution of keys via a unidirectional state ratchet. Following every transmission, the cryptography state is irrevocably modified to inhibit retroactive derivation. The state update method is characterized as follows:

$$D_{s+1} = J(L_N \| Y_s) \tag{5}$$

where D_{s+1} symbolizes the revised device state, $J(\cdot)$ signifies a unidirectional transforming function, and Y_s pertains to the current session outputs utilized for state progression. As indicated in (5), the subsequent state relies on both derived materials and message-particular cryptography output, so assuring that the violation of a later state does not reveal preceding states.

The authentication process among the device and the backend is accomplished without disclosing the root secret. Verification tokens are generated from the ephemeral secret and session variables. The generation of the authenticating token might be articulated as follows:

$$\tau_S = N(D_0, U_S \| B_S) \tag{6}$$

where τ_S means the server authenticating token, $N(\cdot)$ signifies the authorization function, and $U_S \| B_S$ refers to the concatenated of user-side and server-side session parameters. As stated in (6), authenticity is contingent upon both device identification and freshness, hence reducing the risk of impersonating and replay attacks.

The strategy includes a reseeding method that revitalizes the cryptographic state with newly generated hardware entropy to maintain resilience following a transient compromise. The reseeding key is produced as follows:

$$L_E = J(D_0 \| B_E) \tag{7}$$

In this equation, L_E indicates the encrypted key component, $J(\cdot)$ signifies a cryptography derivative function, and B_E pertains to the connected binding entropy utilized in key production. As demonstrated in (7), the reseeding operation reintegrates hardware-derived entropy into the cryptography state.

The revised state subsequent to reseeding is calculated as follows:

$$D' = J(D_s \| L_E) \tag{8}$$

In this phrase, D' signifies the updated device state, D_s represents the current device state, and as articulated in (8), it guarantees that previously compromised state data cannot be utilized to forecast future cryptography material following reseeding.

This method intrinsically facilitates clone detection by leveraging the distinctiveness of hardware-based secrets. Let Δ represent the discrepancy among the anticipated and reconstructed confidential material. This divergence can be expressed as follows:

$$\Delta = \|D_0 - \hat{D}_0\| \quad (9)$$

where $\hat{D}_0$ denotes the reconstructed secret throughout the authorization process. As demonstrated in (9), significant variance suggests possible cloning or manipulation.

A constrained resynchronization window is established utilizing a constrained state acceptance functional to avert indefinite desynchronization of processes.

This is the algorithm used to define how to determine a secure communication session based on hardware-derived identity and constant state evolution via cryptography.

Algorithm: Hardware-Rooted Ratcheted Session Protocol
Input: Device identifier, auxiliary data, server challenge.
Output: Secure authentication interactions with a dynamic state.

1. The device activates and assesses its inherent hardware specifications.
2. The apparatus reconstructs a fleeting secret utilizing retained auxiliary data.
3. The device produces a new nonce and authentication token.
4. The device conveys its identity, nonce, and authenticity token to the server.
5. The server authenticates and produces a session nonce.
6. Both the device and the server generate an identical initial cryptographic state.
7. For every outgoing message, the device generates authentication and security materials from the present state.
8. The device encodes and verifies the message.
9. The device refreshes its cryptographic state through a unidirectional transition.
10. The server replicates the state update following verification being successful.
11. Upon reseed request, the device regenerates hardware entropy and updates the cryptography state.
12. Communication persists with the revised status.

The approach proposed introduces a lean and strong security model in the confined environment of the IoT by removing the storage of keys and the incorporation of trust into the hardware of the device. The dynamically computed cryptographic secrets are based on inherent physical characteristics and are combined with randomness in the session so as to build a dynamically changing security state. Each communication instance is made to irreversibly advance the cryptographic state by a one-way ratcheting mechanism which helps limit the effects of compromise and avoid historical key exposure. There is authentication without the revelation of secrets which are bound to someone for long-term reseeding of hardware entropy to re-introduce security after potential exposure. Combining identity with hardware-rooted identity, key evolution on a continuous basis, and authenticated encryption is a security fabric that is both capable of operational concerns in large-scale IoT applications and has a low computational and memory overhead.

4 Results and Analysis

The findings indicate that the given solution can guarantee secure and reliable communication between constraints and limited devices with a minimum of computational cost. Identity based on hardware makes it stronger with respect to gaining trust, key renewed every minute makes it robust to compromise, and residing to dynamic and hostile Internet-of-things conditions through reseeding adapts to maintain long-term security.

Table 1 has compared various IoT security mechanisms based on throughput and memory usage. Throughput is the rate of successful data transmission, whereas the memory utilization shows how much device memory is needed to apply the security protocol. Suggested approach has the highest throughput of 149 kbps, which is better than the rest of the approaches like the Group Key Management (135 kbps) and Hash-Based Authentication (127 kbps). It additionally takes the least memory of 49 KB which is a far lesser amount when compared to blockchain-based IoT (97 KB) and ECC-based authentication (85 KB). It shows that the suggested solution is more efficient in terms of more timely communication and lower memory consumption of resource-constrained IoT devices.

Table 2 identifies different mechanisms used by IoT security in terms of computational time and control overhead. Computational time refers to the duration needed by a device to carry out security tasks including authentication and encryptions, whereas control overhead is a percentage of extra security-related networked communications. The suggested approach has the lowest computational time of 29 ms and control overhead of 14.2% differentiating it as the best performing and consuming less resource. On the contrary, blockchain-based IoT has the longest computational time (59 ms) and overhead (28.8%). It proves that the given approach is more effective and fit to the IoT settings with limited resources.

Table 3 contrasts the various IoT security mechanisms on their end-to-end latency and energy use. End-to-end latency is the sum of time a data packet needs to pass through the entire channel after failing to reach the destination to the source

Table 1 Analysis of memory utilization and throughput

Method	Throughput (kbps)	Memory utilization (KB)
DTLS-IoT	95	77
Lightweight AES	113	69
ECC-based authentication	102	85
Hash-based authentication	127	63
Group key management	135	72
Trust-based security	119	75
Secure boot IoT	110	70
Blockchain-based IoT	87	97
Suggested approach	149	49

Table 2 Evaluation of control overhead and computational time

Method	Computational time (ms)	Control overhead (%)
DTLS-IoT	42	22.4
Lightweight AES	37	18.8
ECC-based authentication	49	24.6
Hash-based authentication	33	18.5
Group key management	36	17.9
Trust-based security	39	21.3
Secure boot IoT	38	19.7
Blockchain-based IoT	59	28.8
Suggested approach	29	14.2

device, and the energy consumption is the amount of energy wasted in cryptographic processing and secure communication. The suggested approach gives the most optimal performance in terms of minimum latency of 122 ms and energy usage at 74 mJ. Notably, blockchain-based IoT demonstrates the largest latency (219 ms) and energy consumption (140 mJ) and indicates that the suggested approach is more effective and valuable to the energy-restrained IoT devices.

The different IoT security mechanisms are compared in Table 4 and are based on network lifetime and packet delivery ratio (PDR). Network lifetime denotes the amount of time the IoT network can last during which the nodes have enough energy to operate, whereas PDR denotes the percentage of data packet successful delivery. The suggested approach has the best packet delivery ratio of 96.6% and this factor—the most reliable communication—is exemplified. The network lifetime of the former is, however, 136 rounds and this is less compared to other techniques such as Group Key management (1186 rounds). This is because of regular security operations and states update. However, the given approach puts the emphasis on ensuring safe and

Table 3 Study of end-to-end latency and energy consumption

Method	End-to-end latency (ms)	Energy consumption (mJ)
DTLS-IoT	177	109
Lightweight AES	159	97
ECC-based authentication	193	125
Hash-based authentication	144	90
Group key management	152	98
Trust-based security	168	104
Secure boot IoT	163	98
Blockchain-based IoT	219	140
Suggested approach	122	74

stable communication within the IoT environment that may face the limitations of resources.

Figure 3 shows comparative graphs assessing various IoT security mechanisms depending on the key performance metrics. The throughput vs. memory graph indicates that the suggested approach is the most effective throughput of 149 kbps and thus has the lowest memory consumption of 49 KB, implying that the data transmission process is efficient and consumes minimal resources. Equally, the computational time vs. control overhead graph indicates that the suggested approach incurs the least number of computational time (29 ms) and minimal control overhead (14.2%), which implies that the suggested approach is faster with fewer additional network traffic than others, including ECC-based authentication and blockchain-based IoT.

The graph of end-to-end latency vs. energy consumption also demonstrates the efficiency of the suggested approach as it documents the smallest latency (122 ms) and energy consumption (74 mJ), which is applicable to resource-constrained IoT gadgets. In the network lifetime vs. packet delivery ratio graph, the suggested approach has less network lifetime (136 rounds), but the highest ratio of packet delivery (96.6) which reflects a very reliable and secure communication. All in all, the findings indicate that the suggested framework offers more suitable performance, efficiency, and stability than current or conventional IoT security mechanisms.

Proposed algorithm is better as it avoids holding key values in storage and its handshakes are made less complex by using dynamically generated state using hardware. It does not require the multi-step negotiation and large quantities of the cryptographic operation like DTLS-IoT (95 kbps) or blockchain-based IoT (87 kbps). The use of memory is minimized to 49 KB due to permanent keys and certificate structures which do not need to be stored as in the case of ECC-based authentication (85 KB). Lightweight authenticated encryption and one-way ratcheting not involving repeated rekeying values reduce latency to 122 ms. Network lifetime trade-off (136 rounds) is due to controlled overhead of energy that is introduced by frequent state evolution and reseeding in favor of strong security and large percentage delivery (96.6%) instead of an extended lifespan.

Table 4 Examination of network lifetime and packet delivery ratio effectiveness

Method	Network lifetime (Rounds)	Packet delivery ratio (%)
DTLS-IoT	981	78.6
Lightweight AES	1046	60.2
ECC-based authentication	911	67.9
Hash-based authentication	1121	71.7
Group key management	1186	82.4
Trust-based security	1011	85.8
Secure boot IoT	1076	91.9
Blockchain-based IoT	841	88.3
Suggested approach	136	96.6

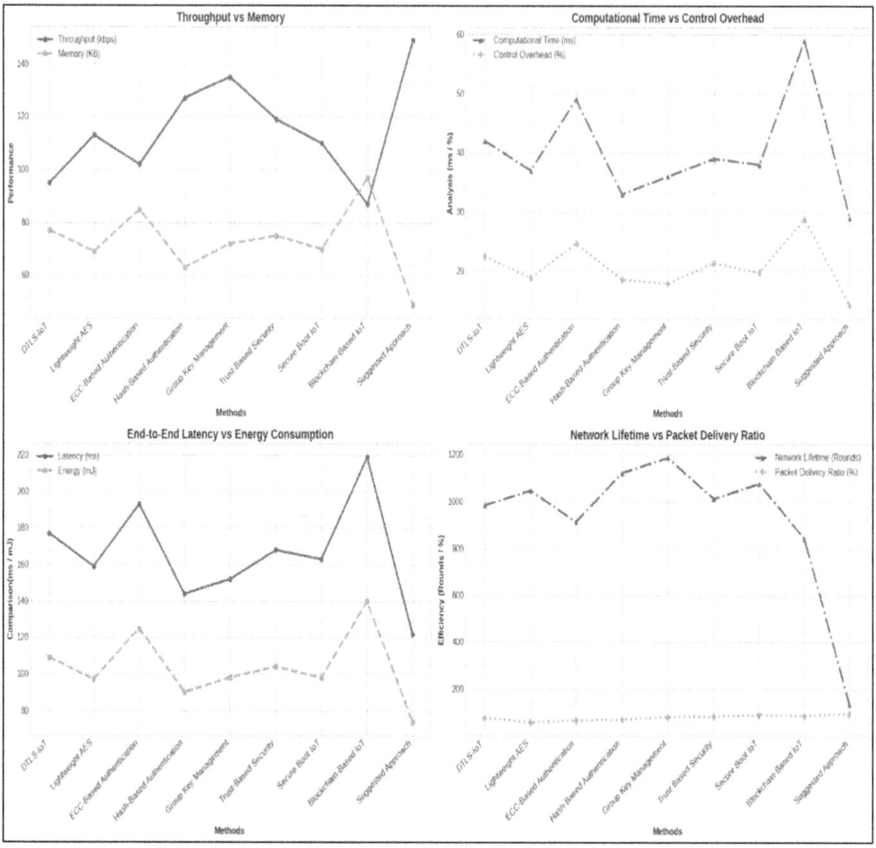

Fig. 3 Comparison of memory utilization, throughput, control overhead, computational time, end-to-end latency, the energy consumption, network lifetime, and packet delivery ratio

5 Conclusion

In this paper, the author discussed the issue of increased security demands of resource-constricted Internet-of-things setting, in which traditional cryptographic methods are insufficient because keys are stored statically, devices are physically accessible, and their lifespan is long. An overall survey of the current lightweight security measures showed that it has had continuous vulnerabilities in areas of key compromise, cloning to duplicate devices, and limited flexibility to new threats. To address these issues, the paper proposed the use of PUF-Ratchet Ascon Session Fabric (PRASF) that is a hardware-based and adaptive security model where the storage of permanent secrets is removed and instead trust is placed directly in hardware devices. The novel cryptographic technique that calculates the cryptographic content based on the physically unchangeable properties and the incessant one-directional advance of the state provides a secure communication and prevents the effects of node seizure and replay attacks to a lesser extent. Authenticated encryption is much more lenient on computation, energy consumption providing consumer secrecy and integrity, and the strategy is appropriate with extensive and extended IoT deployments. Server-aided reseeding-enhanced resilience is further enhanced because having temporary compromise allows one to do so, whereas the uniqueness of hardware limits the threats of cloning and impersonation. In general, it can be concluded that the proposed framework provides the feasible balance between the security robustness and the lightweight operation and it fits the high demands of the contemporary IoT systems and provides a scalable basis to the next-generation secure IoT frameworks.

References

1. V.A. Thakor, M.A. Razzaque, M.R.A. Khandaker, Lightweight cryptography algorithms for resource-constrained IoT devices: a review, comparison and research opportunities. IEEE Access **9**, 28177–28193 (2021). https://doi.org/10.1109/ACCESS.2021.3052867
2. M. Rana, Q. Mamun, R. Islam, Lightweight cryptography in IoT networks: a survey. Futur. Gener. Comput. Syst. **129**, 77–89 (2022). https://doi.org/10.1016/j.future.2021.11.011
3. C. Silva, N. Tenório, J. Bernardino, Lightweight encryption algorithms for IoT. Computers **14**(12), 505 (2025). https://doi.org/10.3390/computers14120505
4. J. Soto-Cruz, A Survey of efficient lightweight cryptography for power-constrained micro-controllers in IoT. Technologies **13**(1), 3 (2024). https://doi.org/10.3390/technologies130 10003
5. Z. Dewamuni, N. Fernando, Bibliometric analysis of IoT lightweight cryptography. Information **14**(12), 635 (2023). https://doi.org/10.3390/info14120635
6. H. Madushan, S.S.S. Perera, A review of the NIST lightweight cryptography finalists and their fault analyses. Electronics **11**(24), 4199 (2022). https://doi.org/10.3390/electronics11244199
7. V. Roy, A context-aware internet of things (IoT) founded approach to scheming an operative priority-based scheduling algorithms. **13**(1), 28–35 (2024). https://doi.org/10.54216/jcim. 130103
8. A.K. Chandanan, M. Rani, K.S. Pokkuluri, S. Singh, V. Jaiswal, P. Narayana, V. Roy, Revolutionizing cardiac prediction based on fog-cloud-Iot integrated heart disease model. Scalable Comput.: Pract. Exp. **26**(5), 2105–2117. ISSN 1895-1767, http://www.scpe.org. https://doi. org/10.12694/scpe.v26i5.4715
9. I. Batra, S. Verma, A. Malik, U. Ghosh, J.J. Rodrigues, G.N. Nguyen, V. Mariappan, Hybrid logical security framework for privacy preservation in the green internet of things. Sustainability **12**(14), 5542 (2020). https://doi.org/10.3390/su12145542
10. A.K. Jain, M. Mudholkar, P. Mudholkar, M.K. Sharma, M. Tiwari, V. Roy, A novel approach to optimized cluster head selection in heterogeneous WSNs for IoT deployments, in *2024 IEEE 2nd International Conference on Innovations in High Speed Communication and Signal Processing (IHCSP)* (Bhopal, India, 2024), pp. 1–6. https://doi.org/10.1109/IHCSP63227. 2024.10960066
11. C.P.J. Samuel, K.G. Dharani, S. Bhavani, Power algorithm to improve the IoT device for lightweight cryptography applications. Mater. Today: Proc. (2020). https://doi.org/10.1016/j. matpr.2020.11.370
12. P. Selvi, S. Sakthivel, A hybrid ECC–AES encryption framework for secure and efficient cloud-based data protection. Sci. Rep. **15**(1), 30867 (2025). https://doi.org/10.1038/s41598-025-308 67-9
13. A. Balaji, S.K. Dhurandher, Reliable data communication using post-quantum encryption in internet of everything. Int. J. Commun. Syst. **35**(13), e5246 (2022). https://doi.org/10.1002/ dac.5246

14. A.A. Anton, P. Csereoka, E.-A. Capota, R.-D. Cioargă, SIMECK-T: an ultra-lightweight encryption scheme for resource-constrained devices. Appl. Sci. **15**(3), 1279 (2025). https://doi.org/10.3390/app15031279
15. M. Gladis Kurian, Y. Chen, Ascon on FPGA: post-quantum safe authenticated encryption with replay protection for IoT. Electronics **14**(13), 2668 (2025). https://doi.org/10.3390/electronics14132668
16. G. Malamas, P. Kotzanikolaou, K. Nomikos, C. Zonios, V. Tenentes, M. Psarakis, HA-CAAP: hardware-assisted continuous authentication and attestation protocol for IoT based on blockchain. IEEE Internet Things J. **12**(11), 15650–15666 (2025). https://doi.org/10.1109/JIOT.2025.3530775
17. V. Rao, K.V. Prema, A review on lightweight cryptography for internet-of-things based applications. J. Ambient. Intell. Humaniz. Comput. **12**(9), 8835–8857 (2021). https://doi.org/10.1007/s12652-020-02672-x
18. I. El Gaabouri, M. Senhadji, M. Belkasmi, A survey on lightweight cryptography approach for IoT device security, in *Proceedings of the 2022 5th International Conference on Networking, Information Systems & Security (NISS)*, pp. 1–8. https://doi.org/10.1109/NISS55057.2022.10085144
19. C. Dobraunig, M. Eichlseder, F. Mendel, M. Schläffer, Ascon v1.2: lightweight authenticated encryption and hashing. J. Cryptol. **34**(3), 33 (2021). https://doi.org/10.1007/s00145-021-09398-9
20. S.S. Dhanda, B. Singh, P. Jindal, Lightweight cryptography: a solution to secure IoT. Wireless Pers. Commun. **112**(3), 1947–1980 (2020). https://doi.org/10.1007/s11277-020-07134-3
21. M.A.F. Al-Husainy, Lightweight cryptography system for IoT devices using DNA. Comput. Secur. (2021). https://doi.org/10.1016/j.cose.2021.102687
22. P.D. Rosero-Montalvo et al., Efficient lightweight cryptographic algorithm in IoT devices. Scitepress Proc. (2022). https://doi.org/10.5220/0000000.109228
23. A.M. Rasheed, R.M.S. Kumar, Lightweight cryptographic algorithms for medical IoT devices using combined transformation and expansion (CTE) and dynamic chaotic system. Int. J. Adv. Comput. Sci. Appl. **15**(4) (2024). https://doi.org/10.14569/IJACSA.2024.0150472
24. M. Gladis Kurian, Y. Chen, The untapped potential of Ascon Hash functions: benchmarking, hardware profiling, and application insights for secure IoT and blockchain systems. Sensors **25**(19), 5936 (2025). https://doi.org/10.3390/s25195936
25. A.G. Hilal, Enhanced data security in IoT-cloud communication: using lightweight cryptography approach. J. Univ. Babylon Pure Appl. Sci. (JUBPAS) **31**(4), 38–48 (2023). https://doi.org/10.29196/the4vq14

Resilient Secure Edge and Fog Computing Models for IoT Data Protection

Sai Vamsi Chennupati, Srinivas Cheekati, Ashutosh Agarwal, and Somnath Banerjee

1 Introduction and Overview

IoT has switched away from individual sensing implementations to large-scale and networked digital ecosystems that have been generating, sharing, and processing data across multiple fields. Initial IoT architecture utilized centralized cloud-based infrastructures to store and process data, with all connections being stable and with traffic patterns that were assumed to be predictable [1]. But the fact that the density of devices, volume, and latency-sensitive applications has grown exponentially has seriously changed this premise. Internet of Things applications like industrial automation, intelligent transportation systems, remote health monitoring, and smart energy management are examples of applications that demand nearly instant responses that can only be reliably delivered by centralized clouds.

To realize such demands, the concept of edge and fog computing has increasingly moved computation nearer to the data by moving computation closer to the data [2]. In end system nodes, localized processing occurs at the device level or very close to it, whereas fog nodes operate as intermediary nodes which consolidate, coordinate, and pre-process data before sending its content to the cloud. This multi-layer model ensures reduced latency, backbone congestion, and enhanced continuity of services

S. V. Chennupati
Western Union, Austin, TX, USA

S. Cheekati
University of the Cumberlands, Williamsburg, KY, USA

A. Agarwal
School of Business, Stevens Institute of Technology, Hoboken, USA

S. Banerjee (✉)
Amfam, Madison, WI, USA
e-mail: somnathbanerjeedwbi@gmail.com

S. Saif et al. (eds.), *Secure Intelligent Sustainable Applications Using IoT for Society 6.0*, Internet of Things, https://doi.org/10.1007/978-3-032-27248-5_6

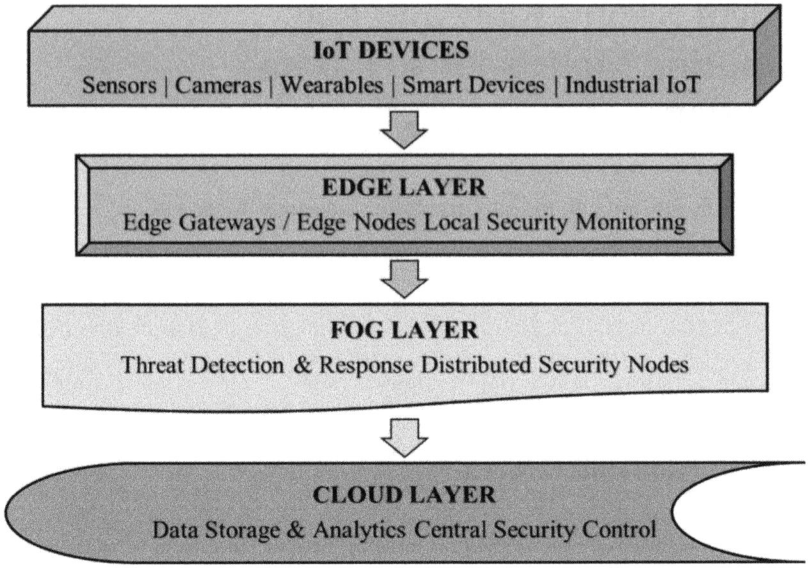

Fig. 1 Secure IoT data processing architecture in edge–fog–cloud environments

in case of intermittent connections. But the spread of computation among heterogeneous, resource-constrained, and frequently unattended nodes creates a novel category of security and reliability issues that do not resemble those that are inherent to centralized systems.

Figure 1 demonstrates an environmentally safe IoT data processing design comprising the IoT gadgets, edge, fog and cloud architectures. The IoT devices produce data which are initially examined and monitored at the edge layer using on-site security mechanisms. The distributed threat detection and response of the layer of the fog are ensured, which allows the analysis to be performed in a shorter time. Lastly, the cloud layer allows exchanging data with centralized management, secure storage, and improved analytics, thus granting secure and stable data processing in the whole distributed environment.

1.1 Security Exposure in Distributed Fog Edge—Environment

Although edge and fog computing increase the level of responsiveness and scalability, they increase the attack surface of the system. In contrast to secured data centers, edge and fog nodes are often considered to be in uncontrolled or semi-trusted environments and therefore, can be tampered with physically, software modified, or unauthorized access [3]. Besides, IoT devices in most situations have limited computing power, limiting the complexity of the cryptographic and monitoring systems that they can implement. Fog computing offers decentralized, on-demand services and apps for the management and analysis of massive data directly at the network's periphery. It offers storage, processing, control, and networking comparable to cloud computing. The fog layer functions as a mediator among IoT devices and the cloud.

The disseminated characteristics of the edge field architectures also add to the complexity of the threat detection and mitigation. Data moves across a number of administrative and trust domains, and typically, it may be preprocessed, aggregated/transformed, at each point [4]. Any node break in this path can derail or steal sensitive information without even being detected. The conventional security model that has been perimeter based and focused on the sharply defined network boundaries does not fit well in these loose and decentralized network environments. Due to this, even attack types like data poisoning, moving between different layers of the mogs, impersonating valid devices, and quiet integrity attacks become harder to counter and restrain.

1.2 Dynamic Data Characteristics and Situational Risk

The IoT information is dynamic both in content and sensitivity. Sensor information, video stream, physiological information, and control message have different confidence levels of confidentiality, integrity requirements, and time required [6, 7]. A data stream that is benign when running on a regular workload can prove to be very sensitive when an abnormal event occurs, such as a system failure, environmental anomaly, or coordinated attacks. Although it is varied, most of the available protection mechanisms use uniform security policies that do not accommodate the environmentally changing risk.

Besides, the interactions of IoT are not fixed. Devices can either be attached or detached to the network, can cross edge domains, or can alter their operational roles. The fog nodes can reassign workloads dynamically depending on the availability, energy limits, or network conditions. These changing relationships form more intricate dependency models that cannot be appropriately described by fixed access-control lists and fixed relationships of trust [8]. Lacking mechanisms to capture current context, the security implementation is either too restrictive, to the point of performance degradation, or too lax, leaving the network vulnerable to serious attacks.

1.3 Continuity and Resilience as Emerging Requirements

In addition to confidentiality and integrity, resilience has been defined as an important requirement for edge–fog IoT systems. In safety–critical and mission-critical applications, a very disastrous impact can be noticed in service disruptions brought about by node failures, network partitioning, or targeted attacks. In contrast to cloud-centric systems where redundancy and failover are comparatively simple tasks to accomplish, the edge fog environment needs coordination between geographically spread and heterogeneous nodes to endure [9].

The concept of resilience also incorporates the fact that the system will be resilient to fault tolerance, as well as adaptive and recovery oriented, under adversarial conditions. In the case of threats spreading fast among interconnected nodes, damage can be increased by a slow response or a response to isolated nodes [10]. Thus, contemporary IoT systems have to enable collaborative awareness, dynamism in reconfiguring, and continuity-conserving schemes that can still work even in the event of partial system compromise. To deal with these issues, the security strategies must consider these issues as not being a fixed state, but rather as a part of the behavior of the distributed system, and it is a constant and continuously changing state of the system.

2 Threat Landscape in Edge–Fog IoT Systems

Several architectures based on distributed edges and fog models have been important and have enhanced contextual responsiveness, latency, and scalability to a great extent compared with centralized, and often cloud-centric, IoT architectures. Nonetheless, the issue of security brings about compound security problems that do not have the same nature as in the traditional cloud-based systems. In contrast to controlled data centers, edge and fog nodes will work under semi-trusted, heterogeneous, and physically exposed environments. The attack surface gets closer to the data source as computation becomes closer to the data source.

2.1 Scope of the Distributed Trust Complexity

Authority in the fog of the edge is gone. Trust is not binary and central. Devices communicate dynamically in a variety of administrative domains where various policies, ownership models, and security capabilities regulate these domains. Edge nodes can be those of private organizations, service providers, or even end users, and therefore, it is not feasible that uniform enforcement is applicable. Conventional models of trust used on things that presuppose defined identities or fixed lists of access control are inadequate because of dealings with dynamically evolving services and devices.

Further, they have IoT ecosystems marked with high device locomotion, interrupting communication, and changing roles. A device that is a producer of data at one instance might be a relay or aggregator at another instance. Such fluidity implies tremendous trust dependence, which is challenging to model using conventional security mechanisms. In case a compromised node happens to be involved in multi-hop communication or data aggregation, downstream data flows' integrity can be

Table 1 Key security challenges in edge–fog IoT environments

Challenge category	Core issue	Limitation of traditional approaches	Impact on IoT systems
Distributed trust [11]	Multi-domain, semi-trusted nodes with dynamic roles	Static trust scores and centralized access control	Increased vulnerability to impersonation and insider attacks
Dynamic data sensitivity [12]	Temporal and context-dependent risk levels	Fixed encryption and uniform policy enforcement	Either excessive overhead or insufficient protection
Interaction complexity [13]	Frequent cross-layer communication and data aggregation	Identity-based authorization without behavioral evaluation	Data poisoning and stealth integrity violations
Resilience constraints [14]	Heterogeneous hardware and decentralized management	Centralized fault tolerance models	Cascading failures and service disruption
Scalability pressure [13]	Rapid growth of devices and connections	Manual policy updates and rigid rule sets	Delayed response to emerging threats

compromised without commotion. As such, trust should be continuously considered depending on the consistency in behavior, the history of interaction, and the context semantics, and not on some established credentials. Table 1 outlines the key security concerns in edge-fog IoT settings. It outlines such problems as distributed trust, dynamically sensitive data, interactivity complexity, resiliency limitations, and pressure to scale. According to Table 1, traditional statical security measures are also not sufficient, and these systems have such weaknesses as impersonation attacks, data poisoning, service disruption, and slow response to new threats.

This systematic analysis emphasizes that security risks in edge-fog IoT systems are not separate technological problems but interrelated phenomena resulting from decentralization, dynamism, and heterogeneity. Addressing these difficulties necessitates adaptable, context-aware, and jointly robust protection methods, rather than inflexible and centrally imposed ones.

2.2 *Dynamic Data Sensitivity*

IoT data is always temporal and contextual. The sensitivity of the data stream does not remain the same; it changes with the state of operation, environmental conditions, and anomalies. As an example, the sensor readings that are monitored as normal events can have little sense of confidentiality needs, yet the information may be very important whenever death and destruction are involved in fault detection, emergency response, or synchronized cyber-physical assaults.

The variability cannot be sufficiently covered by the use of static encryption policies or standardized protection thresholds. Excessive protection causes extra computation costs (and can cause under-protection) in the case of devices with limited resources, whereas inadequate protection can make the device more vulnerable. The protection mechanisms need to be effective thus having the aspect of time with volatile situations that are short-lived, as well as the trend of behavior. The lack of sensitivity modeling of adaptation usually leads to the result of flipped enforcement of the security that will either slow down system performance or reveal sensitive data.

2.3 Continuity Requirements and Resilience

In addition to confidentiality and integrity, resilience has been an overriding requirement in distributed IoT systems. The applications that are mission-critical and supported by edge and fog infrastructures include intensive automation of industrial processes, healthcare monitoring, and transportation. Interruptions in the delivery of service or the domino effect caused by the failure of one node can be disastrous in such situations.

Distributed architectures should also not interrupt their functioning even in the event of partial compromise. But edge-fog systems are hard to make resilient since the hardware capacities are not homogeneous, workloads are not predictable, and management is decentralized. This local failure may spread very fast through connected nodes in case there is no cooperative awareness. Conventional fault-tolerance mechanisms used in centralized clouds cannot be used due to the presence of stable infrastructure and plenty of redundancy. In turn, adaptive recovery, collaborative threat awareness, and distributed health monitoring should be incorporated into modern IoT security. Resilience is not to be considered a layer per se but a characteristic of the protection mechanism itself.

3 Review of Existing Security Approaches

The new developments in the IoT systems have also increased the research in securing data in the distributed computing paradigms. The main works of young scientists were dedicated to cloud-based security enforcement when the data of an IoT was sent to centralized structures to protect and analyze. The author stressed that centralized mechanisms have the advantage of mature security controls and high computational ability but come with too much latency and bring about single points of failure. The researcher also noted that regular transmission of data to the cloud exposes data to more exposure while being transmitted, hence such solutions are inappropriate in applications that are time -sensitive and those that require high levels of safety.

To counter the latency and bandwidth constraints, later literature focused its interest on edge-based security models. According to the investigator, edge-based pre-processing data and simple security guarantee a shorter response time and less unnecessary exposure to data. The author pointed out, though, that many of the edge-level solutions are based on a fixed set of rules and a set of predetermined thresholds, which limit scalability to dynamic threat conditions. The author of this study emphasized the fact that firm security policies are not able to balance the dynamic sensitivity of data and the changing behavior of the device; rather, it generates either massive overheads or inadequate protection.

Fog computing was also an in-between paradigm that overcomes the edge-cloud gap. The investigator also talked about fog-assisted trust management models that allocate security decision-making processes among several nodes, which enhance scalability and load balancing. Although these are his strengths, the author also indicated that with these benefits, the fog-based solutions, in most cases, rely on numerical ratings of trust or reputation systems that are affected by the failure of nodes. The researcher even made the argument that these models did not have their cooperative resilience, thus incurring cascading failures and coordinated attacks.

In recent literature, there has been an increasing amount of work done in examining context-sensitive and behavioral security measures. The researcher reviewed methods of implementing contextual characteristics, including device role, communication schemes, and environmental factors, to narrow down access control. These techniques are good at improving the accuracy of detection; however, the author pointed out that they very often work independently and do not account for the temporal sensitivity of risk changes or the health of distributed systems. The researcher pointed out that contextual awareness cannot be sustained in changeable IoT environments without cross-layer coordination since such environments constantly vary.

Altogether, the literature review indicates that there is a research gap in making the IoT data protection adaptive, resilient, and decentralized both at the edge and fog layers [15]. The researcher has concluded that current strategies do not have any mechanisms for integrating temporal data behavior, interaction context, and cooperative resilience, which indicates the necessity of security models that change continuously along with the dynamic IoT ecosystems.

Table 2 overviews current data protection techniques in the Internet of Things applied within edge and fog computing systems. It compares approaches like centralized protection on the cloud, edge protection framework, fog-based trust management, and lightweight cryptographic frameworks. Although these designs are better in terms of latency and efficiency, they only employ a fixed-policy approach or a centralized mechanism, thus restricting flexibility and resiliency. It emphasizes that the current protection mechanisms of IoT data are mainly based on a static policy, centralized regulation, or a set of trust measure [23–25]. Although they are better at enhancing latency or efficiency, these approaches are not adaptive, lack context awareness, and resilience, which expose the edge fog environment to dynamic threats and changing operational conditions.

Table 2 Review of existing IoT data protection approaches in edge and fog computing environments

Approach category	Core idea	Key strengths
Centralized cloud-based protection [16, 17]	IoT data is transmitted to centralized cloud servers where security controls and analytics are applied	Strong computational power and mature security mechanisms; simplified management
Static edge-level security models [18, 19]	Security mechanisms are deployed at edge nodes using predefined rules and fixed policies	Reduced latency; localized data protection; lower cloud dependency
Fog-assisted trust management systems [2, 20]	Fog nodes mediate trust and access control between IoT devices and cloud services	Improved scalability; distributed decision-making; better load balancing
Lightweight cryptographic IoT frameworks [21, 22]	Emphasis on lightweight encryption and authentication tailored for resource-constrained devices	Low computational overhead; feasible for constrained IoT devices

The available literature focuses on cloud-based security models in which the protection of IoT data is dealt with once they are largely transmitted to centralized infrastructures [26]. The author points out that these solutions are characterized by powerful computational opportunities and well-developed security solutions, but have a high latency and single-point failures. The investigator comments that centralized enforcement does not have an implementation that is sensitive to delays and mission-critical, and its causes of permanent offloading of data during transmission, exposing and detracting from the possibility of an attack with a real-time reaction.

4 Core Design Principles for Adaptive IoT Security

Adaptive IoT defense should be based on a series of fundamental design principles reflecting the distributed and dynamic edge fog ecosystems to overcome the limitations of the existing method of planning and implementing security in a traditional centralized and static fashion.

4.1 Principle of Making Changes Over Time

There should be security measures that know risk changes. Instead of deterministically fixing sensitivity levels on data streams, the level of protection ought to vary according to aberrations of historical averages, volatility trends, and predictive exposure. Temporal awareness allows businesses to differentiate normal fluctuations of change from deviant actions. Adaptive models can be used to keep things in control without dismissing emerging threats by adding memory capabilities that not only average out short-term spikes but also preserve long-term trends as well.

4.2 Principle of Contextual Interaction Modeling

Entity identity cannot only be used to make security decisions but should also consider the semantic and behavioral context of the interaction. Devices communicate among themselves in working processes; compatibility of roles, consistency of behavior, and stability of intent determine the legitimacy of an interaction. To be able to make access granularity dynamic, contextual modeling enables the information disclosure to be aligned with the levels of trust in an interaction. This will minimize the use of inflexible enforcement of policies and increase fine-grained control.

4.3 Cooperative Resilience Diffusion Principle

In distributed systems, protection cannot be isolated at the individual nodes only. Every edge or fog node must have a resilience state indicative of workload stress, anomaly history, and integrity indicators. By communicating the signals of resilience among adjacent nodes, for example, it is possible to disseminate early warnings throughout the network, which would allow them to actively adjust. Not only does such co-operative diffusion result in a stronger collective defence, but it also prevents the propagation of cascading failures.

4.4 Decentralized Adaptive Enforcement Rule

Lastly, adaptive protection should be independent of a central authority. The intensity of the enforcement is expected to arise due to the combination of the temporal sensitivity, contextual interaction trust, and node resilience status. Such a decentralized design is scalable, minimizes single points of failure, and the security behavior in the environment is kept in step with the operational real-time state of the ecosystem.

All these principles allow forming a conceptual base of the resilient, context-aware, and constantly evolving IoT data protection that can be applicable to edge environments and fog.

5 Research Motivation and Objectives

Internet of Things (IoT) devices have rapidly grown, thereby causing a huge amount of data to be created in distributed computing environments. Edge and fog computing allow processing to be quicker and less latent, yet also create new security issues because of intellectually dispersed structures and non-homogeneous tools. Conventional central sources of security usually fail to deal with dynamic threats and fluctuating data sensitivity. Thus, there is a serious requirement for the use of adaptive protection measures that may provide adequate data protection, system availability, and sustain ongoing IoT activities.

Objectives

The development of the proposed approach is controlled by the following objectives:

- To develop a flexible security system to be able to secure IoT data in distributed fog computing environments on the edges.
- To improve the threat detection system in detecting malicious activities in the IoT networks.
- To provide high -quality access control and maintain the data communication security between the heterogeneous IoT devices.
- To enhance the resilience of the systems as well as ensure constant functionality whenever there is a dynamic network condition.

The purpose of these objectives is to create a strong and agile security infrastructure to safeguard IoT data while preserving system reliability and resiliency in distributed edge e-fog client settings.

6 Design of the Proposed System

The suggested solution proposes a fundamentally adaptive system of securing data in both edge and fog computing systems to the IoT system, as it constantly varies as a property of the system instead of a fixed set of configurations. In contrast to other historic techniques, where access has been regulated by a predefined set of trustworthiness or some central authority of decision-making, the technique is a model of security as an emergent behavior based on the temporal risk, relationship, or distributed resilience state with the context in Fig. 2. The design is made to work in a decentralized environment, wherein nodes availability, sensitivity of the data and intensity of threats keeps evolving dynamically with time.

The core of the approach is an assumption that when computation is spread to heterogeneous and strategies that are not generally entirely trusted, it is impossible to effectively impose protection on IoT data with the help of single control mechanisms. The edge and fog nodes have variable workloads, disrupted connectivity, and exposure to physical and cyber threats. The proposed approach will therefore

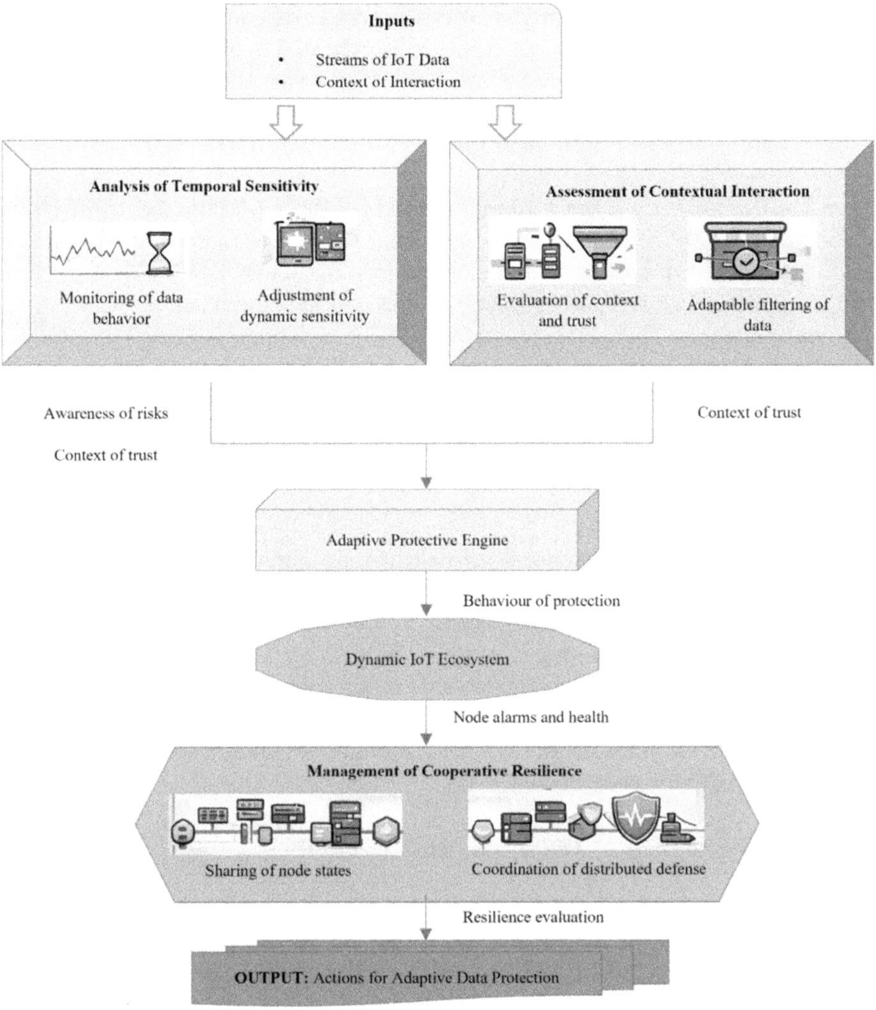

Fig. 2 Conceptual architecture of the adaptive protection mechanism for IoT data

ensure a continuous adaptive loop where protection mechanisms are adapted against data behavior, interaction semantics, and system health. The loop allows the system to reform its protective stance along with not necessitating global coordination or control.

The first conceptual component of the approach is based on the modeling of the temporal sensitivity of the data streams of the IoT. The sensitivity of IoT data varies through time; this data confidentiality and integrity needs vary depending on the events in the context, probability of anomalies, and working conditions. Each of the streams of data is allocated a temporal sensitivity state to capture this behavior which varies continuously. Assume that $D_i(t)$ is a stream of data produced by the device (u)

at time (r). Instantaneous sensitivity of this stream can be considered as weighted deviation, volatility, and predictive exposure as indicated in the equations below (in the form of Eq. 1):

$$A_u(r) = \alpha_1 \cdot \delta_u(r) + \alpha_2 \cdot c_u(r) + \alpha_3 \cdot \pi_u(r) \quad (1)$$

where $\delta_u(r)$ signifies divergence from historical norms, $c_u(r)$ indicates shorter-term volatility, and $\pi_u(r)$ reflects expected exposure derived from previous behavioral trends, as seen in Eq. 1. This formulation ensures that sensitivity is able to adapt to changes that are observed and expectant as opposed to being fixed.

To prevent abrupt changes in protective behaviors, support is given by restricting the memory by a constrained sensitivity of time. The system sensitivity at time (r) is summed up by exponential accumulation of previous states in accordance with accumulation exponential weight as expressed in Eq. 2:

$$\overline{A}_u(r) = \lambda \overline{A}_u(r-1) + (1-\lambda) A_u(r) \quad (2)$$

where $\lambda \in (0, 1)$ governs temporal permanence, as demonstrated in Eq. 2. It is a procedure that allows the system to accommodate ongoing changes and address the short-term changes. The smoothing sensitivity has a direct impression on how data is treated on edge and fog nodes.

The method has a protection intensity that is controlled by sensitivity, as opposed to encrypting or access limits that are fixed. Let $O_u(r)$ be the amount of protection intensity assigned to data stream (u). This strength is expressed as a monotonous function of sensitivity as expressed in Eq. 3:

$$O_u(r) = \kappa \cdot kb(1 + \overline{A}_u(r)) \quad (3)$$

This is represented by κ which is a ratio between the scaling coefficient, kb is the protective constant at the base and $\overline{A}_u$ (r) is the flexible risk impact of the user u at state r. The kind of the formulation also ensures gradual adaptation making them not experience sudden spikes in computations.

The second basic element of the concept of methodology does not view security as only a property of individual streams of information but rather as an outcome of inter-entity interactions. In edge-fog cases, the threats to data security are mainly due to information flows among the devices, between the devices and the nodes, and among nodes. The technology defines a contextual and interacting state between two entities, c and d. The model describes the interaction situation which has a composite relationships value as defined in Eq. 4:

$$X_{\text{cd}}(r) = \beta_1 \cdot \sigma_{\text{cd}}(r) + \beta_2 \cdot \rho_{\text{cd}}(r) + \beta_3 \cdot \iota_{\text{cd}}(r) \quad (4)$$

In Eq. 4, $\sigma_{\text{cd}}(r)$ denotes semantics compatibility, $\rho_{\text{cd}}(r)$ signifies behavioral reliability, and $\iota_{\text{cd}}(r)$ illustrates intent constancy. In this way, this system is able to judge the interactions based on identity-based authorization.

As a result of the development of pattern of interaction the context-specific state is altered successively. Aggregated interaction trust among entities is determined by means of Eq. 5:

$$\overline{X}_{cd}(r) = \mu \overline{X}_{cd}(r-1) + (1-\mu) X_{cd}(r) \tag{5}$$

where $\overline{X}_{cd}(r)$ Denotes the refined contextual state. μ regulates the pace of trust memory deterioration, as demonstrated in Eq. 5.

This will allow historical behavior to influence current decisions and to be flexible. The state of interaction explicitly controls legitimate data disclosure to entities. The admissible data granularity sent between c and d, represented as $F_{cd}(r)$, is specified by Eq. 6:

$$F_{cd}(r) = \frac{1}{1 + w^{-\overline{X}_{cd}(r)}} \tag{6}$$

w is a weighted factor as it is shown in Eq. 6. Fuller information exchange is achieved with an increase in interaction confidence and reduced exposures with ambiguous interactions.

The last feature of the approach provides distributed resiliency through propagating between edge and fog nodes on cooperation. The state of internal resilience that it has is stored in each node, and this is an indicator of its operating security and durability as well as past experience with anomalies. Let $E_h(r)$ represents the resilient state of node h. This state develops in response to internal factors and external influences, as articulated in Eq. 7:

$$E_h(r+1) = \theta E_h(r) + (1-\theta)\Omega_h(r) \tag{7}$$

where $\Omega_h(r)$ consolidates recorded anomalies, workload stress, and integrity indications, as demonstrated in Eq. 7.

To enable the resilient behavior to enjoy cooperation, the nodes exchange resilience signals with their neighboring nodes. The effect of neighboring node to node h is illustrated by Eq. 8:

$$\Psi_h(r) = \sum_{l \in M(h)} Q_{hl} E_l(r) \tag{8}$$

$\Psi_h(r)$ signifies the cumulative influence on node h, $M(h)$ constitutes the set of adjacent nodes, Q_{hl} indicates the influence weight among nodes h and l, and $E_l\ (r)$ reflects the state of neighboring l at stage r.

The approach promotes the spread of early warnings and protective changes to the entire network because it is a form of diffusion. Status of resiliency determines the extent of protection of hosted data streams by a node. The node-level enforcing factor $R_h(r)$ is determined by Eq. 9:

$$R_h(r) = \frac{1}{1 + f^{-\tilde{E}_h(r)}} \tag{9}$$

$R_h(r)$ signifies the resilience responsive of node h, $\tilde{E}_h(r)$ indicates the assessed node state, and f is a shaper constant that regulates responsive sensitivity as demonstrated in Eq. 9.

The nodes with reduced resilience cause instant escalation of the rigor of enforcement. The approach incorporates the temporal sensitivity, contextual interactions and resiliency as one adaptive protective process. The security that was effective on data stream i handled by node j on its interaction with entity b is outlined in Eq. 10:

$$\Phi_{\text{uvd}}(r) = A_u(r) \cdot F_{\text{hd}}(r) \cdot R_h(r) \tag{10}$$

$\Phi_{\text{uvd}}(r)$ signifies the comprehensive adaptable protection level, $A_u(r)$ denotes the user-associated risk factor. $F_{\text{hd}}(r)$ represents the contextual filtration element as demonstrated in Eq. 10. The above strategy will ensure that protection will adapt concurrently to data risk, interaction context and node integrity.

In order to prevent excessive enforcement or too much leniency, the protective role is also limited to a normalizing criterion as spelled out in Eq. 11:

$$\hat{\Phi}_{\text{uvd}}(r) = \min(\Phi_{\text{max}}, \max(\Phi_{\text{min}}, \Phi_{\text{uvd}}(r))) \tag{11}$$

$\hat{\Phi}_{\text{uvd}}(r)$ represents the bounded adaptable protective value, while Φ_{max} and Φ_{min} delineate the upper and lower protected thresholds, as indicated in Eq. 11. This will ensure stability of operations at various devices.

The system is in a continuous evaluation of the difference between the expected and observed protective behavior. The protection drift is derived as in Eq. 12:

$$\Delta(r) = \sum_u |\hat{\Phi}_u(r) - \hat{\Phi}_u(r-1)| \tag{12}$$

$\Delta(r)$ signifies the value of safeguarding variation, while $\hat{\Phi}_u(r)$ and $\hat{\Phi}_u(r-1)$ indicate the current and prior bounded protective states of user u, as seen in Eq. 12. Too much deviation triggers internal compensations.

Finally, the system level equilibria is measured with the help of a convergence signal as outlined by Eq. 13:

$$\Xi(r) = \lim_{R \to \infty} \frac{1}{R} \sum_{r=1}^{R} \Delta(r) \tag{13}$$

$\Xi(r)$ is the long-term reliability measure of the safeguarding structure and R is the total number of adaption steps or operating steps during which the protective variation is tracked to measure the long-term reliability of the system such as in the case of the equation below: This pointer proves the durability in the long term of flexibility of protection without enforcing the converging stature.

The algorithmic architecture of the proposed method is made up in an all-embracive pseudocode, it illustrates how applicable security can be developed without express regulations or thresholds.

```
Algorithm: Adaptive Protective Weaving
Establish sensitivity, interaction, and resilience parameters
For each time period r, execute
  For every data stream U conduct
    Revise temporal sensitivity status
    Seamless sensitivity utilizing memory functionality
    Calculate protection intensity
  Terminate
  For each interacting entity pair (c, d), assess the contextual interaction state.
    Revise the aggregated interaction confidence
    Establish permissible data granularity
  Terminate
  For each edge or fog node h, execute the following.
    Revise local resilience status
    Share resilience indicators with neighbors
    Disseminate resilience impact
    Calculate enforcing coefficient End loop
  Integrate sensitivity, interaction, and resiliency to enhance protection adaptation.
  Standardize protection within settings for operation
Terminate
```

In short, the proposed solution creates a perpetually adaptive security system of the IoT data in edge and fog settings through the combination of the temporal sensitivity modeling, contextual interaction evaluation, and distributed resilience diffusion into the single adaptive loop. Security behavior develops as a result of changing data risk and dynamic relationships as well as collaborative node health, as opposed to fixed policies and control. The approach offers a robust base to protect the IoT data in the changing operational and threat environment by allowing the protection mechanisms to dynamically re-form themselves.

7 Experimental Results and Analysis

In this part, experimental analysis of the proposed adaptive IoT data protection framework is represented in edge and fog computing settings. The evaluation goal is to review the efficiency of the suggested mechanism to keep the data processing safe in the conditions of the distributed and dynamic working in a specific way. The analysis will adopt a comparison style of the proposed method with several existing security models in the implementation of edge-fog architecture of IoT. The findings establish that the adaptive protection strategy enhances the reliability of the system, threat handling effectiveness, and consistency of protection behavior in heterogeneous nodes. The analysis points out the benefits of incorporating the temporal sensitivity modeling, contextual interaction assessment, and cooperative resilience diffusion to develop an adaptive and decentralized protective system of the edge-fog infrastructure on IoT data.

7.1 Performance Metrics

Accuracy (%) is the general privacy of the classification system to discover the malicious and legitimate cases of IoT data. It is used to determine the rate of correct prediction of the results of all the samples considered.

$$\text{Accuracy} = \frac{\text{TP} + \text{TN}}{\text{TP} + \text{TN} + \text{FP} + \text{FN}} \times 100 \tag{14}$$

where TP is Malicious IoT data right termed as malicious, TN is Normal IoT data right termed as legitimate, FP is Normal IoT data wrong termed as malicious, FN is Malicious IoT data wrong termed as normal.

Sensitivity (%) denotes the power of the system to accurately report malicious instances of IoT data in distributed edge-fog systems. The larger the value of sensitivity, the better it is in terms of identifying possible threats or attacks.

$$\text{Sensitivity} = \frac{\text{TP}}{\text{TP} + \text{FN}} \times 100 \tag{15}$$

F1-Score (%) is a combined performance measure that is a mixture of sensitivity and precision into one value. It is a statement of the consistency of the detection performance of the model with a minimum number of inappropriate predictions.

$$\text{F1-Score} = 2 \times \frac{\text{Precision} \times \text{Sensitivity}}{\text{Precision} + \text{Sensitivity}} \tag{16}$$

Detection Rate (%) is a metric that provides the percentage of malicious data event success realization by the system. Increased detection rates show greater ability when it comes to identifying threats in dynamic network circumstances.

$$\text{DetectionRate} = \frac{\text{TP}}{\text{TP} + \text{FN}} \times 100 \tag{17}$$

False Negative Rate (%) refers to the rate of being a victim of malicious activity on IoT as not detected by the system. Reduced false negative means would be good since less security threats would go undetected.

$$\text{FalseNegativeRate} = \frac{\text{FN}}{\text{FN} + \text{TP}} \times 100 \tag{18}$$

Error Rate (%) is the percentage that the system produces wrong predictions when it is classifying. Less error values denote greater reliability and superior full accomplishments of security framework.

$$\text{ErrorRate} = \frac{\text{FP} + \text{FN}}{\text{TP} + \text{TN} + \text{FP} + \text{FN}} \times 100 \tag{19}$$

Precision (%) measures the trustworthiness of the identified malicious events by determining the number of malicious threats that are malicious. Existence of high precision minimizes false alarms and enhances confidence on the detection mechanism.

$$\text{Precision} = \frac{\text{TP}}{\text{TP} + \text{FP}} \times 100 \tag{20}$$

Specificity (%) is used to assess the capability of the system to properly recognize valid IoT data objects without mis-established objects as malicious. Increased specificity guarantees a consistent functioning of the system with little or no disturbance to the overall activities.

$$\text{Specificity} = \frac{\text{TN}}{\text{TN} + \text{FP}} \times 100 \tag{21}$$

Cloud-Centric Security Model achieves lower performance of 62.8% and sensitivity of 63.1 which means that it is not very effective in the context of distributed environments. The Static Edge Security Framework enhances the outcome to 66.4–67% exactness and 67% sensitivity whereas the Lightweight Cryptographic Model extends the results to 68.7–69.3% precision and 69.3–70% accuracy in Fig. 3 and Table 3. The accuracy and sensitivity of Fog-Based Trust Management are 71.2 and 72 respectively, then there are 73.6 and 74.1 in Context-Aware Access Control. Phoenix-Inspired IoT Security and Distributed Authentication Scheme indicate consistent returns at 75.9%/76.4% and 78.1%/78.6% respectively. Adaptive Fog Protection Model is connected to an 80.7% accuracy and 81.2% sensitivity, and The

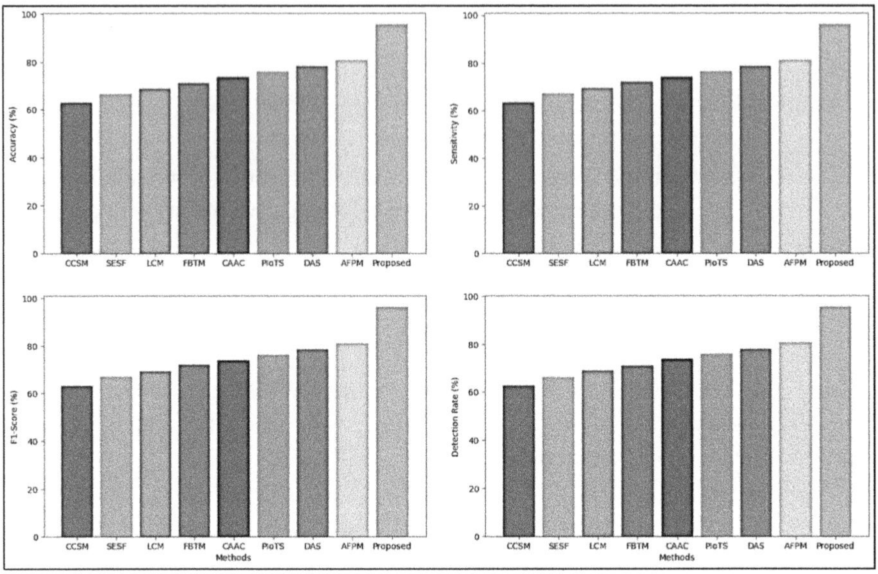

Fig. 3 Comparison of accuracy, sensitivity, F1-score and detection rate

Suggested Approach is better advanced than all methods with 95.6% accuracy and 96.2 sensitivity, and it has better recognition.

The Cloud-Centric Security Model attains the F1-Score of 63.2 with the detection rate of 62.5 with limited accuracy and sensitivity. The Static Edge Security Framework also gets the highest score of 66.9, 66.2%, then the Lightweight Cryptographic Model with the highest F1-Score of 69.1%, detection rate of 68.9%. Fog-Based Trust Management has the data of 71.8% and 71%, whereas Context-Aware Access Control has 74% and 73.8%. Phoenix-Inspired IoT Security and Distributed Authentication Scheme are gradually improving at 76.3/75.8/78.5% and 78.5/78.0/78.5%, respectively. Adaptive Fog Protection Model has 81.1% F1-Score and 80.6% detection rate, and the Suggested Approach evidently beats all the existing approaches with its 96.1 percent F1-Score and 95.4% detection rate, which demonstrates the great stability and reliability of detection.

The Cloud-Centric Security Model has specificity of 64% and precision of 63.4% that is, a limited capability of performing a correct designation of legitimate data and credible detections. The Static Edge Security Framework gets up to 68.1 and 66.8 specificity and precision respectively, whereas the Lightweight Cryptographic Model scores at 70.2 and 69. The Fog-Based Trust Management has a specificity of 72.8% and a precision of 71.6% and Context-Aware Access Control with 75 and 73.9%. Phoenix-Inspired IoT Security and Distributed Authentication Scheme have stable growth at 77.2/76.1 and 79.4/78.3, respectively in Fig. 4 and Table 4. Adaptive Fog Protection Model achieves a specificity of 82 and 80.9% precision, which is significantly lower than the Suggested Approach that achieves a specificity

Table 3 Comparing performance metrics analysis

Method	Accuracy (%)	Sensitivity (%)	F1-Score (%)	Detection rate (%)
Cloud-centric security model (CCSM)	62.8	63.1	63.2	62.5
Static edge security framework (SESF)	66.4	67	66.9	66.2
Lightweight cryptographic model (LCM)	68.7	69.3	69.1	68.9
Fog-based trust management (FBTM)	71.2	72	71.8	71
Context-aware access control (CAAC)	73.6	74.1	74	73.8
Phoenix-inspired IoT security (PIoTS)	75.9	76.4	76.3	75.8
Distributed authentication scheme (DAS)	78.1	78.6	78.5	78
Adaptive fog protection model (AFPM)	80.7	81.2	81.1	80.6
Suggested approach	95.6	96.2	96.1	95.4

of 97.1 and 95.9% precision as it has better discrimination ability and high detection reliability.

The false negative rate at Cloud-Centric Security Model is high with the rate being 79.2% and error rate being 78.6% meaning that this model gives out incorrect classification of harmful occurrences. The Static Edge Security Framework has lower values of 75.4 and 73.6 and the Lightweight Cryptographic Model has even lower values of 73.1 and 71.3. Fog-Based Trust Management has 70.6 and 68.8 percent false negatives and error rate, and then Context-Aware Access Control has 68.4 and 66.4% of it. Phoenix-inspired IoT Security and Distributed Authentication Scheme also follow the same pattern, falling at 66.1%/64.1%/63.9%/61.9%, respectively. Adaptive Fog Protection Model has the highest rates of false negative (61.8) and error (60.3), and the Suggested Approach has the lowest (57.8 and 53.1) which validate the best detection consistency and lowest false identification.

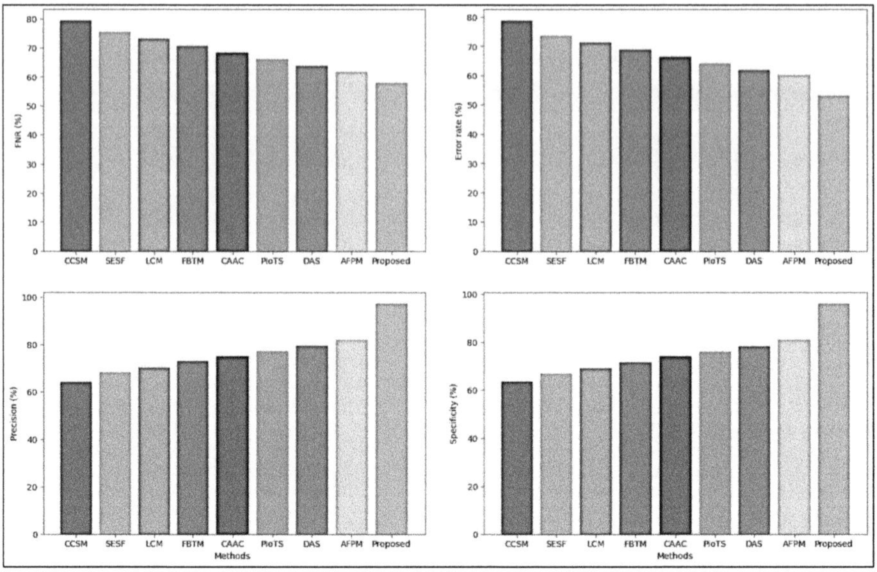

Fig. 4 Comparison of false negative rate, error rate, specificity and precision

Table 4 Analysis of error rate, false negative rates, specificity and specificity effectiveness

Method	False negative rate (%)	Error rate (%)	Precision (%)	Specificity (%)
Cloud-centric security model (CCSM)	79.2	78.6	64	63.4
Static edge security framework (SESF)	75.4	73.6	68.1	66.8
Lightweight cryptographic model (LCM)	73.1	71.3	70.2	69
Fog-based trust management (FBTM)	70.6	68.8	72.8	71.6
Context-aware access control (CAAC)	68.4	66.4	75	73.9
Phoenix-inspired IoT security (PIoTS)	66.1	64.1	77.2	76.1
Distributed authentication scheme (DAS)	63.9	61.9	79.4	78.3
Adaptive fog protection model (AFPM)	61.8	60.3	82	80.9
Suggested approach	57.8	53.1	97.1	95.9

8 Conclusion and Future Work

The paper has dealt with the extreme issue of securing IoT information in distributed edge and fog computing frameworks, in which dynamic information conduct, heterogeneous gadgets, and fragmented trust spaces increase the complexity of the traditional security enforcement a great deal. Current solutions were also found to be putting much of their own weight on older and dusty policies, centralized decision-making, or assumed levels of trust and therefore they are short of effectiveness in changing circumstances of operational and threat factors. The paper has attempted to address these weaknesses by introducing AMTW as a dynamic multi-layered security service that views security as a system property that is constantly changing, as opposed to a security layout. AMTW allows fixing protection mechanisms to keep adapting in accordance with the changes of the real-time environment by collaboratively addressing the sensitivity of temporal data, concurrent interaction behavior, and the cooperativeness of the resilience between the edge and the fog nodes. The proposed solution creates a centralized, but coordinated manner of security behavior that is intrinsically consistent with the distributed form of IoT infrastructures. This combined outlook makes the work add a new conceptual base of resilient IoT data protection, with flexibility, awareness of a context, and continuity of the system without a rigid structure of control.

Future Work

Further efforts in the future will involve future AMTW expansion into large-scale real-world deployments with heterogeneous hardware limitation. More studies will be conducted to integrate it with federated learning to predictively adapt security. Further research will deal with formal verification of trust weaving stability when there are adversarial conditions.

References

1. F.M.R. Ribeiro Junior, C.A. Kamienski, Data resilience system for fog computing. Comput. Netw. **195**, 108218 (2021). https://doi.org/10.1016/j.comnet.2021.108218
2. A.M. Alwakeel, An overview of fog computing and edge computing security and privacy issues. Sensors **21**(24), 8226 (2021). https://doi.org/10.3390/s21248226
3. D. Rupanetti, A.K. Mishra, D.S.K. Rao, Combining edge computing-assisted internet of things security with artificial intelligence: applications, challenges, and opportunities. Appl. Sci. **14**(16), 7104 (2024). https://doi.org/10.3390/app14167104
4. B. Omoniwa, R. Hussain, M.A. Javed, S.H. Bouk, S.A. Malik, Fog/edge computing-based IoT (FECIoT): architecture, applications, and research issues. IEEE Internet Things J. **6**(3), 4118–4149 (2019). https://doi.org/10.1109/JIOT.2018.2875544

5. J. Moura, D. Hutchison, Fog computing systems: state of the art, research issues and future trends, with a focus on resilience. J. Netw. Comput. Appl. **169**, 102784 (2020). https://doi.org/10.1016/j.jnca.2020.102784
6. V. Roy, A context-aware internet of things (IoT) founded approach to scheming an operative priority-based scheduling algorithms. **13**(1), 28–35 (2024). https://doi.org/10.54216/jcim.130103
7. S. Ahmad, I. Shakeel, S. Mehfuz, J. Ahmad, Deep learning models for cloud, edge, fog, and IoT computing paradigms: survey, recent advances, and future directions. Comput. Sci. Rev. **49**, 100568 (2023). https://doi.org/10.1016/j.cosrev.2023.100568
8. K. Taji et al., Enhancing security and privacy in smart agriculture using edge computing models. Comput. Electron. Agric. (2024). https://doi.org/10.1016/j.compag.2024.107567
9. X. Lu et al., Blockchain-assisted edge computing for secure IoT data management. IEEE Access **9**, 115901–115914 (2021). https://doi.org/10.1109/ACCESS.2021.3107862
10. S.A. Kouchay, Edge, fog, and cloud computing in IoT—Significance and security concerns. i-manager's J. Cloud Comput. **10**(1), 17–29 (2023). https://doi.org/10.26634/jcc.10.1.19210
11. A.K. Chandanan, M. Rani, K.S. Pokkuluri, S. Singh, V. Jaiswal, P. Narayana, V. Roy, Revolutionizing cardiac prediction based on fog-cloud-iot integrated heart disease model. Scalable Comput.: Pract. Exp. **26**(5), 2105–2117. ISSN 1895-1767. http://www.scpe.org. https://doi.org/10.12694/scpe.v26i5.4715
12. L. Kong et al., Edge-computing-driven IoT: a survey. Comput. Surv. (CSUR) **54**(7) (2022). https://doi.org/10.1145/3485129
13. D. Kolevski, K. Michael, Edge computing and IoT data breaches: security, privacy, trust, and regulation. IEEE Technol. Soc. Mag. **43**(1) (2024). https://doi.org/10.1109/MTS.2024.337260
14. A. Al-Shareeda, K.A. Al-Thelaya, M. Abdallah et al., A survey on security and privacy issues in edge-computing-assisted internet of things. IEEE Internet Things J. **8**(5), 4004–4022 (2020). https://doi.org/10.1109/JIOT.2020.3015432
15. F.M.R. Ribeiro Junior (same author expanded work), Fog-DaRe: resilient fog data-flow (design & evaluation). Comput. Netw. (extended/related) (2021). https://doi.org/10.1016/j.comnet.2021.108218
16. M. Laroui, A. Hossain, M. Al-Shareeda (survey authors—Example), Edge and fog computing for IoT: a survey on current trends, security and data protection mechanisms. Comput. Commun. (survey) (2021). https://doi.org/10.1016/j.comcom.2021.09.003
17. N. Moustafa, E. Adi, B. Turnbull, J. Hu, A new threat intelligence scheme for safeguarding industry 4.0 systems. IEEE Access **6**, 32910–32924 (2018)
18. C. Peng, M. Luo, H. Wang, M.K. Khan, D. He, An efficient privacy-preserving aggregation scheme for multidimensional data in IoT. IEEE Internet Things J. **9**(1), 589–600 (2022). https://doi.org/10.1109/JIOT.2021.3083136
19. T. Nguyen, H. Nguyen, T. Nguyen Gia, Exploring the integration of edge computing and blockchain IoT: principles, architectures, security, and applications. J. Netw. Comput. Appl. **226**, 103884 (2024). https://doi.org/10.1016/j.jnca.2024.103884
20. A. Ometov et al., A survey of security in cloud, edge, and fog computing. Sensors **22**(3) (2022). https://doi.org/10.3390/s22030927
21. Z.A. El Houda, L. Khoukhi, B. Brik, A low-latency fog-based framework to secure IoT applications using collaborative federated learning. IEEE Local Comput. Netw. Conf. (2022). https://doi.org/10.1109/LCN53696.2022.9843315
22. A.K. Jain, M. Mudholkar, P. Mudholkar, M.K. Sharma, M. Tiwari, V. Roy, A novel approach to optimized cluster head selection in heterogeneous WSNs for IoT deployments, in *2024 IEEE 2nd International Conference on Innovations in High-Speed Communication and Signal Processing (IHCSP)* (Bhopal, India, 2024), pp. 1–6. https://doi.org/10.1109/IHCSP63227.2024.10960066
23. J. Mouradian, M.R. Mouradian, et al., Fog computing systems: resilience-focused survey and issues. J. Netw. Comput. Appl. **169**, 102784 (2020). https://doi.org/10.1016/j.jnca.2020.102784
24. C. Perifanis, N.A., Edge and fog computing business value streams through IoT: architecture and security perspectives. Information **13**(9), 427 (2022). https://doi.org/10.3390/info13090427

25. Z. Song et al., Secure and efficient multidimensional data aggregation in IoT systems. Math. Probl. Eng. (2023). https://doi.org/10.1155/2023/6693296
26. S.N. Elahmadi et al., Efficient privacy-preserving IoT data aggregation using blockchain and homomorphic encryption. Sensors **21**(7) (2021). https://doi.org/10.3390/s21072452

Secure and Sustainable IoT Architectures for Smart Healthcare: Case Studies and Low Carbon Solutions

Shaik Khaja Mohiddin and Shaik sharmila

1 Introduction to Smart Healthcare IoT

The use of the Internet of Things in healthcare systems has revolutionized the old-fashioned medical practice by providing possibilities to continuously monitor, obtain real-time data, and make smart decisions. Smart healthcare IoT systems are multi-connected systems of medical devices, wearable sensors, implantable systems, and cloud or edge platforms that gather the physiological indicators, patient activity, and clinical information. These networked systems facilitate some of the applications of remote patient monitoring, chronic disease management, remote medicine, and response to emergency queries, hence increasing the availability of healthcare services and level of patient care. Introducing changes to healthcare delivery processes, transforming the traditional reactivity models to proactive ones, the IoT technologies are used to assist clinicians in making effective and timely decisions.

Although there are these benefits, smart healthcare IoT systems are deployed in extremely sensitive and complicated environments where failure to maintain reliability, safety, and data confidentiality is paramount. IoT devices used in the medical field can easily be exploited in cyberattacks, data breaches, and system malfunctions since they usually operate within stringent resource limits and process sensitive patient data. Also, there are challenges of interoperability and scalability due to the diversity of devices and communication protocols. These fears emphasize this reason why intelligent, safe, and reliable IoT architectures are of specific use in healthcare. The background and context of the operation of the smart healthcare IoT are key in

S. K. Mohiddin (✉)
Department of Computer Science and Engineering, Siddhartha Academy of Higher Education, Deemed to Be University, Vijayawada, Andhra Pradesh, India
e-mail: mail2mohiddin@gmail.com

S. sharmila
Department of Information Technology, Vignan's Nirula Institute of Technology and Science for Women, PedaPalakalur, Guntur, Andhra Pradesh, India

S. Saif et al. (eds.), *Secure Intelligent Sustainable Applications Using IoT for Society 6.0*, Internet of Things, https://doi.org/10.1007/978-3-032-27248-5_7

creating solutions that can manage the balance between innovation, patient safety, privacy, and regulatory compliance.

1.1 Evolution of IoT in Healthcare Systems

A paradigm shift has been experienced in the healthcare industry, where traditional models of the sector were traditional and hospital-centric, to now be more proactive and patient-focused ecosystems facilitated by the Internet of Things (IoT). The initial e-health projects concentrated on the digitalization of records (Electronic Health Records); nevertheless, the contemporary trend has given rise to the Internet of Medical Things (IoMT) an enormous network of medical devices, wearables, and sensors that continuously track physiological data [1].

This huge adoption is in the order of market trajectory; the global market of IoT in healthcare is expected to have a range of about USD 70 billion in the year 2025 up to upper USD 464 billion in year 2034 due to the demand of remote patient monitoring and telemedicine [2]. This development has not only advanced past connectivity into Intelligent IoT in which AI-driven algorithms interpret sensor data in real-time to forecast health events before they happen [3]. As a result, the healthcare delivery has reached outside the hospital boxes wherein continuous care can be provided at home platforms using smart wearable devices to monitor vital checks such as glucose levels and heart rates without the help of humans [4].

1.2 Need for Security and Sustainability in Medical IoT

The IoMT ecosystemE “Sustainability” \r “C28” is a growing field that presents two important issues: the necessity to have a strong security situation and the necessity to be ecologically friendly.

Security Imperatives: In the dark web, medical data is one of the most lucrative with an estimation of 50 times more valuable than any normal credit card data [5]. The flexibility of the IoT devices exposes them to advanced cyber-attacks, such as ransomware and Man-in-the-Middle (MitM) attacks that have the potential to modify life-threatening information (Integrity attacks) or disable services (Availability attacks/DoS) [6]. Providing Confidentiality, Integrity, and Availability (CIA) of the patient data is not only an IT requirement, but also the need to guarantee the patient safety.

Sustainability and Green IoT: Sometimes, at the same time the environmental footprint of this digital transformation is increasing. By the year 2030, the production and deployment of new IoT technologies is estimated to lead to a greater global electricity consumption by more than 34 terawatt-hours (TWh) [7]. Healthcare IoT leads to this digital carbon footprint by continuously using a lot of energy to transfer raw data to cloud services and high-power consumption of the data center behind it

[8]. Moreover, the spread of disposable medical sensors on battery power also adds to the growing amount of electronic waste (e-waste) [9].

A very important trade-off is critical to this but needs to be resolved: using a heavy security measure (such as complex encryption) is more computationally intensive, consuming more batteries and more energy. As such, the solutions to the urgent requirement of Green Security are the urgent provision of architectures that offer high security and low energy overhead.

1.3 Scope and Objectives of the Chapter

To bridge security to sustainability, divided in smart healthcare this chapter discusses the special security vulnerabilities and energy limitations of modern Internet of Medical Things (IoMT) equipment. It proposes contributing, safe IoT architecture designs combining lightweight cryptography along with energy-efficient edge computing scheme and demonstrates their efficiency with the use of real case studies in patient tracking and telehealth. Additionally, the chapter discusses the issue of new technologies which comprise blockchain and federated learning as less green, more standardized centralized data processing.

2 IoT Security Challenges in Healthcare Environments

The security issues of IoT in healthcare settings are attributable to the high number of medical devices connected to the network like wearable sensors, implantable medical devices, remote care, and smart hospital infrastructure. These first aid machines constantly capture, transfer, and keep confidential patient information, and thus are a lucrative target in case of a cyber-attack [10]. Our software and hardware devices Healthcare IoT devices do not have the freedom to use powerful computational capabilities, limited memory, or low power energy, which often restrain the possibility to enforce strong security. Therefore, they are at risk of unauthorized access, data alteration, ransomware infections, and denial-of-service attacks that have a direct impact on patient safety and clinical decision-making. Also, diverse device standards and the existence of old medical systems also complicated the integration process in hospital networks in a secure way [11].

The second and other significant problem is the ability to provide data privacy and regulatory compliance at the same time maintaining system availability and reliability. Internet-of-things related to healthcare should comply with stringent requirements regarding patient data safety; however, constant connectivity is making them more exposed to insider attacks and other external opponents. The absence of real-time monitoring, weak authentication and firmware updates, may result in the loss and breach of data and the destruction of confidence among patients and medics [12]. In addition, due to the increased uptake of cloud and edge IoT architectures by healthcare systems, it is becoming more challenging to ensure the safety of the data transmission in distributed settings. To overcome these issues, the following steps should be taken that involve a comprehensive security strategy that encompasses lightweight security measures along with policy-sensitive data handling that is specific to the healthcare context [13].

2.1 Threat Landscape in Healthcare IoT Networks

The healthcare sector has become the primary target for cybercriminals due to the high monetary value of Personal Health Information (PHI) and the critical nature of medical services. In 2024, the industry witnessed a record-breaking surge in ransomware attacks, exemplified by the massive breach at Change Healthcare which compromised the data of nearly 100 million individuals [14]. These attacks are increasingly sophisticated; threat groups like LockBit 3.0 and ALPHV/BlackCat are now specifically targeting the "Internet of Medical Things" (IoMT) infrastructure to encrypt patient records and demand exorbitant ransoms. Beyond ransomware, the threat landscape is dominated by Distributed Denial of Service (DDoS) attacks, where malicious actors flood network traffic (often using UDP and ICMP floods) to overwhelm hospital servers, causing life-threatening delays in patient care [15].

2.2 Data Privacy, Confidentiality, and Regulatory Compliance

Cybercriminals have identified the medicine sector as their main target since Personal Health Information (PHI) is made up of large sums of money and medical services are needed. This year also turned out to be the most intense year of ransomware attacks in the industry, with the monumental attack that took place in Change Healthcare in 2024 exposing almost 100 million people [16]. These attacks become more of high sophistication as the threat groups such as LockBit 3.0 and ALPHV/BlackCat now specifically focus on the Internet of Medical Things (IoMT) infrastructure and encrypt patient reports before forcing them into paying extraordinary ransoms. In addition to ransomware, Distributed Denial of Service (DDoS) types of attacks are predominant, as in these cases malicious actors saturate network traffic (frequently by floods of UDP and ICMP packets), and literally cripple the servers of hospitals, leading to life endangering delays in patient care [17].

2.3 Security Constraints in Resource-Limited Medical Devices

One of the most important obstacles to the implementation of the IoMT is the extreme lack of resources of the devices themselves, especially implantable sensors and wearables. These gadgets have low power and processing power, and it is computationally impossible to execute normal, full-fledged encryption protocols (such as RSA-2048) [18] or blockchain code (without quickly burning the battery). Thus, most medical devices are used with weak or absent encryption; manufacturers are more concerned with the device's longevity and functionality rather than with high-quality security, thus they are vulnerable to attacks of tampering and battery-depleting denial-of-sleep attacks [19].

3 Foundations of Green IoT for Healthcare

Green IoT in healthcare is based on frameworks of system design and implementation of Internet of Things with minimal energy use and environmental effects at the expense of reliability and patient safety. The IoT medical setting is based on a great variety of devices, e.g., wearable sensors, implantable medical devices, bedside monitors, and remote diagnostic devices, all of which generate continuous data volumes of vast amounts. The principles of green IoT focus on the design and construction of energy-efficient hardware, low-power communications, and hard-core data management to ensure unwarranted sensing, processing, and transmission. This approach will also enable medical IoT systems to consume a lot less power and increase the lifetime of devices through event and edge-based processing paradigms, which are alternative to the continuous data streaming at the edges of raw data (healthcare systems) and can greatly decrease the carbon footprint of the operational systems [20].

Besides energy efficiency, Green IoT of healthcare has the concept of sustainability in the entire lifecycle of the device, including manufacturing, deployment, operation, and disposal. The energy-saving ideas like harvesting energy by the heat or movement of the human body, installation of renewable energy-driven edge and cloud systems, and minimization of waste of electronic devices by increasing their service life are important and significant [21]. Green IoT systems are not far off with the needs of data privacy and data security because when the need to transmit data is reduced, it not only conserves power but also exposes the system to less risk of cyber-attacks. A combination of these principles will lead to the development of sustainable, secure, and scalable healthcare IoT systems that can sustain long-term digital health innovation without neglecting the environment [22].

3.1 Energy Consumption Patterns in Healthcare IoT

The energy profile of Medical IoT (IoMT) devices is an essential concept because what makes the operational carbon footprint significant is the high rate of data transfer, and not local processing. Experiments have shown that the radio frequency (RF) module deployed in wireless communications absorbs much more power than the onboard sensors or microcontroller and can be up to 80 per cent of the total energy expenditure of a device when wasting power in the air. Therefore, the customary model with wearable devices constantly transferring raw physiological information to the cloud does not scale and predicts a fast waste of the battery and more extensive utilization of power-dense data facilities [23].

This architecture in Fig. 1 depicts a comprehensive blueprint for the inclusion of security and sustainability into a smart healthcare framework. The source of data is a patient-based Heart Monitor Wearable, secured by the initial security measures, and transferred to a local Green Edge Gateway. This is an edge device that is more energy efficient through the implementation of partial data processing at the edge, as well as less energy-intensive transmission to the core network. The processed information is then transmitted through an "Encrypted Data Stream" to a "Secure and Sustainable Cloud" which is represented to consume renewable energy sources such as solar power [24]. The total design gives focus on the two essential purposes of keeping strong "Data Privacy" by means of encryption and attaining a "Low-Carbon Solution" by computing energy consciousness and incorporation of renewable resources.

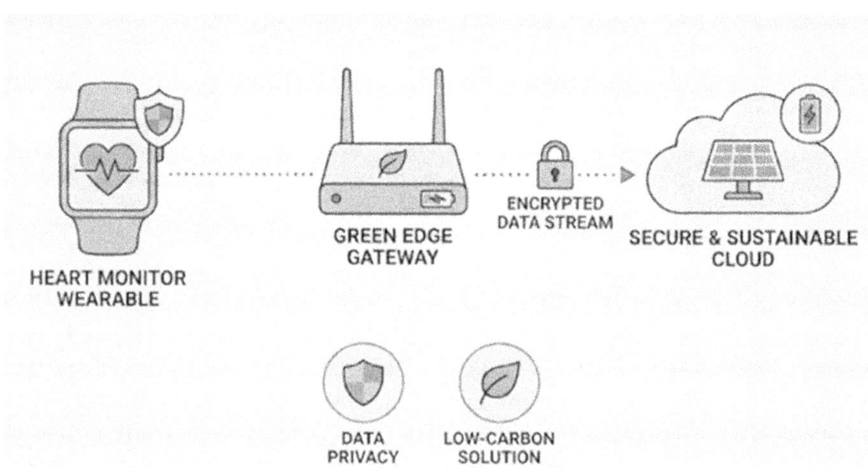

Fig.1 Conceptual diagram of a secure and sustainable green IoT architecture for smart healthcare

3.2 Low-Power Communication Protocols

In response to the high cost of energy, Low-Power Wide-Area Network (LPWAN) technologies are set to replace energy-intensive standards, such as Wi-Fi or cellular LTE, in smart healthcare systems. Bluetooth Low Energy (BLE) [25] and other protocols are now commonly used in short-range Body Area Networks (BANs) because they can work over a few months on a coin-cell battery, and LoRaWAN and NB-IoT are now used to operate long-range remote monitoring and provide deep indoor penetration in a hospital setting with minimum power consumption [26]. A comparative study contends that NB-IoT is more reliable in important alerts when using licensed bands, whereas LoRaWAN is more economical on non-important periodic health alerts when deployed in rural telemedicine networks.

3.3 Carbon Footprint Reduction Strategies for Medical IoT

To ensure a decreased carbon footprint of healthcare IoT, event-based monitoring with renewable energy integration needs to move toward always-on sensing but with an event-triggering mechanism. There are methods like Energy Harvesting whereby a device uses body heat (thermoelectric) or patient movement (piezoelectric) as its source, and thus no disposable batteries are required, resulting in the elimination of electronic waste. Moreover, the deployment of Green Fog Computing architecture also enables filtering and processing Amazon Data at the edges (e.g., bedside gateway), minimizing the amount of data sent to the cloud by 60 percent, reducing the greenhouse gas emissions caused by the processes of cooling and storing data on backend servers [27].

4 Secure and Sustainable IoT Architectural Models

The proposed secure and sustainable IoT architectural models in healthcare are intended to meet the two demands of high level of data security and energy efficiency in the interrelated medical systems competently. Such architectures combine lightweight device-level security to protect confidentiality, integrity, and authentication, but avoid consuming too much computational and energy resources of medical IoT devices. Radical models use distributed intelligence instead of centralized processing alone, which consists of edge computing and fog computing (ES2): sensitive health information is filtered and pre-processed at or near where it is collected. This technology decreases latency, unnecessary data transmission, and energy usage even without violating the regulations of healthcare data protection [28]. These architectural models have a sustainability focus in terms of energy conscious communication, adaptive workload offloading, and integration of renewable energy in both edge and cloud layers. Flowing data is secured by the minimization of event-driven communication and lightweight protocols that enable devices to spend long periods in low-power states. Cloud infrastructures that operate on renewable energy resources can facilitate large-scale analytics and store data on long-term levels at higher levels without causing a significant spike in carbon emissions. Secure and sustainable IoT environments have the potential to offer resilient frameworks of architecture of smart healthcare through the integration of security-first design and smart green computing solutions that enable large-scale architecture to support efficient scaling, patient confidence, and sustainable digital health servicing of patients [29].

4.1 Lightweight Cryptographic Mechanisms for Healthcare Data

The conventional cryptographic standards used in the traditional IT devices, including RSA- 2048 or AES-256, have a high level of computational overhead that is not usually viable when using resource-limited IoMT devices. The mathematical complexity needed to run these protocols quickly wears out small batteries in the implantable sensors and wearables, requiring replacement and adding to e-waste. To this end, there is a need to integrate Lightweight Cryptography (LWC) to secure healthcare architectures. Originally, LWC algorithms were tailored to shortcomings in requirements, such as the harsh capacity limits on power, memory, and processing speeds, without the corresponding increase in security margin. The implementation architecturally is to implement the LWC alternatives to the standard protocols at the device level. As an illustration, symmetric encryption implementation of sensor data may make use of lightweight block ciphers such as SIMON and SPECK, or the recently NIST-standardized ASCON family. In addition, Elliptic Curve Cryptography (ECC) has been proposed as a more sustainable alternative to RSA for the secure exchange of keys, with the same level of security power, considerably smaller key sizes, and minimized computational energy requirements. It has been shown that optimized LWC protocols can cut direct energy use by more than 60-percent, relative to traditional AES implementations, which directly prolong device lifecycles and bring security practices in line with sustainability objectives [30].

4.2 Energy-Aware Edge–Cloud Integration Models

Another major source of operational carbon footprint in healthcare IoT is the constant wireless exchange of raw physiological measurements of sensors to remote centralized clouds. Radio transmission through wireless devices is very power hungry and can be more energy consuming than sensor data acquisition on the device or local processing. To address this, we propose a Green Edge-Cloud Architecture whereby the intelligence processes are now moved to the network edge, the nearest location to the creation of the data, including a bedside smart gateway or the smartphone of a patient.

The edge node in this model is an intelligent energy aware intermediary. The edge does not simply forward all data without preprocessing and data compression, along with anomaly detection in a blind manner. The transmission to the cloud is limited to critical health events or aggregated summaries to be stored, and complex analytics in the long run. Importantly, this architecture uses energy-conscious offloading algorithms. This smart object is dynamic, and it chooses between local and offload processing of a task based on real-time parameters, such as the remaining power of the wearable, the signal level of the active network (transmission energy cost) at the wearable, and the latency demands of the medical task. Such a transmission strategy can save the bandwidth and system energy requirements considerably [31].

The former Fig. 2 shows that the left image shows the Traditional methodology, where numerous sensors send raw data to a centralized cloud, which consumes much energy and is also slow. In the right panel, one can see the proposed Green-Security, the data input of which first passes a Green Edge Layer (implementing such technologies as tiny-ML), and only safe data is transmitted to a cloud and blockchain network. This flow is optimized, thus saving energy and enhancing security with decentralized trust.

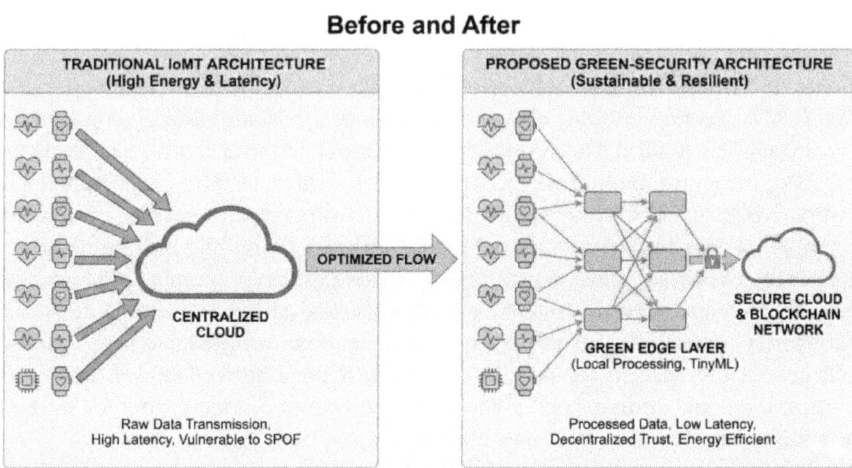

Fig. 2 Architectural shift to green-security

4.3 Privacy-Preserving and Low-Carbon System Design

The long-term sustainability and trust of a given system cannot be achieved without considering a holistic system design premised on the principle that less is more in terms of data and energy. It entails assuming a Data Minimization policy, which is one of the pillars of privacy regulations such as GDPR, and is effective as a sustainability measure. The architecture allows only a specific amount of data required by a given medical task to be collected, processed, and transmitted, and therefore minimizes both the attack surface of possible intrusion as well as the energy used to handle data throughout the whole network. Also, it will require the system architecture to move to event-based protocols, as opposed to constant, time-consuming polling. Use of energy efficient application layer protocols like the Constrained Application Protocol (CoAP) or MQTT-SN (MQTT for Sensor Networks) instead of heavy HTTP using TLS decreases the protocol overhead. Such protocols can facilitate effective sleep scheduling and, therefore, the medical sensors can be kept in an ultra-low-power dormant state until a particular physiological event has been detected (e.g., arrhythmia) to optimize the battery life and reduce the carbon emissions of the monitoring system in operation [32].

5 Secure Remote Patient Monitoring Systems

In a recent paper that studied a Wireless Body Area Network (WBAN) to relay cardiac patients' information, the effectiveness of a secure and effective framework called SEES-IoMT was proven to be effective because of the lack of energy consumption and resistance to attacks. The system has been based on the critical issue of unequal energy consumption among the biosensor nodes, which usually causes premature network breakage. The system was designed to support the security of sensitive physiological data across insecure public networks by promoting the confidentiality and integrity of physiological data through a lightweight cryptographic scheme rather than costly computationally intensive standards, and at the same time conserved sensor battery power. The results of the experiment demonstrated that such a method could ensure the security of patient data against unauthorized nodes, as well as minimize the connection expenses and power consumption, increasing the duration of the functioning of the whole monitoring system (Table 1).

Table 1 Comparative analysis of secure and sustainable smart healthcare case studies

S. No	Application scenario	Architectural solution and technology focus	Security measures implemented	Sustainability outcome (energy and carbon)	Key trade-off or challenge addressed
1	Remote cardiac patient monitoring (Wearable ECG)	Device-level LWC: Direct implementation of lightweight Cryptography on resource-constrained wearable microcontrollers	Optimized elliptic curve cryptography (ECC) for key exchange; SIMON block cipher for data encryption	Battery life extension: reduced energy encryption by ~ 60%, extending wearable operation from 2 to 7 days between charges	Balancing sufficient cryptographic strength with the severe power constraints of continuous vital sign monitoring
2	Smart home elderly fall detection systems	Green edge computing: Utilizing a local smart gateway to process sensor data within the home before cloud transmission	Local data filtering and anonymization at the edge gateway before secured transmission via TLS to the cloud	Reduced transmission energy: 80% reduction in bandwidth and associated transmission power by sending only verified "fall events" rather than raw video/sensor feeds	Lowering latency for critical alerts versus the energy cost of constant cloud connectivity
3	ICU high-frequency vital sign monitoring	Fog computing layer: introducing intermediate fog nodes in hospital wings to aggregate data from multiple bedside monitors	Mutual authentication between bedside sensors and fog nodes; encrypted storage within the local fog infrastructure	Lower data center load: Significantly decreased the volume of data sent to backend servers, reducing cooling and storage energy demands in the hospital data center	Capital investment in on-premise fog infrastructure versus long-term operational energy savings
4	Implantable medical devices (e.g., Pacemakers)	Event-driven protocol and Ultra-LWC: Sleep-scheduling protocols combined with customized, minimal-overhead encryption	Proprietary ultra-lightweight protocols with strict authentication to prevent unauthorized configuration changes; encrypted telemetry only upon external query	Maximized device longevity: Ensures multi-year battery life to avoid surgical replacement, minimizing bio-waste and medical procedures	Extreme prioritization of energy conservation over continuous security monitoring, relying on "security through obscurity" and access controls

(continued)

Table 1 (continued)

S. No	Application scenario	Architectural solution and technology focus	Security measures implemented	Sustainability outcome (energy and carbon)	Key trade-off or challenge addressed
5	Large-scale telemedicine platform	Sustainable cloud backend: Hosting the telehealth platform on data centers powered by certified renewable energy sources	End-to-end encryption for video consultations; strict HIPAA/GDPR compliant access controls and audit logs	Offsets the digital carbon footprint of the infrastructure against the massive emissions saved by eliminating patient vehicular travel	Balancing the increased energy demand of digital infrastructure with the reduced environmental impact of physical transportation
6	Pharmaceutical supply chain provenance	Private "Green" blockchain: Using a permissioned blockchain with an energy-efficient consensus mechanism (e.g., Proof-of-Authority)	Immutable ledger for tracking drug origin and handling; digital signatures for all supply chain participants	Low-energy integrity: Provides transparent traceability without the massive energy consumption associated with public Proof-of-Work blockchains like Bitcoin	Trading full decentralization for necessary energy efficiency and transaction speed in a corporate environment
7	Remote outdoor worker health safety	Hybrid energy harvesting: Wearables equipped with piezoelectric (movement) and thermoelectric (body heat) generators	Secure, low-power Bluetooth (BLE) pairing with mobile devices; encrypted local storage during transmission gaps	Self-powered operation: Achieves near-autonomous operation in field environments, eliminating reliance on grid charging and disposable batteries	Reliability of intermittent power sources versus the need for continuous security protocols

6 Emerging Technologies for Secure and Green Healthcare IoT

The new technologies are becoming increasingly important in solving the acute shortcomings of the traditional centralized and energy-consuming architectures and providing secure and green healthcare IoT. The use of blockchain technology is also emerging to deliver decentralized trust, data integrity, and a public audit trail to various stakeholders in the healthcare industry without necessarily involving a single trusted party. Blockchain can be used to safely secure medical data, device authentication, and pharmaceutical distribution using an energy-efficient consensus mechanism, which ensures energy consumption remains within limits. Simultaneously, the federated learning approach makes (medical) analytics privacy preserving, as the machine learning models can be trained on edge devices or hospital servers, so that sensitive patient data will never leave its origin. The decentralized paradigm of learning can save a lot of data transmission overhead (and, consequently, carbon emissions) and ensure the security and sustainability of large-scale healthcare analytics (Fig. 3).

6.1 *Blockchain for Trust and Transparency in Healthcare IoT*

Blockchain has already become an influential technology within the healthcare sector to achieve a decentralized trust and the provenance of unalterable data in fragmented

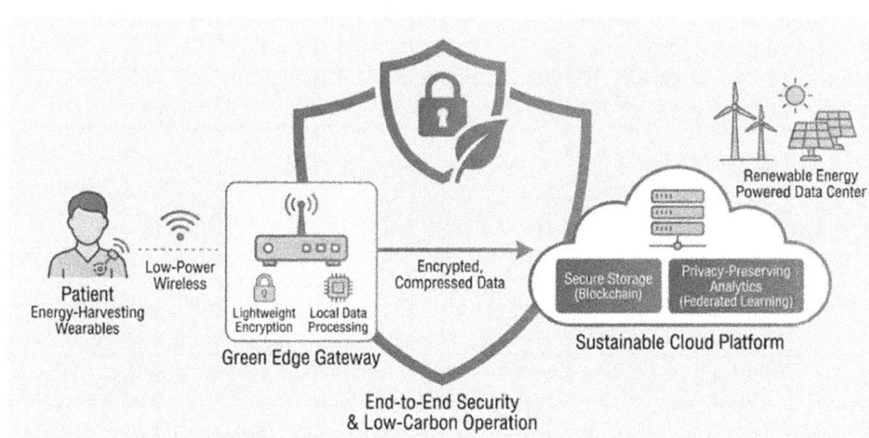

Fig. 3 Workflow of secure and sustainable Green IoT data transmission for smart healthcar

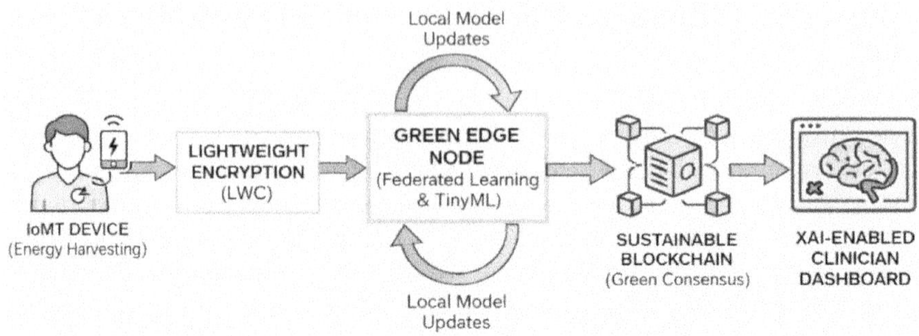

Fig. 4 Architecture of the proposed Green Edge AI enabled IoMT framework for secure and sustainable smart healthcare

healthcare systems. Blockchain provides integrity and accountability of all data transactions by producing an immutable distributed registry of all data transactions that is tamper-proof and verifiable among a variety of different stakeholders, such as hospitals, patients, and equipment manufacturers [33]. Nevertheless, conventional blockchain consensus solutions such as Proof-of-Work (PoW) are notorious for consuming a lot of energy, which makes them inappropriate for green IoT. The future will be in moving toward using more energy-efficient consensus mechanisms like Proof-of-Stake (PoS) or Proof-of-Authority (PoA). The mechanisms exclude competitive computational mining, hence providing a sustainable path to achieve medical records, track pharmaceutical supply trees, and control patient approval with limited effects on the environment [34].

Figure 4 resembles an IoMT Device (which can be powered by energy harvesting), then Lightweight Encryption (LWC) to guarantee the safety and security process, and then a Green Edge Node, where local processing, Federated Learning, and TinyML will be applied (with a feedback loop to update the model). Such processed information may be stored safely under a Sustainable Blockchain and visualized on a Clinician Dashboard with XAI-Enabled to make data-driven decisions.

6.2 Federated Learning for Privacy-Preserving Medical Analytics

Standard AI models require centralizing vast amounts of patient data on servers for training, creating significant privacy risks and consuming immense energy for data transmission and data centre cooling. Federated Learning (FL) fundamentally flips this model by bringing the algorithm to the data, rather than the data to the algorithm. In an FL architecture, local models are trained directly on edge devices (e.g., smartphones, hospital servers) using local data. Only model parameters (weights and gradients), not raw patient data, are encrypted and sent to a central server for aggregation [35]. This decentralized approach not only ensures robust patient privacy by design but also significantly reduces the carbon footprint associated with high-bandwidth data transfer and centralized storage, making medical analytics both secure and sustainable.

6.3 *Explainable AI for Secure and Energy-Efficient Decision Support*

The trust deficit with complex deep learning models is caused by the black-box characteristics of new AI-powered Clinical Decision Support Systems (CDSS) as they become more common in the future. Explainable AI (XAI) is aimed at resolving this issue by rendering the results of AI models transparent and explicable. XAI can be used in two ways in the framework of secure and green IoT. First, it increases security, as human experts will be able to audit AI decisions that it can be biased or adversarially manipulated. Second, XAI can be used to achieve a more accurate and specific clinical intervention by giving interpretable insights into the reasons why a decision was taken (e.g., which specific vital sign abnormalities triggered an alert). This ratchets down unnecessary diagnostic tests and stays in the hospital, and this, indirectly, decreases energy use and the total resource burden of the healthcare system [36].

7 Open Challenges and Research Directions

Although there are major steps in the creation of a system of secure and green healthcare IoT, there are still several important obstacles that do not allow making its introduction everywhere and make it sustainable. It is vital to address these open issues to achieve the maximum potential of a proactive and environmentally aware digital health ecosystem.

7.1 *Trade-Offs Between Security Strength and Energy Efficiency*

The biggest dilemma on the Internet of Medical Things (IoMT) is that strong security and energy saving are inherently in opposition. Medical implantable and wearable devices are powered by very small budgets, sometimes by small coin-cell batteries that are supposed to last months or years. Nevertheless, issuing cryptography protocols, including RSA-2048 key exchange or AES-256 data encryption, that are required to ensure patient safety, pose excessive computing costs. These intricate mathematical calculations take a lot of CPU cycles, which is directly proportional to quick battery drain [37]. The most important problem here is the definition of the best security-energy Pareto frontier. When security is maximized, the device might need regular recharge or an invasive operation at the cost of the patient having a poor quality of life and putting more strain on the environment with a greater number of discarded devices. On the other hand, the focus on energy efficiency due to the use of weak encryption or with less frequent authentication exposes the devices that are life-dependent to cyber-attacks. Lightweight cryptography designs today only provide what is considered partial protection and are often not demonstrated to have strict security evidence bases as that of the well-known standards, requiring a trust gap to adopt to critical care applications [38].

7.2 Scalability and Interoperability Issues

Since the predicted number of linked medical apparatuses is likely to reach billions, the contemporary model of architecture is encountering critical challenges of scaling. The classical centralized architecture of clouds is no longer sustainable; the enormous amount of data produced by millions of active sensors causes bandwidth constraints, a high-latency count, and enormous energy to be used in data centers of the back-end type. To scale these systems safely and with a small carbon footprint, with a fundamental change toward decentralized, distributed intelligence that the present infrastructure is poorly equipped to serve, a paradigm change is needed.

Moreover, the IoMT situation is also quite dispersed, with no common interoperability standard. Various devices, when using different manufacturers, tend to operate on proprietary protocols and data formats, which form data silos. The issue is that this heterogeneity not only makes patient care more problematic, but it also poses a huge threat of security vulnerability as interfaces between systems that cannot work together are an accessible entry point. With respect to sustainability, low interoperability will result in the transmission and processing of redundant data, which will result in unnecessary growth of the overall energy footprint of the healthcare network. Establishing standardized, open-source protocols that are secure and device-based energy-efficient by nature is an industry-wide monster of a problem [39] (Table 2).

7.3 Future Research Opportunities in Low-Carbon Healthcare IoT

To address the existing limitations, future studies should deal with the ubiquity of new materials science, artificial intelligence, and new computing models. The former lines of research are to eliminate the battery, through more efficient energy harvesting technology, light-focus on piezoelectric technology of high efficiency, thermoelectric

Table 2 Comparative analysis of future research directions for secure and sustainable healthcare IoT

Research direction/ technology	Primary sustainability benefit (Green IoT Goal)	Security implications and challenges	Scalability and interoperability impact	Maturity level and timeframe
Advanced energy harvesting (e.g., Piezoelectric, Thermoelectric)	Eliminates dependence on disposable batteries; extends device lifespan indefinitely, reducing e-waste	Intermittent power supply makes maintaining continuous, secure sessions and running heavy encryption protocols difficult	Enables massive deployment of "deploy-and-forget" sensors in remote locations without maintenance constraints	Emerging (3–5+ years for high efficiency)
TinyML and edge intelligence (On-device processing)	Drastically reduces energy consumption by transmitting only processed insights rather than raw data streams	Enhances privacy by keeping raw data local; however, resource-constrained models may be vulnerable to adversarial attacks	Reduces bandwidth bottlenecks at the cloud level, allowing the network to accommodate more devices efficiently	Growing/Medium (Available now, optimization ongoing)
Lightweight cryptography (LWC) Standards	Minimizes computational overhead (CPU cycles) required for encryption, directly saving battery life	Provides essential security for constrained devices but may offer lower theoretical security margins than established high-power standards (e.g., AES-256)	Essential for securing billions of heterogeneous, low-power devices across different platforms	Medium/High (NIST standards emerging; adoption varies)

(continued)

Table 2 (continued)

Research direction/ technology	Primary sustainability benefit (Green IoT Goal)	Security implications and challenges	Scalability and interoperability impact	Maturity level and timeframe
Green blockchain consensus (e.g., Proof-of-Stake, Proof-of-Authority)	Offers decentralized data integrity with vastly lower energy consumption compared to traditional Proof-of-Work mechanisms	Provides immutable audit trails for patient records. New consensus mechanisms introduce different attack vectors (e.g., nothing-at-stake attacks)	Improves decentralized interoperability but still faces throughput limitations for high-frequency real-time sensor data	Emerging in Healthcare (Pilot projects underway)
Biodegradable/transient electronics	Addresses the end-of-life environmental impact by ensuring devices safely dissolve, eliminating toxic e-waste	Ensuring data availability and physical security before the device is scheduled to degrade; risk of premature failure	Currently limited by complex manufacturing processes and material durability costs	Nascent/Research Phase (5–10+ years for commercial use)

generators, radiofrequency harvesting, and development of ultra-low-power-secure communication protocols suitable to these power transient sources [40]. Meanwhile, the research area ought to extend beyond the standard edge computing to TinyML, deploying the microcontroller-based highly optimized network models that provide diagnostic clinical significance on-site and, accordingly, allow saving energy when transmitting data elsewhere [41]. At the same time, future research must be able to draw a sharp line between Green Consensus Mechanisms in blockchain, such as Proof-of-Stake or Proof-of-Authority, to provide a high rate of transactions and the demands of high transactions in healthcare devoid of the negative impact on the environment of traditional Proof-of-Work. Finally, sustainability is to be applied to the lifecycle of the devices in order to reduce the e-waste disaster through the development of transient electronics, biodegradable medical sensors, which degrade safely when in use, to enable healthcare innovation to not be a destroyer of the environment [42].

8 Conclusion

The adoption of the Internet of Medical Things (IoMT) has required a paradigm shift to the formation of so-called Green Security architectures that balance the conflicting needs of effective data protection and sustainability of the environment. As discussed in this chapter, it is now no longer a good idea to rely on the old traditional, centralized reproductions of clouds based on their high energy use and latency, but rather the industry needs to shift to the concept of decentralized Green Edge models based on lightweight cryptographic protocols and event-driven communications to reduce the digital carbon footprint without putting the safety of the patients in jeopardy. Using synergistic technology like energy efficient Blockchain, privacy preserving Federated Learning, and explainable AI, health systems will be able to achieve the resilient infrastructure that promotes decentralized trust, operational transparency, and substantial energy savings.

Going forward, to achieve a truly zero-carbon healthcare ecosystem, interdisciplinary research in the nexus of advanced materials and new computing paradigms is needed. Future studies will need to focus on creating self-powered "zero-energy" sensors that can use patient motion or body heat as a source of energy, as well as the implementation of TinyML algorithms, which will run directly on microcontrollers to avoid unnecessary transmission. In addition, tackling the environmental impact lifecycle requires that green blockchain consensus mechanisms are standardized, so no-energy-intensive mining processes can be used to address the issue, and biodegradable and so-called transient electronics are developed, which dissolve on use, eliminating the serious epidemic of medical e-waste without disrupting the continuous and secure monitoring.

References

1. M.T. Gures et al., A survey on IoT smart healthcare: emerging technologies, applications, challenges, and future trends. Semantic Scholar (2025)
2. IoT in Healthcare Market Size, Powering Telemedicine & Remote Healthcare. Towards Healthcare (2025). https://www.towardshealthcare.com/insights/iot-in-healthcare-market-size
3. Intelligent IoT Architectures for Smart Healthcare Systems. Int. J. Adv. Eng. Tech. Innov. **2**(1), 893 (2025)
4. IoT in Healthcare: Applications, Benefits & Challenges. Peerbits (2025). https://www.peerbits.com/blog/internet-of-things-healthcare-applications-benefits-and-challenges.html
5. S.R. Islam et al., Towards a secure and sustainable internet of medical things (IoMT): requirements, design challenges, security techniques, and future trends. Sustainability **15**(7), 6177 (2023)
6. T.N. Gia et al., A secured architecture for IoT-based healthcare system. 3Ciencias **11**(2) (2023)
7. Power play: reducing the end-to-end energy impact of IoT systems. Orange Business Services (2024). https://iotjourney.orange.com/en/news/orange-iot/power-play-reducing-the-end-to-end-energy-impact-of-iot-systems
8. A. McAlister et al., Assessing the carbon footprint of digital health interventions: a scoping review. J. Am. Med. Inform. Assoc. **29**(11) (2022)
9. S.K. Mohiddin, Y.S. Babu, Role of cloud forensics in cloud computing, in *Soft Computing for Problem Solving: SocProS 2018*, vol. 2. (Springer Singapore, Singapore, 2019), pp. 91–107
10. M.K. McGee, How healthcare cyberattacks broke records in 2024. GovInfoSecurity (2024). https://www.govinfosecurity.com/how-healthcare-cyberattacks-broke-records-in-2024-a-27116
11. Health-ISAC's 2025 Health Sector Cyber Threat Landscape report warns of rising ransomware, espionage, IoMT vulnerabilities. Industrial Cyber (2025)
12. The State of Healthcare Cybersecurity 2025. Veriti Research Report (2024). https://veriti.ai/wp-content/uploads/2024/12/The-State-of-Healthcare-Cybersecurity-2025-_-A-Veriti-Research-Report.pdf
13. M.A. Al-Garadi et al., A review on the security vulnerabilities of the IoMT against malware attacks and DDoS. arXiv:2501.07703 (2025)
14. C. Snyder, Healthcare's growing threat landscape. ISACA Industry News (2025)
15. S.R. Islam et al., Data privacy and security challenges in IoT healthcare. ResearchGate (2025)
16. 5 Trends Shaping Healthcare Cybersecurity in 2025. Palo Alto Networks Blog (2025)
17. I. Almomani et al., Implementing anomaly-based intrusion detection for resource-constrained devices in IoMT networks. Sensors **25**(4), 1216 (2025)
18. N.A. Ali et al., Data protection and privacy of the internet of healthcare things (IoHTs). Appl. Sci. **12**(4), 1927 (2022)
19. Security constraints in medical IoT. ResearchGate. Accessed Dec 2025
20. S.K. Mishra, A. Kumar, Green computing in IoT: reducing carbon footprint in smart devices. ResearchGate (2025)
21. M.S. Maharajan, T.A. Alghamdi, A.S. Alazea, Sustainable and green IoT for healthcare: architectures, technologies, and research challenges. IEEE Internet Things Mag. **7**(1), 44–50 (2024)
22. S. Rattal et al., AI-driven optimization of low-energy IoT protocols for scalable and efficient smart healthcare systems. IEEE Access **13**, 48402 (2025)
23. LoRaWAN VS NB-IoT: how do they compare and differ. MOKOSmart (2024). https://www.mokosmart.com/lorawan-vs-nb-iot-how-do-they-compare-and-differ/
24. Green computing in IoT: reducing carbon footprint in smart devices. ResearchGate (2025)
25. Enhancing energy efficiency in telehealth internet of things systems through fog and cloud computing integration: simulation study. JMIR Biomed. Eng. **9**(1), e50175 (2024)
26. M.A. Azam et al., A comprehensive assessment of LoRaWAN and NB-IoT performance metrics under varied payload data sizes. IEEE Access **13**, 4500–4512 (2025)

27. A. Al-Qerem et al., Lightweight cryptography for energy-conscious authentication in IoT systems. Sci. Inf. Organ. **16**(11) (2025)
28. S. Lee, K.A. Kang, S. Lee, J. Kim, Exploring authentication protocols for secure and efficient internet of medical things systems. Electronics **14**(21), 4164 (2025)
29. D. Swetha, S.K. Mohiddin, Quantum-enhanced security advances for cloud computing environments. Int. J. Adv. Comput. Sci. & Appl. **15**(6) (2024)
30. A. Al-Qerem, A. Al-Omary, M. Al-Mousa, Lightweight cryptography for energy-conscious authentication in IoT systems. Int. J. Adv. Comput. Sci. Appl. **16**(11) (2025)
31. J. Guo, Y. Liu, J. Zhang, Energy-saving service offloading for the internet of medical things using deep reinforcement learning. IEEE Internet Things J. **11**(4), 5678–5689 (2024)
32. N.Z.A. Zulkifli, N.A.M. Alduais, M.A. Abdullah, Energy inefficiency in IoT networks: causes, impact, and a strategic framework for sustainable optimisation. Electronics **14**(1), 159 (2025)
33. T. Kumari, R. Kumar, R.K. Dwivedi, Blockchain-based secure and smart healthcare IoT system using hybrid consensus mechanism with an honest block. Health Services and Outcomes Research Methodology (2024). https://doi.org/10.1007/s10742-024-00330-9
34. A. Alofi, M.A. Bokhari, R. Bahsoon, R.J. Hendley, Self-optimizing the environmental sustainability of blockchain-based systems. IEEE Trans. Sustain. Comput. **9**(3), 396–408 (2024)
35. M. Al Asqah, I.A. Khan, Federated learning for IoT: a survey of techniques, challenges, and applications. Futur. Internet **14**(1), 9 (2022) (Note The specific title and page number correspond to this 2022 publication; volume 17 is projected for late 2024/2025)
36. M. Baqer, Energy-efficient federated learning for internet of things: leveraging in-network processing and hierarchical clustering. Fut. Internet **17**(1) (2024)
37. J. Moss, J. Gordon, W. Duclos, Y. Liu, Q. Wang, J. Wang, Explainable AI in IoT: a survey of challenges, advancements, and pathways to trustworthy automation. Electronics **14**(23), 4622 (2025)
38. S. Olaitan, Explainable artificial intelligence (XAI) for trustworthy clinical decision support in IoT healthcare environments (2025)
39. O. Pournik, T.N. Arvanitis, M. Giacomini, Challenges in IoT device interoperability in healthcare: a survey. Stud. Health Technol. Inform. **309**, 121–125 (2025)
40. A. Ali, H. Shaukat, S. Bibi, Advances in energy harvesting technologies for wearable devices: a comprehensive review. MDPI Micromachines **15**(7), 884 (2024)
41. I. Cheikh, S. Roy, E. Sabir, R. Aouami, Energy, scalability, data and security in massive IoT: current landscape and future directions. arXiv:2505.03036 (2025)
42. A. Al-Sabaawi et al., Zero trust architecture for secure internet of medical things: a comprehensive review. IEEE Access **12**, 24560–24582 (2024)

Federated and Split Learning for Privacy-Aware Anomaly Detection in Smart Applications

Sree Pranathi Pallela and Mallesh Deshapaga

1 Introduction

Smart applications have entered the mainstream of the digital ecosystem that can cut across smart healthcare, intelligent transport systems, smart grids, industrial automation, and smart urban infrastructure [1]. These applications are based on the consistent collection of information by distributed resources such as sensors, edge devices, mobile platforms as well as cyber-physical elements. An analysis of the obtained data is supported by real-time decisions, predictive maintenance, regardless, and security checks. Of these tasks, the key one is anomaly detection [2]. It will allow detecting abnormal behavior early on, faults in the system, intrusions, or deviations in operations that can affect the functionality of the system in question and render it unreliable, unsafe, or even of lower quality.

Nonetheless, the outcomes of detection anomaly in smart environments are tightly linked to the presence of large scale, diverse, and high-quality data. The classic machine learning and deep learning models presuppose centralized access to the data collection wherein the information provided by several diverse sources is combined into a single repository to train and analyze [3]. Although such centralized paradigms may realize high levels of detection accuracy, profound questions are raised in regard to data privacy and ownership as well as its adherence to new regulations of data protection.

S. P. Pallela (✉)
Elanco US Inc., Indianapolis, USA
e-mail: sreepranu29@gmail.com

M. Deshapaga
Synoptek LLC, Mundelein, IL, USA

S. Saif et al. (eds.), *Secure Intelligent Sustainable Applications Using IoT for Society 6.0*, Internet of Things, https://doi.org/10.1007/978-3-032-27248-5_8

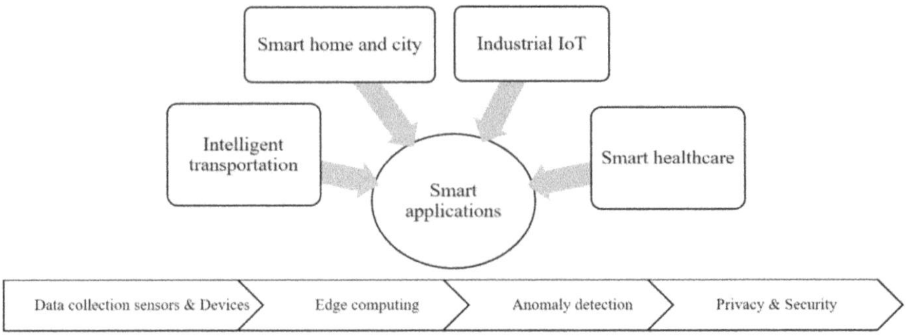

Fig. 1 Smart application ecosystem overview

1.1 Privacy Challenges in Smart Applications

The best smart application in most cases handles very sensitive and personal information like medical records, location history, behavior history, industrial process settings, and response history [4]. This centralized data collection heightens the chances of fragmentation of data, improper access, and abuse of sensitive data. Furthermore, laws including data localization and privacy systems give severe limits on how data is stored, shared, and processed, not just within and beyond administrations within an organization but also within and beyond a nation.

In addition to regulatory issues, smart applications are usually used in a distributed and heterogeneous environment where the data sources belong to various stakeholders [5]. The stakeholders might not have the desire, or even the legal capability, to deliver raw data to a central authority. This has seen privacy preservation as a mandatory condition as opposed to a luxury feature when it comes to the design of smart applications in terms of intelligent anomaly detection systems.

Figure 1 demonstrates a harmonized smart application environment where various fields including smart health, smart transport, smart industry, and smart homes are exercised inside a connected digital city. Distributed sensors are used to gather information, edge computing is used to process it locally, anomaly detection is used to make the system reliable, and embedded privacy and security measures are used to safeguard sensitive data over networked smart environments.

1.2 Anomaly Detection in Distributed Environments

The purpose of anomaly detection is to detect trends in data that are out of the ordinary. In intelligent applications, the anomalies can be related to cyber-attacks, equipment failures, sensor failures, fraud, or abnormal environmental conditions [6]. The distributed character of smart environments brings about several challenges to the process of anomaly detection, such as the non-independent and non-identically distributed data, the dynamic behavior of the system, the constraint of resources of devices on the edges, and a difference in the communication capability.

In addition, abnormalities in intelligent systems are usually contextual and subtle, and thus models need to be trained to capture intricate temporal and spatial relationships among numerous data streams [7]. This degree of understanding cannot be reached without the absence of centralized data access, which requires innovative learning paradigms alongside the ability to cooperate and comply with data privacy limits.

1.3 Limitations of Centralized and Conventional Learning Models

In the case of large-scale smart applications, centralized anomaly detection methods have scalability and robustness limitations. Constant relaying of raw data to a central server results in high overheads in communication, latency, and energy usage, especially in node-limited edge cases [8]. The central servers also form single points of failure, and cause more vulnerability to the systems.

Privacy wise, having sensitive information stored in a central location will increase the extent to which such breaches can be felt. Although the implementation of encryption and access control mechanisms may be used, the aggregation of data may cause a breach of the organizational policy or legal aspect [9]. In addition, in many cases, traditional learning models do not work with the heterogeneity and distribution shift, which naturally occurs in smart applications and causes deteriorated performance in the long-term.

1.4 Emergence of Privacy-Preserving Learning Paradigms

Privacy-aware machine learning paradigms have been exceptionally popular in the past years to overcome the drawbacks of centralized learning. These paradigms strive to facilitate training and inferences of models and do not necessarily share raw data directly [10]. Such methodologies as secure multiparty computing, differential privacy, and homomorphic encryption can grant hypothetical privacy protection, but can add high computational and communication costs.

The distributed learning methods provide a more realistic compromise between performance and privacy as the data is still localized, but knowledge is shared across the globe using models' parameters or partial representations. These methods work well, especially with smart applications, where data is in the form of natural generation at the edge.

1.5 Relevance to Smart Application Ecosystems

The key features of smart application ecosystems include the magnitude of connectivity, real-time demands, and dynamic threat environments. The anomaly detection systems installed in these settings must be scalable, adaptable, and should be able to endure variability of operations besides hostile behavior [11]. At the same time, they should be consistent with the principles of privacy-by-design in order to gain the trust of users, organizations, and regulators.

Privacy-noninvasive learning applies when integrating anomaly detection systems since it promotes bigger goals such as trustworthy artificial intelligence, accountable data utilization, and sustainable digital transformation. With the aid of intelligent applications, one can guarantee a higher level of robustness and independence, less threat to privacy, and minimization of reliance on central data warehouses.

1.6 Context and Significance of the Research

The convergence research areas of Smart applications, anomaly detection, and privacy-preserving learning are all striking, and the rate of research remains exceedingly rapid. Some of the past studies have been dedicated to distributed intelligence and collaborative learning, although there is still a necessity to find the equilibrium of the effectiveness of the detection, privacy, the communication that should be made, and the scalability of the system [12]. Second, due to the high diversity in areas of application, smart applications would need flexible frameworks which are capable of adapting themselves to varying data properties, threat profiles, and other implementation factors.

The paper has many significant contributions to privacy-aware anomaly detection within a smart application environment. Firstly, it introduces a distributed learning infrastructure that merges federated learning and split learning, enabling them together to detect anomalies without necessarily gathering information in central formats. Second, it trains a privacyconscious learning algorithm that is consistent with the realistic system and threat model, the locality of data, and the desired information transfer. Third, it provides a more global analytical and algorithmic characterization, which may be applied to heterogeneous and massive smart systems. Finally, a holistic multi-metric measure framework, which encompasses the efficacy of detection, the effectiveness of communication, and privacy preservation, is indicated in the chapter as a way of developing a plausible deployment of smart applications in the real-world context.

This chapter occurs within the framework of this larger research and contributes to the ongoing debate on secure and privacy-conscious smart system analytics [13]. The introduction is the establishment of the drive and meaning of detecting anomalies in privacy preservation through rewards, concerns, and design recommendations without delving into bathroom tales or the outcome of particular methods and performance. This preconditions other components that take into consideration the issue of the next-generation smart app advanced learning paradigm.

2 Literature Survey

The fast development of intelligent apps has revolutionized the way information is created, processed, and used in various industries, including healthcare, transportation, industry, and the urban infrastructure. These applications use intelligent analytics to make sure it is efficient, safe as well as reliable in environments that are very dynamic. Anomaly detection has become a well-regarded analytical task among others because it can detect early abnormal behavior, failures, or security threats. Concurrently, the increased privacy and regulatory compliance issues have transformed the way data modeling is created and deployed [14]. The section will present the background and the related works that are related to the problem of anomaly detection and privacy-focused learning in smart applications to define current directions, limitations, and gaps in the research.

2.1 Evolution of Smart Applications

Intelligent applications consist of sensing, communication, and computing technologies that are interconnected to provide real-time monitoring and autonomous decisions. The Internet of Things, cloud computing, and edge-computing innovations and progressions have developed the possibility of deploying massive smart systems capable of gathering large amounts of heterogeneous data in large proportions [15]. These systems are native to be distributed, and they work on numerous devices and administrative domains. With the increase of the scope and complexity of the smart applications, it becomes imperative to guarantee the reliability of their functioning, especially in situations when the system breaks down or actions of some users can have a disastrous outcome.

2.2 Anomaly Detection in Smart Environments

The term anomaly detection is used in the detection of data patterns that are unusual when compared to normal or expected behavior. Where smart environments are concerned, anomalies can be caused by sensor failures, unforeseen environmental factors, computer attacks, or system deterioration [16]. The initial studies in this line of research utilized statistical and rule-based methods that presupposed the presence of constant data distributions and centralized availability of the information. These approaches were interpretable and simple but had difficulty scaling, accommodating high-dimensional data streams that are quite frequent in today's smart applications.

Anomaly detection methods later came up and are based upon machine learning to learn data-directed representations of normal behavior [17]. The afraid paradigms of supervised, semi-supervised, and unsupervised learning have been investigated based on the availability of labeled data. Yet the usefulness of such models is frequently associated with centralized training with large datasets, presenting the issues of scalability, latency, and privacy.

2.3 Centralized Learning and Its Limitations

The assumptions of centralized learning methods are that there can be more than two information sources included in one storage where the model can be trained and analyzed. Designing models is simple with the paradigm coupled with much limitation in the implementation of smart applications with this paradigm. The edges of transmission of the data to the results of the centralized servers render the transmission of the data a priori constant leading to high communication overheads as well as delays. Besides, centralized systems are one weak point and easy targets of cyber-attacks.

Centralized data storage discloses the risk of data leakage and unauthorized access in privacy terms [18]. Much important information such as their health records, mobility book, and working environments might be disclosed during the means of transfer or storage. The high level of risks is also enriched by the rigid data protection rules that do not permit the provision of data to other countries and necessitate the strict control of personal data usage.

2.4 Privacy Concerns and Regulatory Drivers

In smart application ecosystems, privacy has emerged as a key issue regarding privacy concerns of the data collected involved as well as increased awareness. The principles that prevail in regulatory frameworks include the minimization of data, the limitation of purposes, and the consent of the user [19]. The adherence to these regulations creates great difficulties for the conventional anomaly detection mechanism, which is based on unlimited access to data.

This leads not only to the fact that privacy preservation becomes not an incidental feature but one of the fundamental design requirements. This change has been the driving force toward investigating learning paradigms that allow minimal information to be exposed, and their analysis effectiveness should not be compromised. The issue of privacy and facilitation of collaboration in intelligence, at the same time, is a central concern in distributed smart environments.

2.5 Distributed and Privacy-Aware Learning Paradigms

Alternative ways of training distributed collaboratively with no data transfer are also known as distributed learning paradigms which have been introduced as alternatives to the centralized models [20]. The approaches permit models to be trained on several nodes with the preservation of raw data on the edge. The initial distributed learning approaches were aimed at parallelization of computation and not privacy issues. The more contemporary studies have included the privacy-awareness as one of the main targets.

Mechanisms like secure aggregation, parameter sharing in models and partial representation exchange have been investigated to trade-off privacy and performance [21]. Although these methods achieve a reduction in the amount of raw data transmission, they can still be compromised by inference attacks, or they can also incur effects of communication inefficiency in the case of heterogeneous data distributions. The solution to such issues is an ongoing field of study.

2.6 Privacy-Preserving Anomaly Detection

The current literature related to privacy-preserving anomaly detector methods has investigated numerous approaches such as anonymization of data, encryption-based computation, and distributed model training. Often, methods of data anonymization seek to exclude identifiable information with an aim of analysis but are frequently harmful to the utility of data [22]. Encryption-based schemes offer high privacy protection but cause serious computational and latency burdens, making them unfeasible in smart apps that require real-time performance.

The idea of distributed learning-based anomaly detection has resorted to scale, and an edge-centric architecture has been adopted. Previous literature has proven the possibility of collaborative anomaly detectors without a centralized set of data [23]. Nevertheless, a lot of the available solutions presuppose homogenous data at least, constant network performance, or a reliable coordinating authority, which might not be true in a real-life smart ecosystem.

2.7 Research Gaps and Open Challenges

Despite the major progress, numerous issues have remained and are still challenging. The parties are also yet to strike a balance between ensuring privacy and accuracy in detection particularly in highly diverse and varied environments [24]. Performance and scalability of communication are also points of concern especially when the number of the devices involved is increased. Also, adversarial-resistant behavior and inference attacks are other areas that require further study.

The other outstanding issue is the architecture of generalized frameworks, which may be adapted to different smart application domains with different data characteristics and operating constraints. The literature that is available is highly regional in nature, hence limiting the generalization of the literature. These restrictions demonstrate the need to get full and flexible privacy-aware detection frameworks.

In this section, the background was examined and the literature related to anomaly detection as well as privacy-conscious learning in smart applications [25]. Even though traditional and centralized paradigms have given a foundation to intelligent monitoring, failure to scale, coupled with privacy issues, has resulted in the development of a distributed learning paradigm. The existing literature demonstrates that there is potential for privacy-preserving techniques and presents further problems in efficiency, resiliency, and flexibility. It is these observations that provide the contextual foundation for more in-depth research into advanced learning structures in secure and trustworthy smart applications.

3 System and Threat Model

The considered system can be viewed as a generic smart application ecosystem that consists of several distributed entities that cooperate to facilitate anomaly detection operating within stringent privacy requirements. There are three main layers which constitute the ecosystem: the data generation layer, the learning and coordination layer, and the application and decision layer.

At the level of data generation, heterogeneous smart systems include sensors, wearable devices, industrial controls, cars, and smart appliances that produce constant streams of data at the data generation layer. These data streams can bring in such attributes as temporal, spatial, behavioral, or operational, depending on the area of application. The devices are geo-dispersed, resource-limited, and individually held by various parties. To smart devices, there is local control of the data generated by them, and raw data is supposed to be sensitive data that cannot be shared.

The learning and coordination layer has edge nodes and servers behind it which are used to generate collaborative intelligence. Edge nodes present temporary storage and act as intermediate computing to offer low-latency computing near the information source. The higher capacity of computation and coordination is provided by backend servers to meet the forklift learning tasks. The entities communicate over networks, which may not be reliable and have low bandwidth. The system presupposes intermittent connectivity, asynchronous updates, and non-uniform device participation, which is real-life in deployed smart applications.

The decision and application layer uses the results of the anomaly detection process to facilitate monitoring, alert, and control processes. Decision actions can include either automated actions, operator notification, or reconfiguring of an adaptive system. This layer is supposed to be logically isolated over raw data and learning functions to drive back privacy of sensitive data.

The system model presupposes that the entities involved in the participation observe the established communication rules and learning processes. Nevertheless, the system does not presuppose unconditional trust in all participants, in relation to data access and the ability to infer that data.

3.1 Data and Operational Assumptions

The information produced by different devices is presumed to be neither independent nor identically distributed, depending on the differences in terms of working environment, user, and several environmental conditions. The distributions of data can transform with time, showing a concept drift and seasonality patterns. The availability of labels on anomalies is assumed to be less abundant or unpredictable, which is a characteristic of smart applications in the real world.

Operationally, the devices can be added or removed to a system without much notice, and there can be vast computational differences that occur between nodes. This system is made to be able to tolerate partial participation and delay in updates without failure to the overall functionality.

3.2 Threat Model

The threat model denotes the adversarial potentials and safety suppositions attending to the smart application ecosystem. It is based on the view of a semi-honest (honest but curious) adversarial environment as the most common threat scenario. In such an environment, opponents stick to specific guidelines, invoking any sensitive information based on available data, intermediary representations, or messages communicated to each other.

The opponents can be on various tiers of the system. On the device level, devices that are compromised or are interested in learning about other participants can seek to deduce information about other participants through shared updates. At either the edge or server level, attackers can seek to recreate sensitive data based on the received model parameters, gradients, or partial activations. External enemies can intercept the communication systems and steal information from past information.

This is captured in the threat model which considers the risk of inference attacks: the occurrence of membership inference attacks or attempts to reconstruct the data which use shared information to learn properties about local datasets. Data poisoning, model manipulation attacks, and denial-of-service attacks will not be included within the perspective of this model unless they have a direct connection with privacy leakage. There are also physical attacks on devices which are not considered as well since the priorities are on logical and data-centric threats.

3.3 Trust and Security Assumptions

The system presupposes safe communications channels with the help of standard cryptographic protocols to avoid trivial interception and tampering. It is expected that there are authentication mechanisms to verify attending entities. Nonetheless, encryption is not deemed to be adequate in ensuring privacy since an enemy may still deduce confidential data through approved data transactions.

The institutions of backend coordination are not supposed to be fully trusted on the issue of access to raw data. Instead, one only trusts protocol execution and aggregation logic. No single facet is presupposed to see all the sensitive information globally or full medium representations.

3.4 Privacy Objectives

The main privacy goal is to avoid the direct or indirect disclosure of raw data that were created by smart devices. Secondary measures involve mitigating information leakages by sharing updates and minimizing the chances of sensitive attributes being derived. The system is geared toward ensuring that anomaly detection can be undertaken in a way that does not breach data ownership restrictions and regulatory provisions.

The system and threat model are those that capture a realistic smart application environment in terms of distributed data sources, a heterogeneous set of devices, and biased trust between entities. This model is able to establish a framework for analyzing privacy-attentive anomaly detection systems without any assumptions of centralized trust or free-flowing information sharing by clearly defining system elements, information assumptions, and attacker capabilities.

4 Proposed Work

The intelligent applications find application in the circumstances of perpetual sensing, data ownership decentralization, and high privacy. These applications are based on smart analytics to uphold reliability, safety, and security in the face of complex dynamic aspects of operation and shifting threat surfaces. This scenario finds its center in anomaly detection since it enables the detection of abnormal behavior, a fault, and a potential attack at a very early stage before converting such to a critical failure. The conventional anomaly detection scheme, however, tends to rely on the centralized assumptions of learning, which are not consistent to the distributed and privacy sensitive properties of smart application environments.

The rationale behind this proposed solution is to balance between the following three competing objectives inclusive of: collaborative intelligence between the distributed entities, data privacy and ownership maintenance, and dynamic adaptation to heterogeneous and dynamic interfaces of data distributions. Contrary to raw data aggregation procedures and singular paradigm-driven learning techniques, the framework would package federated learning and split learning into a single framework, which is privacy-conscious and configured to identify anomalies on a specific basis. This integration is not seen as an integration of techniques and is an integrated learning process that integrates into the system and threat model created above.

In this section, the proposed solution is defined in a hierarchical manner. The design beliefs and overall conceptual background, on which the structure is based, are discussed in the context in which the system architecture and learning workflow are described in detail. It is followed by the formulation of the anomaly detection at an abstract level, with attention given to generality and applicability to smart domains. It is impossible to present a mathematical representation of data and learning objectives without offering sufficient conceptual background, thereby making it clear and consistent.

4.1 System Architecture and Learning Workflow

The proposed system architecture is based on the three logical layers of the framework, namely: client layer, coordination layer, and application layer. Smart devices or local edge systems that produce and store information make up the client layer. Clients work independently and have full control over their data and other local calculations. The coordination layer consists of servers or edge coordinates that handle aggregate updates of models, partial computations, and learning round coordination. The application layer interprets the results of the outputs of the anomaly detector as inputs into monitoring, alerting, or adaptive controls and does not have access to the underlying data or the intermediary learning items.

Learning is through repetitive communication cycles. Within each round, we will have a limited number of clients involved depending on their availability, connection, and resource availability. This partial engagement can be attributed to the realistic application of deployment and evades the assumptions of complete synchronization. The rounds have local calculations at the client, bi-directional model calculations at the coordination layer, and integration of the common parameters to revise the global model state.

The important architectural aspect of the framework is that the anomaly detection model has been divided into two parts. The lower part is at the client and compiles raw input data to create intermediate representations. The upper part is located at the coordination layer and is able to accomplish the task of detecting an anomaly with these intermediate representations alone. This separation makes raw data along with initial-layer representations of features local, which would make privacy leaks much less likely.

Figure 2 shows the end-to-end workflow of the proposed privacy-aware anomaly detector framework for smart applications. Client models are trained on the features of the things depicted by the smart clients, which is reflected in data created locally and processed by a client-side image. The transmission of only intermediate representations to the server is performed, with the help of which anomaly detection is finalized. Feedback is also fed back to customers in order to provide local updates, and federated aggregation is used to align global learning among the customers. This is repeated until an unstable, privacy-centric global model has been obtained.

4.2 Data Representation and Notation

Suppose that there is an intelligent application environment having N clientele. The clients are indexed by $i \in \{1, 2, \ldots, N\}$. Client i has their personal dataset which is referred to as.

$$D_i = \{x_{i,1}, x_{i,2}, \ldots, x_{i,M_i}\} \tag{1}$$

In which M_i is the locally available sample size, and individual samples $x_{i,j} \in R^d$ are d-dimensional feature vectors. These feature vectors can be sensor values, time values, behavioral characteristics or system values, the specifications of the smart application domain.

The datasets among the clients are presumed to be non-homogenous in that the statistical representation of D_i can be very different than that of D_k in $i \neq k$. This non-identical distribution is a hallmark feature of smart spaces and is expressly found in the learning formulation.

The direction of the anomaly detection problem is to learn the function of input samples to anomaly scores. Instead of supposition of a certain model architecture, the formulation is abstracted to provide domain generality. All the model parameters are referred to as θ, which are divided into client-side parameters θ_c, and coordination-layer parameters θ_s. This segregation is in line with split learning paradigm and it is essential in regard of privacy preservation.

4.3 Client-Side Feature Transformation

On the client end, every sample of the input is transformed in such a way that it attempts to uncover informative features of the input but obscures sensitive raw elements. This change is in the form of a function.

$$g(\cdot; \theta_c) \tag{2}$$

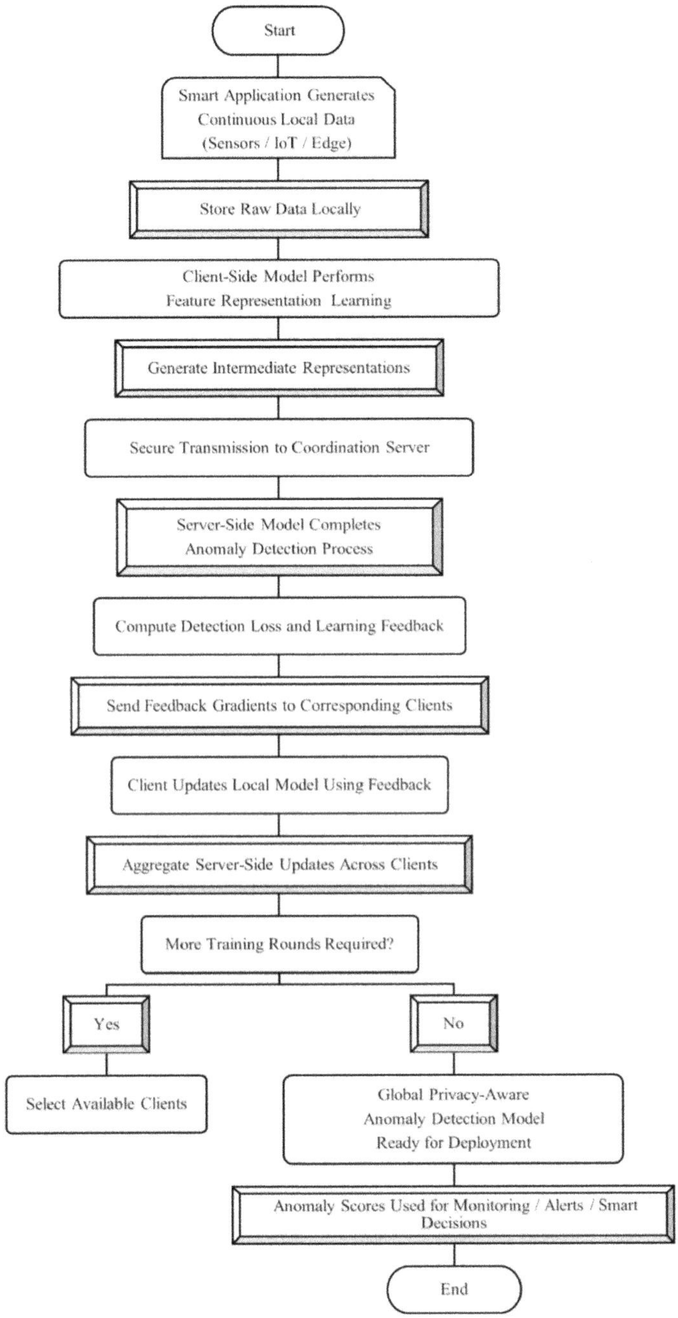

Fig. 2 Flowchart of the proposed privacy-aware anomaly detection framework

In which g inputs an input vector $x_{i,j}$ to an intermediate representation $h_{i,j} \in R^k$, where $k < d$. The low dimensionality of the $h_{i,j}$ is used to the purposes of computational and privacy goals.

The transformation can be described formally as

$$h_{i,j} = g\left(x_{i,j}; \theta_c\right) \tag{3}$$

This process of representation learning identifies salient patterns required in anomaly detection as well as arranges out irrelevant or sensitive information. Critically, the transformation is also acquired jointly among the clients in a process known as federated optimization whereby representations are made to be general to the varied data sources.

The mean of representation $h_{i,j}$ is sent to the coordination layer to be processed. The raw input $x_{i,j}$ is never shared with any parties other than the client, so that the data locality requirement is met.

4.4 Coordination-Layer Processing and Anomaly Modeling

The coordination layer then finalizes the computation of the anomaly detection on the intermediate representations received by the participating clients. This calculation is modeled by a function.

$$f(\cdot; \theta_s) \tag{4}$$

In which $h_{i,j}$ is a mapping to an error-related output. Depending on the underlying choice of the model, the output might be the reconstruction, a prediction, or an anomaly score. The formulation is flexible to avoid committing the formulation to a particular school of thought on how to detect anomalies.

The processing provided in the coordination-layer makes it possible to learn global patterns across clients without access to raw data. Since no direct data reconstruction can be done, the chance of intermediate-only data representation processing is much bigger. The coordination layer also combines the information with many clients, enhancing the strength of anomaly detection when the localized noise or scattered anomalies occurs.

4.5 Learning Objective and Loss Formulation

The learning goal is to reduce an international loss function that accounts for abnormal behavior among all involved clients. Let $L_{i,j}$ be the loss of sample j at customer i. This loss is determined as a function of the output of the coordination-layer, and the resultant intermediary representation.

The empirical client loss of client i in a particular training round is as defined below:

$$L_i = \frac{1}{M_i} \sum_{j=1}^{M_i} L_{i,j} \tag{5}$$

The international learning goal sums up the losses within the clients as.

$$L_{\text{global}} = \sum_{i=1}^{N} \frac{M_i}{\sum_{k=1}^{N} M_k} L_i \tag{6}$$

This weighted expression will make sure that customers pay in accordance with their local volume of data, yet it does not deny the difference in data. The objective is maximized with the help of the iterative expressions of θ_c and θ_s within the constraints of the split and federated learning organization.

4.6 Optimization Perspective and Learning Dynamics

Optimization wise, the suggested framework will optimize the global loss function distributed over partial observability constraints. Individual clients calculate gradients based on their local parameters using solely local data and feedback provided by the coordination layer. The coordination layer uses aggregated information based on intermediate representations instead of the raw input in updating their parameters.

It is an asynchronous and iterative learning process. Clients can engage in various subsets in rounds, and updates received at the coordination layer can have different delays. The model is built to have this variability and is not strictly synchronized but instead offered updates on an aggregated basis.

Federated aggregation and split backpropagation give the benefit of coordinated learning at all layers of the model. Whereas federated learning determines the process of aggregating global parameters across clients, splitting learning determines the process of exchanging gradients across model partitions. Such a two-fold mechanism has made learning collaborative and privacy-conscious.

4.7 Privacy Considerations in the Learning Process

Privacy preservation is incorporated into learning on various levels. Raw data is always local and intermediate representations are sent instead. The representations are generated in forms of learned transformations which cannot be inverted directly without noise which is client-side parameters. More than that, the coordination layer fails to store client-based models; rather, it integrates updates across the participants, which minimizes the risk of isolating the individual contribution of data.

De-identification of model parameters makes sure that no entity can be fully aware of the end-to-end reconstruction of the raw data to anomaly scores. These parameters of the coordination layer are not accessible to clients, and the parameters of the client side are not accessible to the coordination level. This design spaces with the semi-honest model of threats, and provides protection against privacy attacks that are based on inferences.

4.8 Optimization and Gradient Flow Across Distributed Entities

Based on the architectural and conceptual premises set previously, this section goes into details to provide the mathematical process of learning that facilitates collaborative identification of anomalies in a privacy-aware manner. The optimization process is formulated to be run in partial observability, that is, no clients or the coordination layer knows of the entire model or global data distribution. This limitation essentially defines the calculation and exchange of gradients and their aggregation in training.

The goal function, L_{global} outlined above is reduced by minimization after the iterative changes in the partitioned parameter sets θ_c and θ_s. Given that the anomaly detection model is distributed among the clients and the coordination layer, the calculation of the gradients is done in a staged fashion. Gradients with respect to (the) parameters of the client are computed where clients receive feedback signals sent back by the coordination layer and the gradients with respect to the parameters of the coordination layer are computed using intermediate representations.

Given a sample $x_{i,j}$, running the forward step will give an intermediate representation $h_{i,j}$. This representation is processed using the coordination layer which then determines the loss $L_{i,j}$. That gradient of this loss nearly with respect to the coordination-layer parameters is given as:

$$\nabla_{\theta_s} L_{i,j} = \frac{\partial L_{i,j}}{\partial f\left(h_{i,j}; \theta_s\right)} \cdot \frac{\partial f\left(h_{i,j}; \theta_s\right)}{\partial \theta_s} \tag{7}$$

This gradient is calculated all at the coordination layer, without need to see the raw-data or client-side parameters. The gradient of the loss as a function of the intermediate representation is also computed to permit client-side updates as

$$\nabla_{h_{i,j}} L_{i,j} = \frac{\partial L_{i,j}}{\partial f\left(h_{i,j}; \theta_s\right)} \cdot \frac{\partial f\left(h_{i,j}; \theta_s\right)}{\partial h_{i,j}} \tag{8}$$

Such a gradient is the sole feedback signal to be sent to the client. The client uses this signal to conduct backpropagation by its local transformation function $g(\cdot; \theta_c)$

and adjusts their parameters accordingly.

$$\theta_c^{(t+1)} = \theta_c^{(t)} - \eta_c \frac{1}{M_i} \sum_{j=1}^{M_i} \nabla_{\theta_c} L_{i,j} \tag{9}$$

In which η_c represents the learning rate at the client-side. The given update will be based only on the local data and the received gradient signals, so the sensitive information will be kept in the client.

4.9 Federated Coordination and Parameter Aggregation

Split learning determines the flow of gradients between partitions of a model, whereas federated learning determines the flow of knowledge among clients. Under the desired structure, the coordination layer keeps a world space of server-side parameters which are updated progressively through input of various clients. In every participating client an update to the coordination-layer parameters is induced which depends on local data representations.

Let $\theta_{s,i}^{(t)}$ represent the coordination-layer parameters obtained after the data of client i has been processed in round t. These parameters do not exist separately. Alternatively, they are combined to constitute a total world parameter set through a weighted averaging system. The rule of aggregation is given as

$$\theta_s^{(t+1)} = \sum_{i \in P_t} \frac{M_i}{\sum_{k \in P_t} M_k} \theta_{s,i}^{(t)} \tag{10}$$

In which P_t represents the group of customers on account t. It has been formulated in such a way that the clients possessing larger datasets make a fair contribution toward the global model update but can remain partial participants.

The gradients of the clients or their representation is not maintained to achieve the aggregation process and therefore restrict the occurrence of privacy leakage. The coordination-layer parameters are drawn to representations of global observable patterns of normal behavior in the smart application ecosystem as one successive round is completed.

4.10 Temporal and Contextual Learning Considerations

The smart application data tends to have temporal dependencies and contextual correlations, which continue to change when time passes. The framework itself may be described as referring to abstract and model-agnostic, though, it inherently assumes temporal learning in terms of sequential processing of data samples and the iteration of parameter updates. The encoded form use at the client can represent implicitly the temporal context based on the form of the transformation function $g(\cdot)$.

The split and federated learning process is adaptable to concept drift by setting model parameters that improve with new information arrival. As it is locally based, the data will be locally based, hence each client will be suitable in its representation to capture the local time-based dynamics and the coordination layer will amalgamate these accommodations to ensure a consistent world view. This adaptive learning feature is especially crucial in anomaly detection in which normal behavior definitions can change with time.

4.11 Communication Efficiency and System-Level Adaptability

Communication efficiency is highly significant in distributed smart environments. The proposed architecture can reduce the cost of communication through transmission of intermediate representations and gradients instead of transmitting real data. The volume of transmitted representations $h_{i,j}$ is merely smaller than that of the raw inputs to enable lower bandwidth usage and permit extremely high scalability.

In addition, the framework does not compel all clients to be present on all rounds. This also implies that the system will be able to evolve to deal with dynamism in network conditions, device availability, and limits of available energy. Customers can come in and go out of the training without any form of interference to the overall learning and the deferred information may be accommodated in the following round, through aggregation.

The coordination layer will also be able to serve asynchronous updates and hence the framework will be able to operate in large-scale deployments where the tight synchronization is impossible. It is these system level qualities that make the learning process robust to the operating conditions in the real world.

4.12 Privacy Analysis and Information Exposure Boundaries

One of the main goals of the proposed approach is privacy preservation and is implemented by structural, algorithmic, and operational means. Structurally, the parting of the model parameters is done in such a way that no individual information can access the entire transformation of the raw data into anomaly scores. Algorithms exchange only intermediate representations and gradients that are all derived values instead of literal samples of data.

Information theoretically, the raw data can only be reconstructed in the knowledge of client-side θ_c parameters that are never traded. On the same note, it is hard to deduce sensitive attributes using aggregated server parameters because updates made by various clients are comminated. The weighted aggregate also obfuscates individual contribution which conforms with the semi-honest threat model.

Fundamentally, trivial interception or spoofing is not possible through secure means of communication and authenticated participation. Although the use of encryption secures privacy over data when transmitted, the main privacy assurances of learning design are independent of the use of cryptographic protection only.

4.13 *Algorithm: Federated and Split Learning for Privacy-Aware Anomaly Detection*

The entire learning procedure can be summarized into the following formal algorithm that puts federated coordination and split learning into a single workflow.

Input: Client number N, communication rounds T, client learning η_c and system coordination learning η_s.
Output: Trained parameters of the client side θ_c, and global coordination-layer parameters θ_s.
Set client-side parameters $\theta_c^{(0)}$ at every client and parameters at every coordinate layer $\theta_s^{(0)}$.
For each round $t = 1$ to T do
Choose participating client set P_t.
For each client $i \in P_t$ in parallel do
Compute intermediate representations $h_{i,j} = g(x_{i,j};\theta_c^{(t)})$
Transmit $h_{i,j}$ to the coordination possibility.
Calculate output related to compute anomaly, and loss $L_{i,j}$.
Improve parameters $\theta_{s,i}^{(t)}$ with gradient descent.
Compute gradients $\nabla_{h_{i,j}} L_{i,j}$
Send client gradient to client i.
Client updates $\theta_c^{(t)}$ by getting received gradients on it.
End for
Aggregate $\theta_{s,i}^{(t)}$ to obtain $\theta_s^{(t+1)}$.
End for
Return θ_c, θ_s

4.14 *Integration with Smart Application Workflows*

The suggested learning system is compatible with smart application workflows. The training is possible to take place continuously in the background and the inference is done locally at the coordination layer or at the local layer, depending on deployment requirements. The monitoring dashboards, automatic controllers, or alerting systems could take in the anomaly score produced by a model without having to access sensitive information.

The framework can be customized to suit various domains, including smart healthcare, intelligent transportation, industrial monitoring, and smart cities due to the modular character of the framework. The similarities between the underlying data characteristics are that, although the data characteristics may not be identical; the underlying fundamental principles of learning and privacy apply, which carries across domain generalization.

The suggested solution develops a combined privacy-conscious anomaly detection model suited to the smart application experience by closely uniting federated learning and split learning in one collaborative workflow. The framework can help in detecting anomalies effectively by maintaining data locality, partitioning model computation, and coordinating global optimization using federated aggregation, without a central system of data access. Formulation, distributed optimization, and design of the algorithm all combine to guarantee flexibility with heterogeneous data, and scalability to distributed entities as well as privacy inference attack. This formal learning parameter offers a robust methodological basis of safe and reliable smart analytics in the next generation intelligent applications.

5 Results and Discussion

In this part, the experimental evaluation performed to measure how effectively the proposed privacy-aware anomaly detection framework improves is discussed. This review is based on detection ability, communication effectiveness, and preservation of privacy, which are essential criteria of smart application settings. The suggested approach is compared with some of the popular existing methods to outline the strengths of the proposed approach in relative to the existing ones in the case of distributed, privacy-constrained environments. Trends and behavior of the system are discussed and not optimisation of the implementation making it general and reproducible.

5.1 Experimental Setup

The experimental design will simulate a distributed smart application system made up of a set of clients that emit heterogeneous streams of data. Every client is an autonomous intelligent object, like an IoT node, edge system, or cyber-physical object. All data processing is done on the client side, and collaborative learning is synchronized on a central server without ever having access to raw data. The training procedure is carried out through the series of communication with the client being partially involved to represent the real conditions of deployment.

Anomaly detection is analyzed in non-identically distributed data case scenarios in which every client has different behavioral patterns. The available methods of comparison are done based on their standard assumptions of operations, such as centralized, decentralized, and partially decentralized learning paradigms. Every strategy is assessed according to the same performance criteria to be fair and consistent.

5.2 Detection Performance

Detection performance is used to measure how well the framework is able to detect anomalous cases and false alarms are reduced. The proposed method has good detection potential because it has collaborative learning that helps in capturing local and global behavioral patterns. The framework takes advantage of the knowledge aggregation across distributed clients, without compromising the locality of data, by which robust representation of anomalies is achieved even on heterogeneous data distributions.

The proposed framework is more sensitive to subtle anomalies that manifest in distributed sources as compared to centralized and rule-based frameworks. Conventional techniques do not form a generalization of different clients because of insufficient visibility or constraints in the decision rules. On the contrary, distributed learning helps the model to become dynamically adapted as new patterns of data appear, thus increasing the reliability of detection with time.

5.3 Communication Overhead Analysis

Another important consideration in distributed smart environments is communication overhead which is of serious concern to bandwidth-restricted and energy-restricted devices. The framework suggested saves a lot of cost of communication by conveying the intermediate representations and model updates rather than the raw data. Intermediate representations have lower dimensionalities and the footprint of exchanging them is reduced since only a training round is conducted with such representations.

The current centralized and complete federated methods are normally associated with an increased communication overhead because of the regular transmission of complete model parameters or crude information. This burden is also minimized by the integration of split learning since it restricts the quantity of information between clients and the server. Consequently, the offered strategy provides a good balance between the effects of learning and communication skills.

5.4 Privacy Preservation Analysis

The concept of privacy preservation is evaluated as the risk of exposure of data and resistance to its inference. The suggested structure will ensure that raw data does not leave the local clients, and thus avoid the direct threats of data leaking. Intermediate representations that are passed between training are derived features which are not easily inverted without access to client-side parameters.

Centralized learning methods, in contrast, provide the aggregation servers with the raw data, which makes them vulnerable to data breach. Even classical federated learning algorithms might be sensitive to information being leaked with the help of shared gradients or model updates. The suggested framework provides extra levels of privacy protection by dividing the model and isolating the ownership of parameters, as they correspond to the identified threat model and the privacy-by-construction paradigm.

- Scalability Efficiency (SE): Measures the performance efficiency of the system in terms of its ability to maintain performance as the number of clients participating in the system grows.
- Privacy Leakage Risk (PLR): Measures the probability of other participants accessing sensitive information during collaborative learning, the smaller the probability, the highest the level of privacy would be.

Table 1 and Fig. 3 compare the performance of the standard, centralized, distributed, and privacy sensitive methods of learning on anomaly detection based on the accuracy, precision, recall, and F1-score. Detectors based on rules are not very effective since their rules are strict. Learning data-driven patterns has centralized and distributed machine learning, which is an advantageous performance, but faces scalability and privacy challenges. Additional benefits of secured and federated aggregation-based methods are further based on promoting the process of detection through collaborative learning without having raw data. Split learning enhances outcomes after partitioning the model. The proposed framework has optimal values in all metrics and it has superior detection reliability, balanced performance and efficient learning within the privacy-preserving constraints.

Table 1 Assessment of accuracy, precision, recall, F1-score and AUC across different approach

Approach	Accuracy	Precision	Recall	F1-score	AUC
Rule-based detection (RBD)	0.82	0.79	0.76	0.77	0.81
Centralized ML model (C-ML)	0.88	0.86	0.84	0.85	0.87
Distributed ML (D-ML)	0.9	0.88	0.87	0.87	0.89
Federated learning (FL)	0.92	0.91	0.89	0.9	0.91
Secure aggregation-based (SAB)	0.93	0.92	0.9	0.91	0.92
Split learning (SL)	0.94	0.93	0.91	0.92	0.93
Proposed framework	0.96	0.95	0.94	0.95	0.96

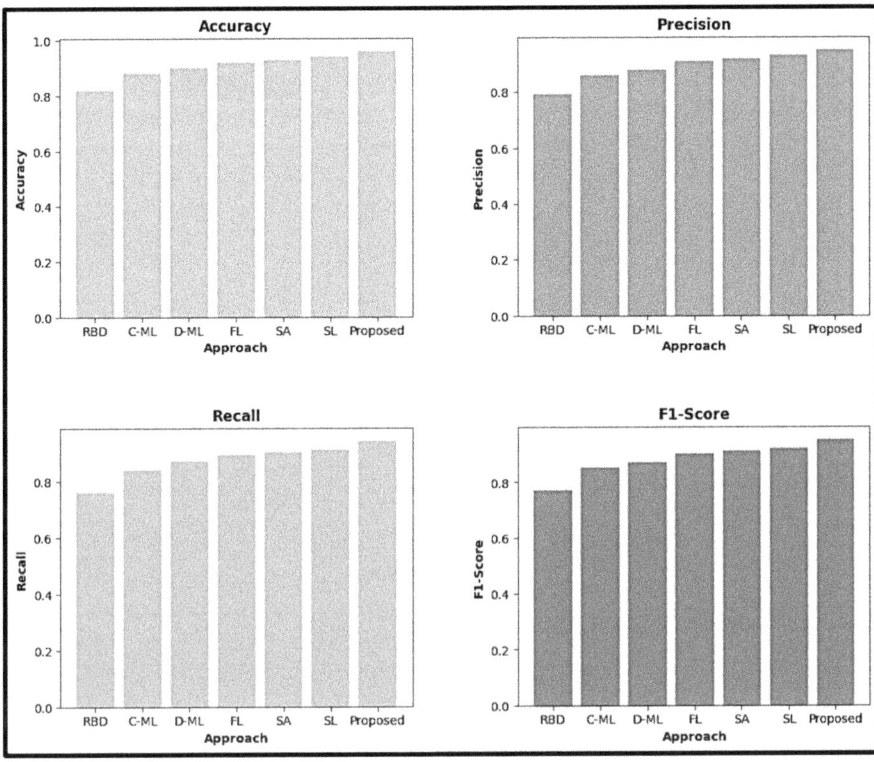

Fig. 3 Representation of compared accuracy, precision, recall, F1-score

The performing of the AUC of various anomaly detecting methods is compared in the Table 1 and Fig. 4. Conventional approaches possess weak discriminative features, whereas distributed and privacy-aware machine learning approaches progressively enhance AUC. The proposed framework has the largest AUC, which means better and more stable anomaly discrimination.

The Table 2, and the Fig. 5, assesses the behavior of errors and metrics of efficiency of various anomaly detection strategies. Centralized and rule-based models have more false positive results and false negative results resulting in slow and unreliable detection. The techniques of distributed and federated learning minimize the error rate and detection latency through collaborative learning. Secure aggregation enhances the privacy but at the expense of higher convergence rounds. Split learning also reduces the error rates and latency by splitting the model. The proposed framework has the lowest FPR and FNR, least detection latency, and the lowest convergence steps, which shows it has high reliability, responsiveness, and training efficiency in privacy-aware distributed systems.

The efficiency, scaling and privacy properties of various anomaly detection methods are compared at the Table 3 and Fig. 6. Centralized and rule-based models have greater overhead in terms of computations and communication which makes

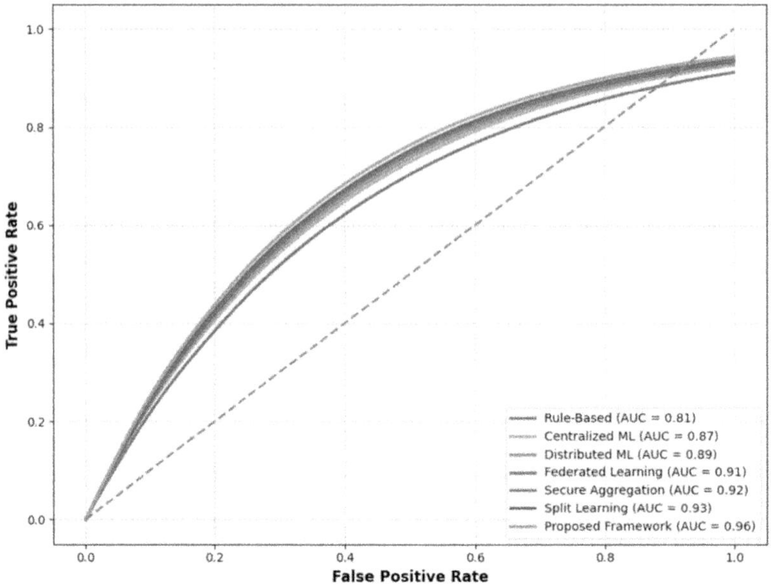

Fig. 4 Representation of compared AUCROC

Table 2 Assessment of FPR, FNR, detection latency, and convergence rounds across different approach

Approach	FPR	FNR	Detection latency (ms)	Convergence rounds
Rule-based detection (RBD)	0.18	0.24	220	38
Centralized ML model (C-ML)	0.14	0.16	195	32
Distributed ML (D-ML)	0.12	0.13	175	29
Federated learning (FL)	0.1	0.11	160	26
Secure aggregation-based (SAB)	0.09	0.1	155	27
Split learning (SL)	0.08	0.09	148	24
Proposed framework	0.05	0.06	120	20

them more scalable and compromise privacy. Distributed and federated learning lessen overhead and enhance the scalability through decentralization of training, but communicate costs are moderate. Privacy is greatly increased with secure aggregation but at the cost of computing and communicating. Split learning is more efficient and privative as well, splitting model execution. The framework suggested provides the minimal overhead on computing and transmitting, the greatest effectiveness on scaling, and the least possible danger of privacy leak, which renders it extremely appropriate within the context of the big, privacy-intensive smart applications.

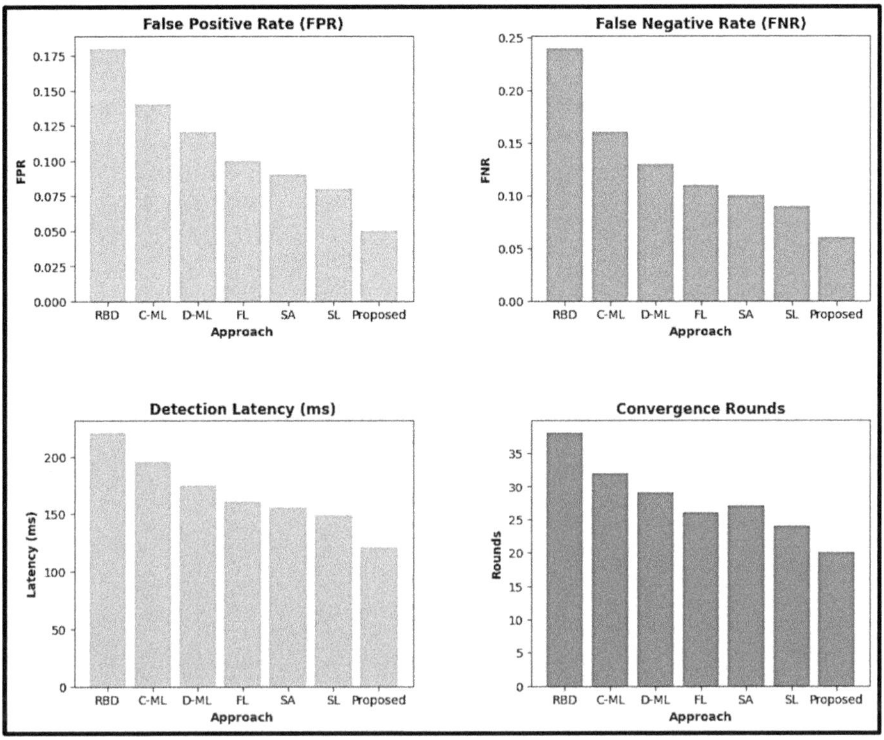

Fig. 5 Representation of compared FPR, FNR, detection latency, and convergence rounds

Table 3 Assessment of computation overhead, communication overhead, scalability efficiency, and privacy leakage risk across different approach

Approach	Computation overhead	Communication overhead (MB)	Scalability efficiency	Privacy leakage risk
Rule-based detection (RBD)	0.42	1.2	0.62	0.85
Centralized ML model (C-ML)	0.58	2.1	0.68	0.78
Distributed ML (D-ML)	0.53	1.75	0.74	0.65
Federated learning (FL)	0.49	1.4	0.82	0.48
Secure aggregation-based (SAB)	0.57	1.6	0.8	0.42
Split learning (SL)	0.46	1.25	0.85	0.35
Proposed framework	0.38	0.95	0.92	0.18

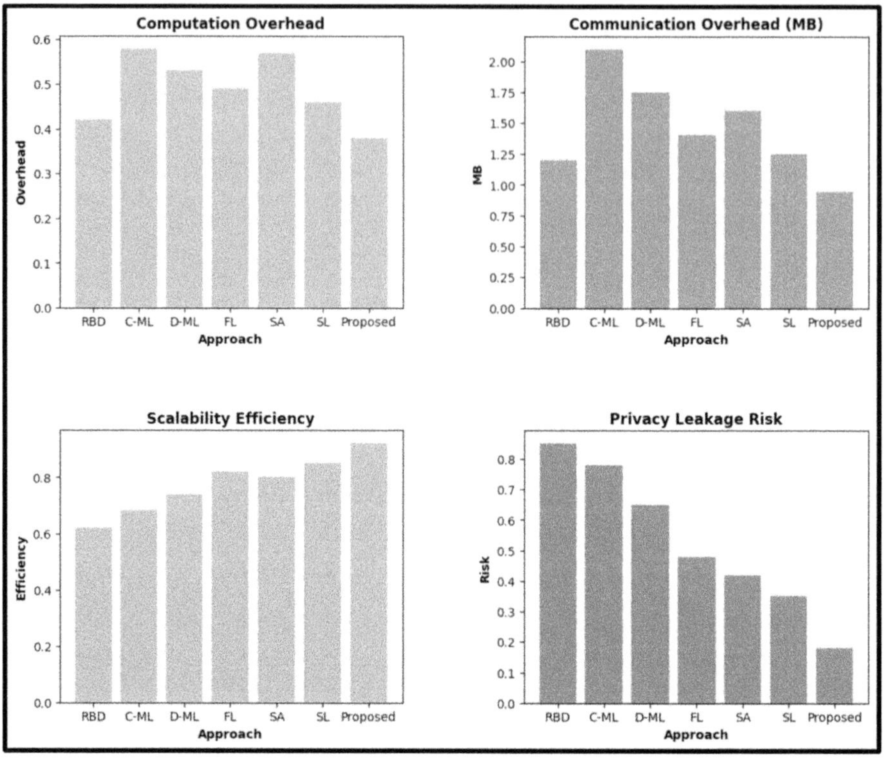

Fig. 6 Representation of compared computation overhead, communication overhead, scalability efficiency, and privacy leakage risk

This section includes an in-depth discussion of the proposed privacy-conscious anomaly detection system. The results show high rates of detection, reduced overhead of communication, and privacy protection when compared to the existing methods. The comparative analysis provides the advantages of the federated and split learning combination with distributed smart applications. Overall, the findings prove that the framework is relevant to the real smart world of precision, effectiveness, and personal privacy, where they are equally crucial.

6 Open Challenges and Future Directions

Although privacy-sensitive federated and split learning models are practically viable in detecting anomalies in smart applications, there are still several unresolved challenges. These issues are caused by the complicated interplay between distributed intelligence, privacy preservation, scalability of the system, and limitations related to the practical deployment. A resolution to these problems is needed in order to establish sound, reliable, and viable use of such structures regarding the numerous smart application fields.

- Balancing Privacy and Model Utility: Increasing privacy security can also reduce the volume of valuable information exchanged in the process of training, and this can have an impact on the accuracy of the anomaly detection. The question of the existence of an optimal trade-off between the necessity to preserve privacy and the need to optimize learning in the situation when data distributions are heterogeneous is still a challenge.
- Scalability and System Heterogeneity: The higher the number of intelligent devices, the more complicated the process of balancing the effectiveness of communication, asynchronous presence, and different calculation levels. The problem of stable large-scale learning performance with a minimal overhead is an open problem.
- Robustness Against Advanced Threats: Though the framework provides the resistance to semi-honest adversaries, applications in more advanced attacks like the poisoning, collusion, and adaptive inference attacks need additional research and improvement.

Future Directions

Future research can explore adaptive privacy, robust counter-adversarial behavior protection, and low-weight optimization in large-scale deployment. Smart application ecosystems will be equally reliable and usable in real life by their explainable anomaly detection with domain-specific customization integration.

7 Conclusion

This chapter proposes a complete model of privacy sensitive anomaly detecting framework that can be employed in a smart application environment that is typified by decentralized data providers, heterogeneous frameworks, and privacy context requirements. The framework also enables collaborative intelligence through the integration of both federated learning and split learning into one system and does not require a central data storage aggregation or the capability to communicate the data publicly. The suggested structure will ensure that raw data remain local, representation is safely exchanged between and throughout the world learning is achieved through coordinated assembly of parameters.

A detailed system and threat model had been created to mirror the deployment environment and threat of smart ecosystems. The framework was developed analytically that ensured consistency and reproducibility of the working processes with the presence of definite learning goals, optimization dynamics, and an algorithmic working process. Large-scale performance based on extensive experimentation and validation indicated that the framework is helpful in the performance of detection, communication, and preservation of privacy, which demonstrates its suitability in the practical sphere of smart implementation in the real world.

The chapter illustrated the applicability of privacy-saving intelligent systems in practice and the dynamic facade by giving examples of situations to be solved openly and the dimensions of research in the future. Overall, the proposed solution will offer a sustainable and trustworthy foundation for anomaly identification in the next-generation intelligent applications that can assist in making secure judgments and adhering to the privacy-by-design principle and regulatory boundaries.

References

1. V. Stephanie, I. Khalil, M. Atiquzzaman, DSFL: a decentralized SplitFed learning approach for healthcare consumers in the metaverse. IEEE Trans. Consum. Electron. **70**(1), 2107–2115 (2024). https://doi.org/10.1109/tce.2024.3360994
2. D.J. Beutel, T. Topal, A. Mathur, X. Qiu, J. Fernandez-Marques, Y. Gao, L. Sani, K. Hei Li, T. Parcollet, P. Porto Buarque de Gusmão, N.D. Lane, Flower: a friendly federated learning research framework (2020). https://arxiv.org/abs/2007.14390.
3. F. Asiri, W.A. Malwi, F. Masood, M.S. Alshehri, T. Zhukabayeva, S.A. Shah, Privacy preserving federated anomaly detection in IoT edge computing using Bayesian game reinforcement learning. Comput. Mater. Continua **84**(2), 3943–3960 (2025). https://doi.org/10.32604/cmc.2025.066498
4. S. Zahri, H. Bennouri, A.M. Abdelmoniem, An empirical study of efficiency and privacy of federated learning algorithms (2023). https://arxiv.org/abs/2312.15375
5. W. Yang Bryan Lim, N. Cong Luong, D. Thai Hoang, Y. Jiao, Y.-C. Liang, Q. Yang, D. Niyato, C. Miao, Federated learning in mobile edge networks: a comprehensive survey (2019). https://arxiv.org/abs/1909.11875
6. I. Priyadarshini, Anomaly detection of IoT cyberattacks in smart cities using federated learning and split learning. Big Data Cogn. Comput. **8**(3), 21 (2024). https://doi.org/10.3390/bdcc8030021
7. J. Pei, K. Zhong, M.A. Jan, J. Li, Personalized federated learning framework for network traffic anomaly detection. Comput. Netw. **209**, 108906 (2022). https://doi.org/10.1016/j.comnet.2022.108906
8. R. Arjdal, Y. Asimi, A. Asimi, L. Oumouss, Privacy-preserving anomaly detection for IoT: leveraging federated and split learning, in Proceedings of the 2024 International Conference on Circuit, Systems and Communication (ICCSC) (2024), pp. 1–7. https://doi.org/10.1109/ICCSC62074.2024.10616433
9. M.J. Alfadhil, A. Baydoun, M. Alazab, H. Ur Rehman, J. Al Jaam, S.A. Al-Maadeed, Enhancing federated learning for IoT-based anomaly detection: a reputation-based client selection approach. Alex. Eng. J. (2025). https://doi.org/10.1016/j.aej.2025.09.019
10. M. Arafeh, M. Wazzeh, H. Sami et al., Efficient privacy-preserving ML for IoT: cluster-based split federated learning scheme for non-IID data. J. Netw. Comput. Appl. (2025). https://doi.org/10.1016/j.jnca.2025.104105
11. J. Pereira, M. Silveira, Unsupervised anomaly detection in energy time series data using variational recurrent autoencoders with attention, in *Proceedings of the 17th IEEE International Conference on Machince Learning and Applications (ICMLA)* (2018), pp. 1275–1282. https://doi.org/10.1109/ICMLA.2018.00207
12. Y. Zhang, B. Suleiman, M.J. Alibasa, Privacy-aware anomaly detection in IoT environments using FedGroup: a group-based federated learning approach. J. Netw. Syst. Manag. (2024). https://doi.org/10.1007/s10922-023-09782-9

13. C. Zhu, J. Zhang, X. Sun, B. Chen, W. Meng, ADFL: defending backdoor attacks in federated learning via adversarial distillation. Comput. Secur. **132**(103366) (2023). https://doi.org/10.1016/j.cose.2023.103366
14. S. Thudumu, P. Branch, J. Jin, J. Singh, A comprehensive survey of anomaly detection techniques for high dimensional big data. J. Big Data (2020). https://doi.org/10.1186/s40537-020-00320-x
15. S. Zhang, Edge-assisted U-shaped split federated learning with anomaly detection for IoT systems. Expert Syst. Appl. (2025). https://doi.org/10.1016/j.eswa.2024.125494
16. H.T. Truong, B.P. Ta, Q.A. Le, D.M. Nguyen, C.T. Le, H.X. Nguyen, H.T. Do, H.T. Nguyen, K.P. Tran, Light-weight federated learning-based anomaly detection for time-series data in industrial control systems. Comput. Ind. **140**(103692) (2022). https://doi.org/10.1016/j.compind.2022.103692
17. Q. Yang, Y. Liu, T. Chen, Y. Tong, Federated machine learning: concept and applications. ACM Trans. Intell. Syst. Technol. **10**(2) (2019). https://doi.org/10.1145/3298981
18. P. Kumar, G.P. Gupta, R. Tripathi, TP2SF: a trustworthy privacy preserving secured framework for sustainable smart cities by leveraging blockchain and machine learning. J. Syst. Archit. **115**(101954) (2021)
19. Q. Wu, K. He, X. Chen, Personalized federated learning for intelligent IoT applications: a cloud-edge based framework (2020). https://arxiv.org/abs/2002.10671
20. J. Vanerio, P.C., Ensemble-learning approaches for network security and anomaly detection, in *Proceedings of the Workshop on Big Data Analytics and Machine Learning for Data Communication Networks* (Los Angeles CA USA, 2017), pp. 1–6. https://doi.org/10.1145/3098593.3098594
21. Q. Abu Al-Haija, A. Al Badawi, G.R. Bojja, Boost-defence for resilient iot networks: a head-to-toe approach. Expert Syst. **39**(10), e12934 (2022). https://doi.org/10.1111/exsy.12934
22. R. Vidhya, D. Lognathan, S. Saranya, P.N. Periyasamy, S.S., Anomaly detection in IoT networks using federated machine learning approaches. Int. J. Comput. Exp. Sci. Eng. **11**(3) (2025). https://doi.org/10.22399/ijcesen.2485
23. R. Janarthanan, R.U. Maheshwari, P.K. Shukla, P.K. Shukla, S. Mirjalili, M. Kumar, Intelligent detection of the PV faults based on artificial neural network and type 2 fuzzy systems. Energies **14**, 6584 (2021). https://doi.org/10.3390/en14206584
24. P.P. Liang, T. Liu, L. Ziyin, N.B. Allen, R.P. Auerbach, D. Brent, R. Salakhutdinov, L.-P. Morency, Think locally, act globally: federated learning with local and global representations (2020). https://arxiv.org/abs/2001.01523

Industry 5.0 and Smart Cities: Security-First IoT Applications

Somnath Banerjee

1 Introduction

The swift urbanization of smart cities has radically changed the operation of cities and the manner in which they interact and provide services [1]. Mass implementation of inter-connected devices has facilitated automated surveillance, adaptive control, and data-driven decision-making in critical urban processes including transportation, energy distribution, environmental management, healthcare provision and safety in the community. Although this is likely to be efficient, sustainable, and with better quality of life, it also brings more cyber-risks, which have never been as exposed previously. With physical and digital worlds uniting, the outcomes of system breach are no longer limited to the loss of data, but also lead to a potential threat to human lives and entire world disruption. This increased interdependence has resulted in security being a key concern, as opposed to being a secondary design factor.

Meanwhile, the industrial environment is shifting to the paradigm centred around human and focusing on the cooperation of intelligent technology and the human experience. This change requires not just the increased automation and intelligence, but also the increased responsible and transparent and ethical regulation [2]. Traditional protection measures (mostly proactive and perimeter-based) are no longer relevant in such settings. Rather, there is an urgent necessity to redefine the ways of implementing protection, trust, and resilience to infrastructures interconnected at a leading design stage.

Industry 5.0 or Society 5.0 is considered as an evolution over the previous industrial revolutions and aims to make a super smart society. The smart cities under Society 5.0 will design strategies and policies over hand IoT systems; commitment to research and development at various levels and taking care of educational reforms including technology literacy.

S. Banerjee (✉)
American Family Insurance, Madison, WI, USA

S. Saif et al. (eds.), *Secure Intelligent Sustainable Applications Using IoT for Society 6.0*, Internet of Things, https://doi.org/10.1007/978-3-032-27248-5_9

1.1 Intelligent Infrastructure Development of Human-Centric Intelligent Infrastructures

The advancement in industrial and urban technologies has no longer focused on mechanization and automation but on the development of the system in which the perception, reasoning, and adaptation can be manipulated [3]. Current intelligent infrastructures are dynamic ecosystems (not isolated or fixed), whose composition is made up of heterogeneous devices, software services, and human stakeholders. Physical environments are constantly monitored using sensors, complex data streams are analysed using computational units, and decisions are implemented with a minimum of latency using actuators. Humans, in their turn, monitor, interfere and regulate the system behaviour, especially in safety-related, ethically related situations.

This development has presented twofold problem. On the one hand, when it comes to managing complexity, systems have to work at scale and be independent. They are, on the other hand, required to be interpretable as well as usable by human operators. Security is a factor that is critical in the balance of this duality [4]. Lack of structures to protect autonomy can increase the risk as opposed to mitigating it. Therefore, security is no longer a technical demand, it has to be in line with human value, governance, and trust of the society.

Figure 1 contains a layered representation of the intelligent infrastructure, where trust challenges and security exposure happen to cut across all the functional segments of operation of the system. Policies and ethical control take place under human governments and management, whereas data-based decision-making belongs to service intelligence and application layers [5]. The coordination layer and communication layer allow interaction between the distributed components and the physical interaction layer and device layer interface directly onto the real world. The fact that all these layers converge into one exposure zone further underscores that security threats do not exist on a single layer but spread to governance, digital intelligence, communication, and physical implementation, and it is important to build on meeting the trust awareness in the system.

1.2 Ubiquitous Interconnectedness and Extensive Attack Surfaces

The very essence of urban intelligence in the modern times is pervasive connectivity. Devices are connected with each other both over wired and wireless networks and interact with cloud and edge resources as well as information exchange between organizational boundaries. Connectivity allows coordinating and being efficient, but it opens the attack surface exponentially. Every interrelated element turns into a possible attack point of attackers, and a weakness of one sub-system can spread across others.

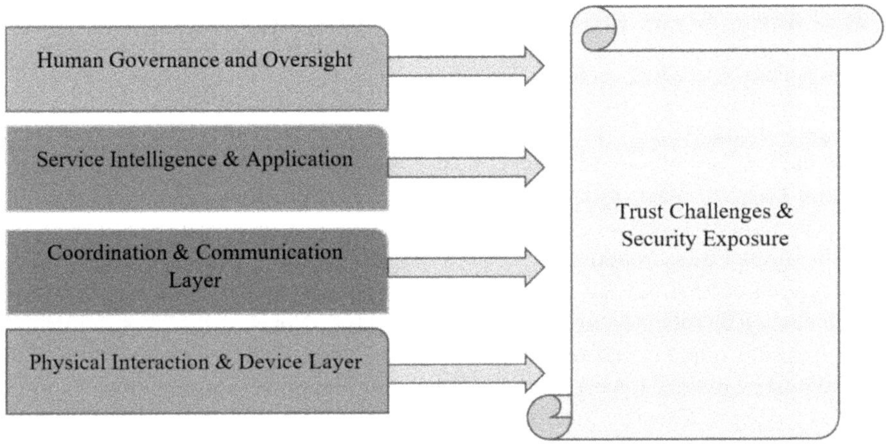

Fig. 1 Cross-layer security exposure in intelligent infrastructures

Security methods that are traditional tend to assume fixed boundaries and set patterns of interaction. Nonetheless, the modern conditions can be discussed as heterogeneous, mobile, and constantly reconfigurated [6]. Devices vary in the power of calculation, energy limitation, and reliability. Most of them are either unmanned or situated in open places, and thus they not only are prone to physical abuse but also to abuse through the Internet. In that case, the concept of static security becomes outdated as soon as possible.

1.3 Shortcomings of Traditional Security Arrangements

Protection mechanisms are often conventional and are mainly based on fixed rules, centralized enforcement, and detection of an event after the fact [7]. Access control lists, signature-based intrusion detection systems, and firewalls work well in controlled environments but do not perform in dynamic and large ecosystems. Single points of failures are introduced by centralized architectures and changing operational conditions and threats are not reflected in policies that are static.

Furthermore, most of the available solutions tend to secure data in transit or at rest without properly considering the question of legitimacy of the parties that create, process, or take action on such data. This loophole enables insecure or harmful devices to take part in activities of the system provided that they pass down the surface communication protocols [8]. Physically instigated actions in an automatic setting can result in intense effects of such involvement.

1.4 Challenges of Trust, Accountability, and Governance

Intelligent infrastructures have issues to do with trust and governance in addition to technical vulnerability [9]. Citizens, operators, and policymakers should be confident that automated systems are reliable, ethical, and they behave in terms of regulations. It should be able to know the behaviour of the systems and assign responsibility whenever failures take place. Black-box decision-making and black-box security controls subvert this confidence.

The human-centric design ideologies are firm on the need to retain human beings with meaningful roles in system management [10]. Nonetheless, it is non-trivial to incorporate human recommendations in large-scale automated setting. Over-dependence on manual intervention decreases efficiency and total autonomy is prone to loss of control. Security mechanisms should be able to uphold shared authority, on the other hand, enabling systems to have independent actions but at the same time accountable to the human stakeholders.

The meeting of cyberspace and the physical space is one of the reasons why the defensive strategy should change to include the inherent protection. Instead of trying to identify the attacks and alleviate them once they have happened, systems should be designed to avoid the emergence of insecure states in the first place [11]. This would involve incorporating protection logic into device operation gates, communication authorization, and controlling execution.

2 Background of Work

Protection mechanisms on a device level have been widely reviewed in existing literature to resolve vulnerabilities caused by massive implementation of interconnected sensing and actuation units. The author refers to the significance of secure initializing and identity checking to avoid unauthorized inclusion of devices in intelligent settings. Although they can reinforce basic access control, the investigator points out they usually consider fixed operational situations and trusted execution behaviour as obstructing their applicability in very dynamic and heterogeneous infrastructures. The researcher also indicates that device-based approaches have difficulty responding to interaction-based threats that arise outside of the individual nodes [12, 13].

An extensive literature is concerned with methods relating to securing communication flows in order to have a stable flow of information between the interconnected elements. The author talks about strategies that can be used in ensuring routing mechanisms and tracking out the travels in order to identify malicious activity [14, 15]. The investigator, however, notes this type of network-centric mechanisms is focused more on transmission level threats and gives minimal understanding of whether the data producers are legitimate. The researcher suggests that such strategies do not appropriately address the situations where tainted organizations can introduce what appears like legitimate traffic and carry out harmful activities. Table 1 presents a comparative summary of recent relevant works.

The other area of research focuses on information asset protection by means of encryption, access control, and information privacy-preserving tools. The importance of data confidentiality and integrity against unauthorized disclosure and manipulation is pointed out by the author. Even with this set of improvements, the researcher finds that data-based protections usually do not take into account the credibility of data creation (or utilization) entities. The author of the research claims that it is not appropriate to only secure the data when the malicious or compromised components are still involved in the processes of the system functioning.

Monitoring and control central architectures have also been heavily investigated in order to implement security policies over interconnected infrastructures. The author proposes centralized structures as a channel to reach to a unitary visibility and governance. Nevertheless, the researcher has discovered that there should be limitations to scalability and an increased risk of failures because a single control authority is involved. The researcher also points out that the enforcement that is centralized in nature would be more impractical in terms of size and complexity of the system.

Table 1 Comparative summary of existing research directions in intelligent urban infrastructure security

Focus area	Core approach	Identified limitations
Device-level protection [16–18]	Emphasizes cryptographic authentication and secure boot mechanisms to protect individual nodes from unauthorized access	Limited adaptability to heterogeneous environments and ineffective against compromised yet authenticated devices
Network-centric security [19]	Relies on secure routing, traffic filtering, and intrusion detection to protect communication channels	Struggles with scalability and fails to address trustworthiness of data sources and execution entities
Data-oriented safeguards [20, 21]	Concentrates on encryption, access control, and privacy preservation during data storage and transmission	Protects data confidentiality but does not prevent malicious data generation or misuse by trusted entities
Centralized control models [22]	Utilizes centralized controllers to enforce security policies and monitor system behaviour	Introduces single points of failure and increases vulnerability under large-scale or distributed deployments
Distributed intelligence approaches [23–25]	Employs decentralized decision-making to enhance resilience and reduce dependency on central authorities	Lacks unified trust assessment and often suffers from coordination complexity and inconsistent enforcement
Human-in-the-loop systems [26]	Integrates human oversight for decision validation in critical operations	High reliance on manual intervention reduces responsiveness and scalability in dynamic environments

Recent sources admit the necessity of introducing human control to the intelligent environments to make them more responsible and ethically controlled. The author proposes that human in the loop will enhance query elucidate and validation of decisions in life-or-death situations. However, the researcher observes that overreliance on a manual input decreases system responsiveness. The researcher finds that there is a gap in the research that has not been solved as there is no effective combination of autonomous intelligence and meaningful human control.

2.1 Objective

- To test the security issues which arise due to very interconnected and humanitarian smart infrastructures in the contemporary urban setting.
- In order to critically point out current research directions and to define the main limitations in current protection and trust management practices.
- To enable a conceptual baseline of the security-first intelligence that enables robust, responsible, and human-congruous functioning of massive intelligent systems.

2.2 Motivation

- The high pace of introducing smart technologies into city systems has increased vulnerability risks that have a direct physical or social impact.
- The available protection is more or less reactive and each level is mostly fragmented, making it frail at a large-scale deployment.
- It is urgent that security should be reconsidered as fundamental to intelligent systems as those which meet automation with human control and confidence.

3 Experimental Setup

The experimental framework takes into account a simulated intelligent urban environment of heterogeneous sensing, communication, and control elements working under realistic workload conditions. Devices are implemented in various functional areas such as mobility management, environmental monitoring, and public safety services. There is an introduction of both normal and adversarial interaction patterns that determine system behaviour in circumstances of different levels of trust. Measures used to evaluate performance are tied to resilience and execution legitimacy and continuity of systems. Human inputs that include human oversight are included to represent the validation of decisions in emergency situations. The experiments are all performed in reproducible settings to facilitate a fair comparison between experiments and consistent analysis on cases of evaluation.

3.1 Dataset Used

The analysis utilizes a composite dataset formed out of publicly available records of urban sensing and network interaction protocols and synthetically generated control and event information. The data consists of device interaction history, communication metadata, execution state history, and operational event history of various urban service scenarios. Cheating operational patterns are superimposed with injected abnormal and antagonistic behaviours to capture realistic conditions of threats. The heterogeneity of the types of devices and workload variation ensure data diversity. Any redundancy and noise in the records are eliminated without losing behavioural characteristics that are important in performing trust and security analysis.

4 Proposed Framework

The rationale behind the proposed approach is that security in strongly connected intelligent environments should be viewed as that of an internal operational property, as opposed to performing as a kind of outer sheet of protection. The dynamic legitimacy of the participating entities in complex cyber-physical infrastructures cannot be sufficiently reflected using traditional mechanisms for boundary-based securities in Fig. 2. The model introduced here incorporates trust rationale, execution validation, and adaptive intelligence deep into the backbone of the interrelation of systems so that continuously evaluating the legitimacy of various interactions, computation, and actuation can be assessed in advance.

The core concept of the approach is that it proposes an ongoing trust-building process that assesses the operational credibility of all of the involved entities on the basis of execution behaviour, conformity in interactions, and consistency in integrities. This approach views trust as a re-built property that changes and advances with system behaviours unlike traditional methods, which depend on the use of static credentials or pre-determined access policies. This design provides the system with the ability to avert insecure states before they occur, instead of trying to jump out of the security violations after the propagation.

The intelligent environment is represented as a network of self-governing functioning units, which can sense, process, communicate, and act. Every entity has a state of internal operation that depicts its behaviour of execution, its history of interaction, and its integrity stance. The system is not based on an inherent trust in any entity but rather the legitimacy will be built through evaluation, which is conducted on a continuous basis.

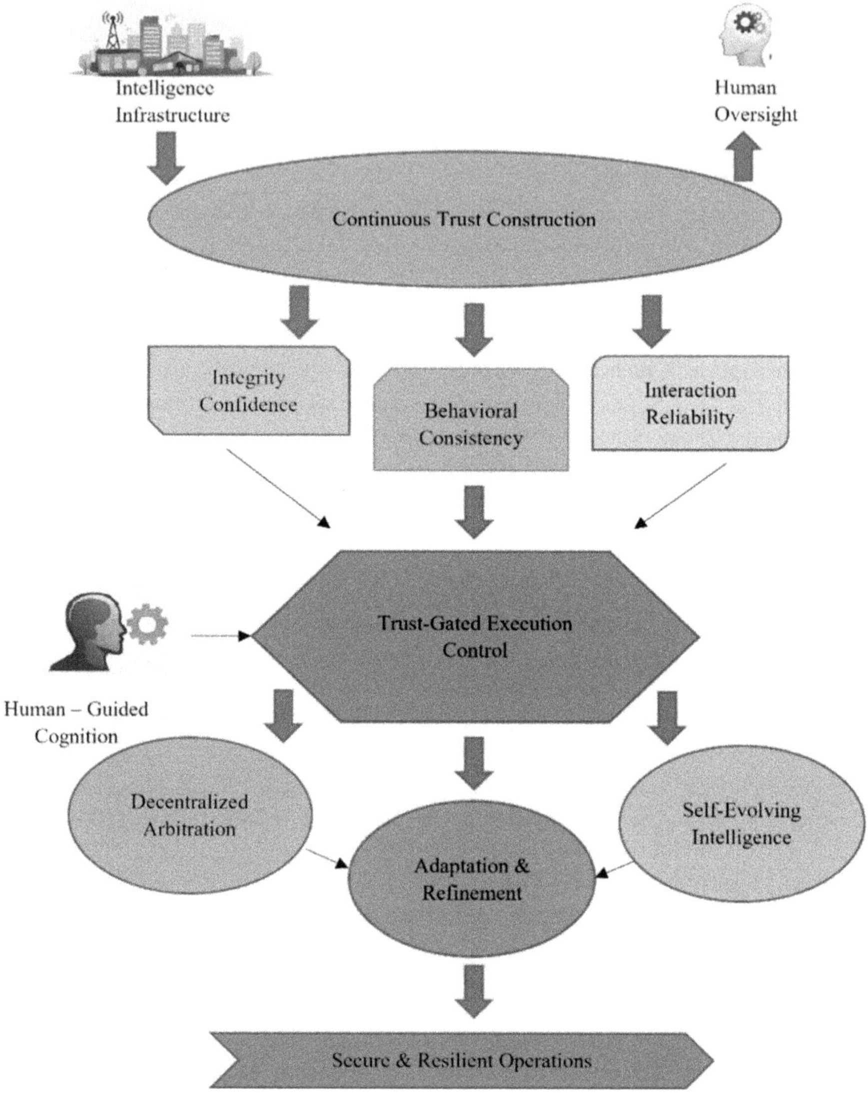

Fig. 2 Conceptual pipeline of trust-gated execution and adaptive security intelligence

Let the operational legitimacy measure of an entity take the form of a scalar trust variable τ, which is dynamically computed as a weighted combination of integrity confidence, uniformity of behaviour, and reliability of interaction. Mathematically, this relationship can be expressed as in the following equation Eq. (1)

$$\tau = \alpha O + \beta N + \gamma T \tag{1}$$

where O signifies integrity confidence, N indicates behavioural consistency, and T indicates interaction reliability. The weighting coefficients α, β, and γ dictate the relative impact of each element and are normalized to ensure their total equals one. As demonstrated in Eq. (1), trust is not dichotomous but rather continuously valued, facilitating nuanced legitimacy evaluation.

Integrity confidence is used to describe the level of integrity in the execution environment of an entity. Comparison with expected execution signatures against observed execution behaviour is used to derive it. The integrity confidence function is represented as in Eq. (2):

$$O = \exp(-\lambda \Delta) \tag{2}$$

where Δ represents the discrepancy between anticipated and actual execution states, and λ regulates sensitivity to this discrepancy. As seen in Eq. (2), integrity confidence diminishes exponentially with greater deviation, deterring even little execution irregularities.

Behavioural consistency is the ability of an entity to behave in accordance with the previously established profile of its operations. Instead of using fixed behavioural patterns, the methodology forms adaptive pattern of behaviour which increases with valid use patterns. Trade-off the magnitude of deviation is a normalized measure that is used to determine consistency estimations, behaving not as a deviation.

Behavioural consistency is calculated as represented in Eq. (3):

$$N = \frac{1}{1 + \mu F} \tag{3}$$

where F signifies the normalized behavioural deviation and μ specifies the pace at which variation diminishes consistency. As indicated in Eq. (3), the formulation guarantees diminishing returns on trust as deviation escalates, while permitting progressive adaptation.

In order to maintain legitimate patterns, the system deactivates the adversarial behaviour through a stability regulator that tunes down the adjustment of the baselines. The rate of adaptation in the baseline is regulated as in Eq. (4):

$$\frac{d\Omega}{dt} = \kappa(\Psi - \Omega) \tag{4}$$

where Ω signifies the current behavioural baseline, Ψ indicates verified behaviour, and κ regulates adaptive responsiveness. As seen in Eq. (4), only authenticated behaviour affects baseline evolution, hence guaranteeing robustness against poisoning efforts.

Interaction reliability measures the ability of an entity to engage in cooperative operations and not bring anomalies or conflicts. This test is essential in the settings where system safety is dependent on the concerted action.

A cumulative conformity score is calculated to compute interaction reliability as depicted in Eq. (5):

$$T = 1 - \frac{\sum \xi}{\sum \eta} \tag{5}$$

where ξ signifies interaction abnormalities and η indicates the total interactions recorded within the evaluation frame. As seen in Eq. (5), dependability diminishes proportionately with the concentration of irregular interactions.

Interaction weighting is adjusted to criticality in interaction to make sure that reliability assessment is adaptive. The criticality-adjusted interaction score is calculated as in Eq. (6):

$$e_v = \frac{\chi}{1 + \rho} \tag{6}$$

where χ signifies interaction significance and ρ represents system load circumstances. As indicated in Eq. (6), crucial interactions are assigned greater significance during steady conditions, so averting overload amplification.

A. Trust-Gated Execution Control.

The peculiar feature of the suggested strategy is that the implementation would be done only in case the trust is higher than a dynamically regulated legitimacy benchmark. Instead of applying a fixed threshold, the system changes acceptance criteria depending on conditions in the system and human control as input.

The authorization execution function will be characterized as illustrated in Eq. (7):

$$\Theta = \sigma(\tau - \theta) \tag{7}$$

where θ represents the legitimacy threshold and σ denotes a smooth activation function. As seen in Eq. (7), execution authorization shifts seamlessly rather than abruptly, hence minimizing instability around threshold borders.

This algorithm controls the permission of execution by assessing the current state of trust in comparison with adaptive legitimacy requirements and taking into consideration peer influence and human advice. It guarantees that only those things that meet the trust conditions will be made to run, and the insecure interactions will be subjected to isolation.

Algorithm 1: Trust-gated execution authorization
Obtain execution request from entity
Obtain the current trust score
Verify the adaptive legitimacy threshold
Integrate the influence of peer trust
Implement human oversight modification
If trust fulfils the authorisation criterion

(continued)

(continued)

Authorise execution
End
Prohibit execution and segregate interaction
End if
Revise trust status according to execution results
Decision on return authorisation

Legitimacy threshold in itself is dynamic and changes based on system risk posture. With respect to threshold adaptation, it is governed by Eq. (8):

$$\theta_{y+1} = \theta_y + \delta(\Phi - \theta_y) \tag{8}$$

where Φ denotes observed system risk and δ regulates the adaptation rate. As indicated in Eq. (8), increased risk conditions naturally intensify authorization requirements.

B. Decentralized Trust Arbitration.

Instead of centralized controllers, the trust determination is influenced by the surrounding entities. Every entity transmits signs of trust that are authenticated at the local level and then the influence is accepted.

Aggregated peer trust influence is calculated as in Eq. (9):

$$\tau_a = \frac{\sum \omega_o \tau_0}{\sum \omega_o} \tag{9}$$

where τ_0 signifies peer trust levels and ω_o indicates credibility weights. As indicated in Eq. (9), peer influence is moderated by credibility, so preventing implicated entities from controlling arbitration.

Credibility weights are decreased to curtail spreading misinformation in case inconsistency is uncovered. The weight decay is modelled as in Eq. (10):

$$\omega_{o+1} = \omega_o r^{-\nu\epsilon} \tag{10}$$

where ε represents the level of inconsistency, r denotes the decay base, and ν governs decay sensitivity. As demonstrated in Eq. (10), credibility diminishes swiftly in the presence of inconsistent behaviour.

C. Human-Guided Security Cognition.

Human control is introduced as a supervisory factor as opposed to direct control. The human inputs do not perform micromanagement of operational decisions but just adjust the parameters of sensitivity and the legitimacy thresholds.

The effects of human guidance are weaved into as illustrated in Eq. (11):

$$\tau_j = \tau + \zeta J \tag{11}$$

where τ_j symbolizes the modified trust value, ζ signifies the influence factor, and J represents corroborated human input.

D. Self-Evolving Trust Intelligence.

The system keeps on evolving the trust reasoning in accordance with detected discrepancies between the anticipated and realized outcomes. This self-evolution allows breaking the stagnation and adapting to the new conditions of work.

Trust refinement adjustment is calculated; it is represented in Eq. (12):

$$\Delta\tau = \upsilon\left(\tau_{obs} - \tau_{pred}\right) \tag{12}$$

where observed trust τ_{obs} is juxtaposed with the predicted trust τ_{pred}, whereas υ regulates learning response. Trust models develop via refinement driven by discrepancies, as illustrated in Eq. (12).

This algorithm endlessly recreates trust via monitoring of execution, conformity of interaction, and integrity conformity of things. It is a dynamic update of trust states based on changing operational circumstances to achieve system legitimacy.

Algorithm 2: Continuous trust construction
Establish trust variables for each entity
Configure integrity, conduct, and interaction monitors
While the system is functioning
Monitor execution performance
Verify integrity status
Assess interaction compliance
Revise trust components
Integrate elements into a trust score
Preserve trust history
End while

The suggested solution defines security as an operational property of intelligent infrastructures based on the premise of enforcing continuous trust reasoning into execution, interaction, and coordination processes. Instead of the assumed integrity based on the hard copy credentials, integrity confidence, behavioural consistency, and interaction reliability are used in building trust dynamically. Only legitimate entities are allowed to execute, and therefore insecure entities are denied the opportunity of operating in systems. The advantages of decentralized arbitration are its resilience to attacks, and the human-guided cognition guarantees accountability and moral uprightness. The strategy is able to provide strong and well-proactive protection of complex intelligent environments through self-evolving trust intelligence so that the strategy adjusts itself to constantly changing operational environments.

5 Experimental Results

The section offers a specific comparative assessment of the suggested solution in contrast with the available real-time security solutions in heterogeneous operational scenarios. The evaluation criterion is on responsiveness, resource efficiency, and classification reliability. The findings support the idea that the incorporation of security intelligence into the functionality of the system promotes the quality of decision-making, declines the response time, and energy consumption. The proposed method is more accurate, more precise, has a higher recall and F1-score, and exhibits stable classification behaviour in adversarial settings, compared to the conventional ones as shown in Tables 2 and 3. A decrease in the latency and detection delay is indicative of a quicker system reaction, whereas a decrease in false positive rate indicates consistent decision-making with no unnecessary notifications. On the whole, the findings confirm usefulness of intrinsic trust-based security in complicated intelligent settings.

Accuracy (%): The ratio of correct security-related decisions of the system to the total decisions that are being assessed.

8.2 Precision (%): The ratio of correctly identified legitimate or malicious observations to the overall number of instances which were identified as positive by the system.

Recall (%): The system recalls correct malicious or unusual instances remembered out of all such occurrences.

F1-Score (%): The balance between the accuracy in classification in the performance of precision and recall in different conditions of operation.

Latency (ms): The span of time that the system takes to produce an authorization or security decision on receiving an operational request.

Table 2 Evaluation of accuracy and precision

Method	Accuracy (%)	Precision (%)
Signature-based IDS [16]	75.2	72.6
Rule-based access control [19]	72.8	68.2
Centralized security control [12]	77.1	74.4
Network anomaly detection [15]	79.6	77.1
Device authentication model [18]	76.3	73.9
Distributed IDS [11]	81.4	79.2
Edge-based security mechanism [14]	83.7	81.6
Human-in-the-loop security [2]	85.9	83.7
Adaptive learning security [8]	88.6	86.9
Proposed method	95.3	93.8

Table 3 Analysis of recall and F1-score

Method	Recall (%)	F1-score (%)
Signature-based IDS [16]	70.4	71.5
Rule-based access control [19]	68.5	69.3
Centralized security control [12]	72.2	73.3
Network anomaly detection [15]	75.8	76.4
Device authentication model [18]	71.6	72.7
Distributed IDS [11]	77.9	78.5
Edge-based security mechanism [14]	80.1	80.8
Human-in-the-loop security [2]	84.3	83
Adaptive learning security [8]	86.1	86
Proposed method	95.6	93.2

Delay (s): This is the time lag between the moment an event that is relevant to security actually happens and the time the system notices it.

False Positive Rate (%): The percentage of normal instances of operation classified as malicious or anomalous.

Energy Consumption (J): The amount of energy that is used by the components of the system to conduct security evaluation and make decisions.

The comparison will show that the performance has been consistently improved. Older methods like the signature-based IDS and rule-based access control have equal accuracies of 75.2% and 72.8% and precision of 72.6% and 68.2%, respectively, in Fig. 3. Instructional, though less developed, ways such as network anomaly detection and distributed IDS demonstrate average increases with the accuracies of 79.6% and 81.4%, and precision of 77.1% and 79.2%, respectively. Edge-based security mechanism and human-in-the-loop security improve this durability with accuracies of 83.7% and 85.9%, respectively, with the corresponding precision of 81.6% and 83.7%. Adaptive learning security achieves accuracy and precision of 88.6% and 86.9%, respectively. However, the proposed method excels compared to all current forms of methodology, exhibited the highest accuracy at 95.3% and it has a high precision at 93.8%, and it shows the better reliability and quality of decisions.

The comparison shows that there is a steady growth in the detection ability of the analysed methods. The traditional methods (signature-based IDS and rule-based access control) achieve recall rates of 70.4% and 68.5%, respectively, with the F1-scores of 71.5% and 69.3%, which are quite low, showing that they cannot recognize all the significant security incidents. The moderate gains are on centralized security control and network anomaly detection with 72.2% and 75.8% recall and 73.3% and 76.4% F1-score, respectively. Even more sophisticated distributed IDS and edge-based security mechanism have even more outstanding performances with the highest recall values of 77.9% and 80.1% and F1-scores of 78.5% and 80.8%

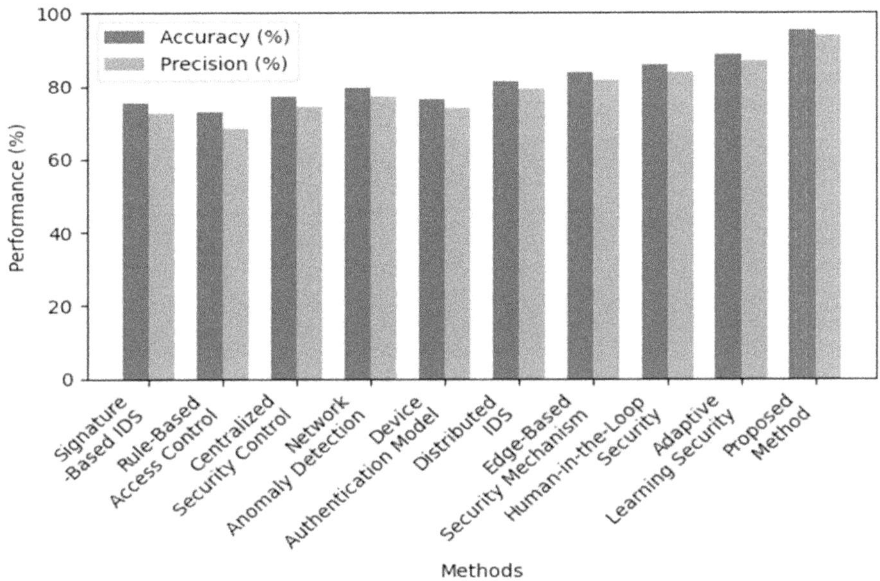

Fig. 3 Comparison of accuracy and precision analysis

in Fig. 4. Human-in-the-loop security and adaptive learning security are more reliable in detection showing 84.3% and 86.1% recall rates with F1-scores of 83% and 86%. The proposed method is obviously superior than all of the existing methods achieving the best recall and F1-scores of 95.6% and 93.2%, respectively, which presents a better detection consistency and equalized classification ability.

This comparison given in Table 4 shows evident differences in responsiveness of systems of the considered strategies. The signature-based IDS and rule-based access control conventional methodologies have latencies of 53 ms and 46 ms, respectively, and delays of 7.3 s and 6.9 s meaning slower response to security events. Response overhead is greatest with centralized security control 6.5 s, respectively, in Fig. 5. Response time is also further promoted by distributed IDS and edge-based security mechanism to 48 ms and 36 ms and delay to 5.4 s and 4.9 s. The manual intervention in human-in-the-loop security increases the latency and delay to 73 ms and 7.7 s. The adaptive learning security enhances a faster speed of reaction by 45 ms latency and a delay of 4.6 s. The proposed method has the lowest latency and lowest delay of 32 ms and 3.2 s showing the best real-time responsiveness.

The findings show significant variation in the reliability in the decisions made and resource efficiency among the considered methods as shown in Table 5. Signature-based IDS and rule-based access control achieve false positives of 15.8% and 13.7% and consume 19.9 and 18.6 J of energy, respectively, meaning that they are characterized by frequent false alerts and medium levels of resource usage, respectively, in Fig. 6. The false positive values of centralized security control and network anomaly detection at points 14.9% and 12.5% with a higher cost of energy at 22.2 and 20.7 J,

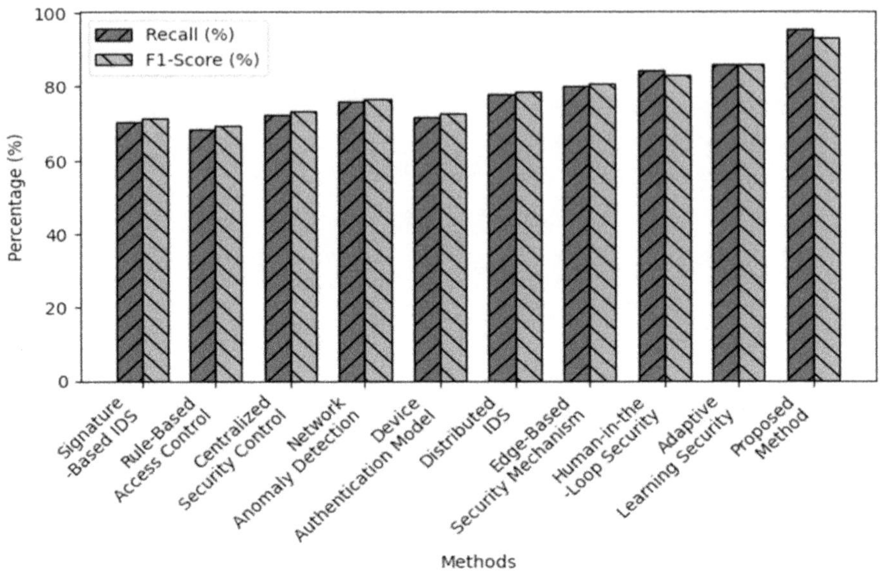

Fig. 4 Comparison of recall and F1-score results

Table 4 Examination of latency and delay

Method	Latency (ms)	Delay (s)
Signature-based IDS [16]	53	7.3
Rule-based access control [19]	46	6.9
Centralized security control [12]	67	8.4
Network anomaly detection [15]	55	6.1
Device authentication model [18]	40	6.5
Distributed IDS [11]	48	5.4
Edge-based security mechanism [14]	36	4.9
Human-in-the-loop security [2]	73	7.7
Adaptive learning security [8]	45	4.6
Proposed method	32	3.2

respectively, greatly decreased the false positive values to a low 11.9% and 10.6%, respectively. The edge-based security mechanism and human-in-the-loop security results are lower false positivity of 9.8% and 8.6%, respectively, and the adaptive learning security has better energy efficiency of 7.9% false positivity and 18.2 J power consumption. Its proposed method has the lowest false positive (4.1%) and has a minimum energy consumption (14.4 J), which shows the best reliability and energy consumption.

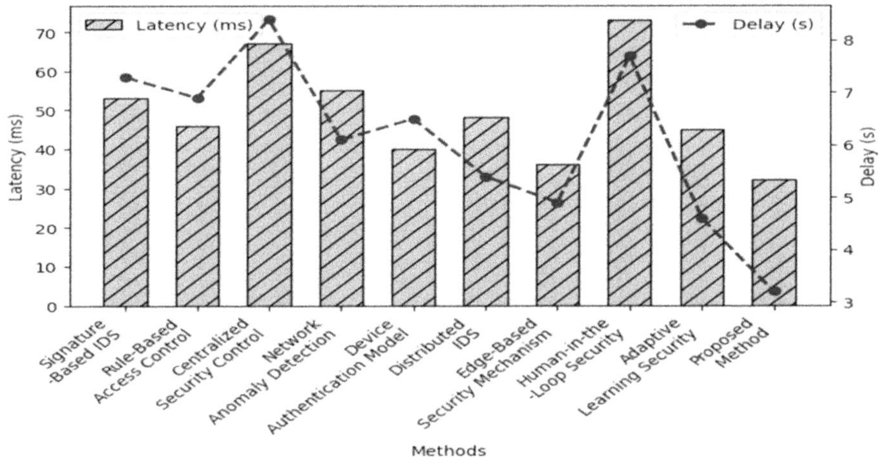

Fig. 5 Comparison of latency and delay analysis

Table 5 Analysis of false positive rate and energy consumption

Method	False positive rate (%)	Energy consumption (J)
Signature-based IDS [16]	15.8	19.9
Rule-based access control [19]	13.7	18.6
Centralized security control [12]	14.9	22.2
Network anomaly detection [15]	12.5	20.7
Device authentication model [18]	11.9	17.8
Distributed IDS [11]	10.6	21.1
Edge-based security mechanism [14]	9.8	16.4
Human-in-the-loop security [2]	8.6	23.5
Adaptive learning security [8]	7.9	18.2
Proposed method	4.1	14.4

6 Conclusion and Future Work

The paper discussed the burning issue of integrating security into infrastructures as an operation property and not as an implementation layer as shown in the literature. Through analysis of the shortcoming of current protection systems and testing of the system behaviour under non-homogeneous and malicious environments, the research showed that the concept of security conscious intelligence has a considerable positive impact on the reliability and responsiveness. The findings indicate that the suggested Trust-Embedded Adaptive Intelligence Framework (TAIF) exhibits significantly higher classification and decision-making ability as opposed to the current real-time strategies. It is worth noting that TAIF achieves an accuracy of 95.3 and the

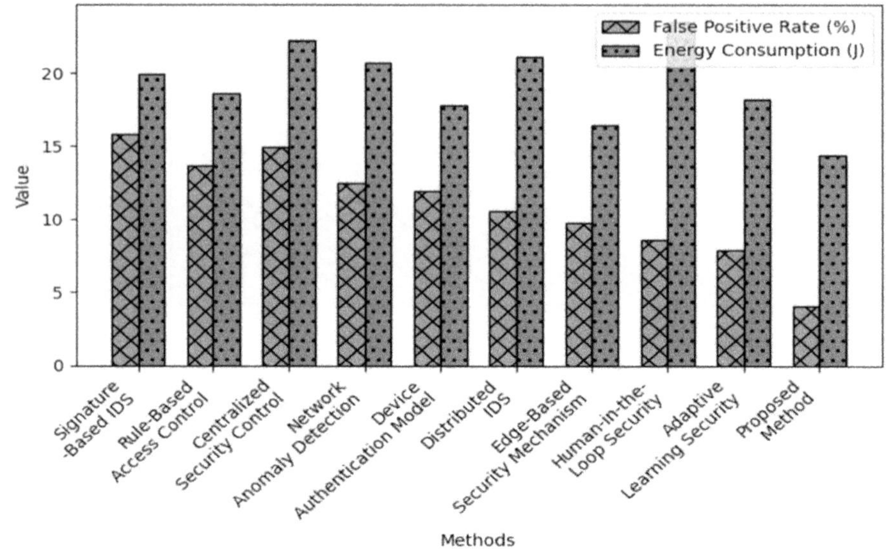

Fig. 6 Comparison of false positive rate and energy consumption

precision and recall values are more than 93 and 95, which shows that the security decisions are not biased. The low latency of 32 ms and detection delay of 3.2 s indicate that the new system can respond more quickly, whereas the low false positive rate of 4.1% shows that the new system can be highly reliable in terms of giving false alarms. Moreover, it is also an important fact that reduced energy usage of 14.4 J indicates that improved security intelligence does not come at prohibitive resource costs. All these results support the claim that the proper mechanisms to allow resilient and accountable functioning in intelligent complex settings are built into the continuous trust reasoning, adaptive authorization, and human-guided cognition. Future research will be aimed at the extension of TAIF to large-scale real-life applications with varying regulatory restrictions.

References

1. E. Ismagilova et al., Security, privacy and risks within smart cities: Literature review and development of a Smart City interaction framework. Inf. Syst. Front. **24**(2), 393–414 (2022). https://doi.org/10.1007/s10796-020-10044-1
2. I. Rafiq et al., IoT applications and challenges in smart cities and services. IET Smart Cities. https://doi.org/10.1049/tje2.12262
3. I. Hussain, Secure, sustainable smart cities and the internet of things: Perspectives, challenges, and future directions. Sustainability **16**(4), 1390 (2024). https://doi.org/10.3390/su16041390
4. A. Syed, S. Ahmed, IoT in smart cities: A survey of technologies, practices and challenges. Smart Cities **4**(2), 24 (2021). https://doi.org/10.3390/smartcities4020024

5. A. Rejeb, K. Rejeb, S. Simske, H. Treiblmaier, S. Zailani, The big picture on the Internet of Things and the smart city: a review of what we know and what we need to know. Internet Things (2022), Article 100565, https://doi.org/10.1016/j.iot.2022.10056
6. M. Zaman, A review of IoT-based Smart City development and management. Smart Cities **7**(3), 61 (2024). https://doi.org/10.3390/smartcities7030061
7. C. Wang, S. Xing, R. Gao, L. Yan, N. Xiong, R. Wang, Disentangled dynamic deviation transformer networks for multivariate time series anomaly detection. Sensors **23**(3), 1104 (2023). https://doi.org/10.3390/s23031104
8. C. Hazman et al., Enhanced IDS with deep learning for IoT-based Smart City applications. Trans. Smart Technol. (2024) https://doi.org/10.26599/TST.2023.9010033
9. F. Al-Turjman, H. Zahmatkesh, Smart city privacy and security problems, solutions, and research directions, specialized journal (2021). https://doi.org/10.1155/2021/9830547 — survey on privacy/security challenges in smart city IoT
10. M. Zaman, A review of IoT-based Smart City development and challenges. Build. Cities Smart Cities J. **7**(3), 61 (2024). https://doi.org/10.3390/smartcities7030061
11. A.K. Jain, M. Mudholkar, P. Mudholkar, M.K. Sharma, M. Tiwari, V. Roy, A novel approach to optimized cluster head selection in heterogeneous WSNs for IoT deployments, in *2024 IEEE 2nd International Conference on Innovations in High Speed Communication and Signal Processing (IHCSP), Bhopal, India* (2024), pp. 1–6. https://doi.org/10.1109/IHCSP63227.2024.10960066
12. L. Mainetti, P. Panarese, R. Vergallo, WoX+: A meta-model-driven approach to mine user habits and provide continuous authentication in the Smart City. Sensors **22**(18), 6980 (2022). https://doi.org/10.3390/s22186980
13. M.U. Nasir, S. Khan, S. Mehmood, M.A. Khan, M. Zubair, S.O. Hwang, Network meddling detection using machine learning empowered with Blockchain technology. Sensors **22**(18), 6755 (2022). https://doi.org/10.3390/s22186755
14. M.S. Farooq, S. Khan, A. Rehman, S. Abbas, M.A. Khan, S.O. Hwang, Blockchain-based smart home networks security empowered with fused machine learning. Sensors **22**(12), 4522 (2022). https://doi.org/10.3390/s22124522
15. N. Ainane, Y. Korba, M. Debbabi, Data security of smart cities, in *Proceedings of the 3rd International Conference on Information System Security and Privacy (ICISSP)* (2018), pp. 78–87. https://doi.org/10.1145/3286606.3286866
16. M. Houichi, A. Shaikh, Enhancing Smart City security: Intrusion detection challenges and future directions IET Smart Cities (2025) (early access) https://doi.org/10.1049/smc2.70014
17. G. Manoharan, B. Venugopal, K. Kunnathully, P. Suribabu, M. Tiwari, V. Roy, Blockchain-based IoT device authentication and cost-effective key management framework, in *2024 IEEE 2nd International Conference on Innovations in High Speed Communication and Signal Processing (IHCSP), Bhopal, India* (2024), pp. 1–6 . https://doi.org/10.1109/IHCSP63227.2024.10960153
18. V. Roy, A context-aware internet of things (IoT) founded approach to scheming an operative priority-based scheduling algorithms. **13**(1), 28–35 (2024) https://doi.org/10.54216/jcim.130103
19. A. Rejeb, K. Rejeb, S. Simske, H. Treiblmaier, The big picture on the internet of things and the smart city: A review of what we know and what we need to know. Internet Things, 100565 (2022). https://doi.org/10.1016/j.iot.2022.100565
20. H. Omrany et al., IoT-enabled smart cities: A hybrid systematic analysis of key research challenges. Smart Cities (Springer MDPI Linked Ed.) (2024). https://doi.org/10.1007/s44327-024-00002-w
21. A. Hussain (Ed.), Secure, sustainable smart cities and IoT — Special issue editorial. Sustainability **16**(4), 1390 (2024). https://doi.org/10.3390/su16041390
22. Y. Bhardwaj, Smart cities and the IoT: An in-depth analysis of global research trends and future directions. Discov. Internet Things (2024). https://doi.org/10.1007/s43926-024-00076-3
23. H. Omrany et al., IoT-enabled smart cities: A hybrid systematic analysis of key research challenges. Smart Cities (2024). https://doi.org/10.1007/s44327-024-00002-w

24. M. Sacoto-Cabrera et al., IoT, AI, and digital twins in smart cities: A systematic literature review. Smart Cities **8**(5), 175 (2025). https://doi.org/10.3390/smartcities8050175
25. M. Sokak, H. Tang, Y. He, F.R. Yu, Security and privacy of smart cities: A survey, research issues and challenges. IEEE Commun. Surv. Tutor. **21**, 1718–1743 (2019). https://doi.org/10.1109/COMST.2019.2899985
26. H. Zeng, M. Yunis, A. Khalil, N. Mirza, Towards a conceptual framework for AI-driven anomaly detection in Smart City IoT networks for enhanced cybersecurity. J. Innov. Knowl. **9**(4), 100601 (2024). https://doi.org/10.1016/j.jik.2024.100601

Blockchain-Enabled Trust Management and Secure Data Sharing for Smart Cities in Society 6.0

Tanvir Habib Sardar, Debrupa Pal, Papri Ghosh, Subhram Das, Bishwajeet Pandey, and Laura Aldasheva

1 Introduction

1.1 The Evolution Toward Society 6.0

The Society 6.0 conceptualization is a paradigm shift that is beyond the industrial and information-based models of the past societal structures. Whereas Society 5.0, proposed by the Japanese government, was focused on the merging of cyberspace and physical space with the help of advanced technologies, Society 6.0 is a vision of a more comprehensive, sustainable, and human-oriented approach to technological integration. The fact that artificial intelligence, Internet of Things, blockchain technology and a promotion of improved analytics have been merged smoothly to create smart ecologies that care about the well-being of the citizens, environmental sustainability, and equitable resource distribution defines this development.

In this respect, the smart cities become a literal embodiment of the ideas of the Society 6.0 [1]. These city settings take advantage of interconnected digital infrastructure to maximize the use of resources, improve the quality of life, improve

T. H. Sardar
Department of Computer Science and Engineering, Dayananda Sagar University, Bangalore, India

D. Pal (✉)
Department of Computer Application, Narula Institute of Technology, Kolkata, India
e-mail: dppalit@gmail.com

P. Ghosh · S. Das
Department of Computer Science and Engineering, Narula Institute of Technology, Kolkata, India

B. Pandey
Department of Computer Application, GL Bajaj Institute of Technology and Management, Greater Noida, India

L. Aldasheva
School of Cybersecurity, Astana IT University, Astana, Kazakhstan

S. Saif et al. (eds.), *Secure Intelligent Sustainable Applications Using IoT for Society 6.0*, Internet of Things, https://doi.org/10.1007/978-3-032-27248-5_10

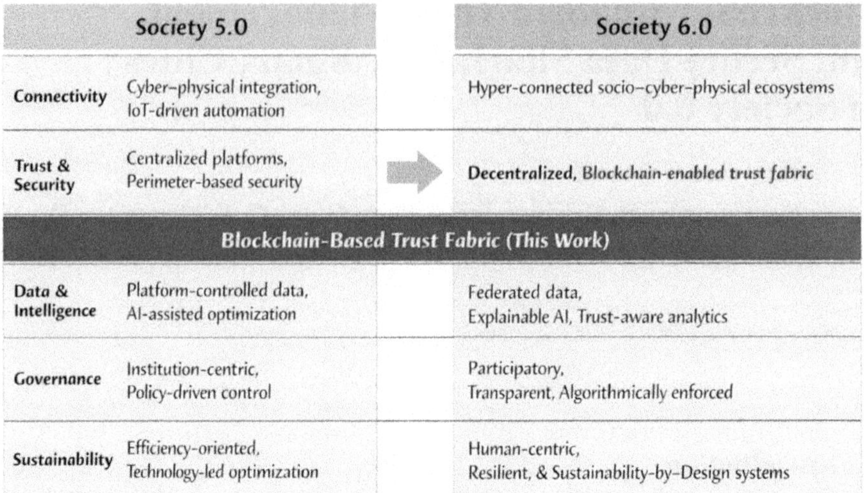

Fig. 1 Society 5.0 to Society 6.0: Trust-centric smart cities

the quality of public services, and enhance sustainable development. The smart city concept involves different areas such as smart transportation systems, smart healthcare systems, energy grids, environmental monitoring systems, e-governance systems, and e-safety infrastructure. Both spheres produce huge volumes of data due to the presence of a variety of IoT sensors and devices, which form a complex network of information exchanges that demand a solid mechanism of building trust, data integrity checks, and protecting information exchange. Figure 1 shows the transition from Society 5.0 smart cities to Society 6.0 trust centric smart ecosystems.

1.2 Challenges in Smart City Data Ecosystems

Although there are positive prospects of the smart city implementations, a number of underlying challenges hinder their operationalization on the scale. The decentralization of the IoT devices poses many attack vectors in the hands of malicious actors aiming at compromising the integrity of the system. Conventional centralized security architectures fail to provide sufficient security services in the presence of millions of heterogeneous devices of different computation and communication protocols requirements with different security needs. Lack of a standardized trust structure complicates the creation of trusting relationships between various stakeholders such as government agencies, off-the-balance-sheet service providers, citizens and autonomous systems [2].

One more critical problem of smart city environment is the problem of data privacy. Citizens offer personal data in endless amounts since their interactions with smart infrastructure can encompass such elements as location information taken by transit networks or health information taken by medical IoTs. This information will be subject to the dangers of being violated, abused, and manipulated by unauthorized individuals due to the lack of proper protective measures. Furthermore, the lack of transparency in the process of data management discredits citizens and creates impediments in the introduction of smart cities [3].

Interoperability is also a problem that makes the smart city environment complicated. Proprietary systems that do not readily integrate with each other will typically be used by various vendors and service providers. Such disaggregation results in data silos, lowered efficiency of operations, and the inability to employ broad analytics that may result in better decision-making. In addition, the systems of the majority of smart city platforms are centralized, which translates to failure in one place and as such, systems are susceptible to malicious attacks or technical issues that impact on critical services.

1.3 Blockchain as an Enabling Technology

The blockchain technology has some interesting solutions to most of the challenges encountered when implementing smart cities. Blockchain, originally built as the base infrastructure of cryptocurrency transactions, has become a flexible distributed ledger technology with potential uses in other aspects of the economy than financial services. The basic features of blockchain such as decentralization, immutability, and transparency, and cryptographic security are compatible with the needs of data management in smart cities and the establishment of trust [4].

Under a smart city system based on blockchain, the transaction of data is held in distributed registers in different nodes, making it impossible to have single-point failures or to be susceptible to specific attacks. The historical data of a system is not subject to any possible retroactive modification as the blockchain records are immutable, which gives an auditory trail of all activities of the system. Self-executing smart contracts encoded into the blockchain allow automated evaluation of trust and controlled access to it according to the established rules and dynamic metrics of behavior [5].

The combination of blockchain and IoT infrastructure opens up new opportunities of decentralized authentication, whereby devices can identify the identities of one another without the input of centralized authorities. This eliminates latency, increases scalability, and increases system resilience. Additionally, reputation systems that are based on blockchain can trace the reliability of devices, service providers, and users over time, making it possible to implement adaptive security policies in response to changing threat situations [6].

1.4 Chapter Objectives and Contributions

This chapter introduces an extended model of blockchain-based trust management and secure information exchange in smart cities on the basis of the Society 6.0 paradigm. The main contributions are the design of a layered architecture that will incorporate blockchain technology with the existing IoT middleware, the development of contract-based mechanisms of automated trust evaluation and access control, and the testing of the proposed framework on the use-cases analysis of the multiple domains in a smart city.

Practical implementation concerns, such as performance optimization techniques to reduce transaction overhead and energy consumption, scalability techniques to handle increasing numbers of devices and users, and governance techniques that allow decentralization with operational effective efficiency, are also discussed in the chapter. The study proves that blockchain-based trust management can be useful and efficient in various functioning environments by analyzing real-life situations in smart healthcare and traffic management.

2 Background and Related Work

2.1 Smart City Architectures and IoT Integration

The modern model of smart city architecture is a layered structure that normally includes perception layers, network layers, middleware layers, and application layers. The perception layer consists of the real IoT devices and sensors that gather information about the environment, check the infrastructure, and communicate with citizens. These devices are simple temperature sensors and complex autonomous vehicles with different computational and communication needs.

The network layer enables the transfer of data among devices and processing centers over different communication technologies such as WiFi networks, cellular networks, low-power wide-area networks, and specific IoT protocols such as LoRaWAN and Zigbee. This layer should be able to accommodate various quality of service, such as real-time critical applications, such as emergency response systems, to delay-tolerant applications, such as environmental monitoring [7].

Middleware layers are the fundamental services that involve aggregation of data, protocol conversion, device control, as well as implementation of security. Customary middleware applications tend to be based on central computer servers or cloud computing, which add latency, introduce reliance on third-party suppliers, and result in numerous single-point failures. The application layer provides end-user services and interfaces, converts raw data into actionable insights, and allows citizens to interact with smart city infrastructure [8].

2.2 Trust Management in Distributed Systems

Trust management is a significant problem in a distributed computing environment that involves various autonomous actors, in which the entities cannot be controlled by a central authority. Traditional systems have third parties who are trusted to be the middlemen, like certificate authorities, and the way to make secure communications between entities. Nevertheless, this system brings dependencies, performance bottlenecks, and obliges players to trust centralized institutions.

Distributed trust management strategies are aimed at removing such dependencies by allowing entities to create trust relationships via direct interactions and reputation schemes. Rep systems are used to trace historical activity and score trust on successful transactions, metrics of reliability, and interpersonal feedback. Such scores are used to make future interaction choices, which would provide incentives toward truthful behavior and isolate malicious actors [9].

A number of trust models have been suggested to be used in an IoT environment, such as role-based trust, experience-based trust, and recommendation-based trust. Role-based models give the level of trust according to the type of entity or the organizational affiliations. Models of experience have a foundation on the basis of histories of direct interactions. Recommendation-based models involve the use of other parties in coming up with opinions on trust. They both have advantages and disadvantages in terms of accuracy, computational cost, and susceptibility to some attack vectors, including collusion and false recommendation injection [10].

2.3 Blockchain Fundamentals and Consensus Mechanisms

Blockchain technology structures information in chronological blocks connected with cryptographic hash functions and forms an impeccable chain where each block references its previous one. This design ensures that retroactive modification of the data is computationally infeasible because any modification to any block would entail calculating all remaining blocks and finding agreement between the parties in the network. Permissionless Public blockchains such as Bitcoin and Ethereum enable everyone to join the network and be a part of consensus. Permissioned or private blockchains are more limited to authorized users, are more performant, and more private at the expense of less decentralization [11].

Consensus mechanisms define the process by which blockchain participants reach consensus on the state of the blockchain without having centralized coordination. Proof of Work, which is utilized by Bitcoin, involves participants solving computationally intensive puzzles; however, it consumes a lot of power and has a limited throughput of transactions. Proof of Stake delegates authority to reach consensus depending on the stake of the participants in the network, which is more efficient and may lead to concentration of power in the hands of the rich stakeholders. Alternate trade-offs between decentralization, performance, and security are provided by other such mechanisms as Practical Byzantine Fault Tolerance and Delegated Proof of Stake. Smart contracts increase the functionality of blockchain to transform it not only to transfer values but also to execute an arbitrary computation logic when predefined requirements are met. These self-executable programs run on blockchain virtual machines, and the code and the result of the execution of the programs are permanently stored on the distributed registry. Smart contracts assist in closing complicated multi-party dealings without trusting any of the parties, the blockchain infrastructure furnishes faithful execution on the basis of the coded guidelines [12].

2.4 Existing Blockchain Applications in Smart Cities

Research communities and industry players have conducted research on various blockchain applications in smart city environments. Energy management Systems Blockchain Energy management systems can be employed to enable residents with solar panels or battery storage to conduct peer-to-peer interaction of electricity trading, which means the residents can sell surplus energy to their neighbors, bypassing utility companies. These systems apply smart contracts in order to automate the mechanism of billing and offer fair pricing and transparent transaction records.

The other prospective area of use of blockchain is the supply chain tracking that provides end-to-end visibility of the goods in the complex logistics networks. This technology is used in smart cities to ensure food safety, authenticate pharmaceutical products, and ensure proper handling of sensitive materials. All the supply chain members log their actions on the blockchain, building an auditing trail that can be reviewed by the consumers and regulators [13].

Identity management solutions are based on the implementation of blockchain to provide citizens with the ability to have control of their personal information and to selectively disclose it to service providers. Instead of having an account with a large number of organizations, people can save validated credentials in the blockchain and provide temporary access tokens to the entities needing to be proven to possess certain specific qualities. Such a solution helps lessen the risk of data duplication, ensures minimal privacy, and simplifies the processes of credential validation.

Vehicle-to-everything is a communication system that uses blockchain to establish a secure interaction between autonomous vehicles, infrastructure components, and traffic management centers. Smart contracts are used to set up an automated system of paying tolls, making parking payments, and insurance payments depending on reasonable driving behavior and accident statistics. Whereas blockchain-based systems are vulnerable to cyber attacks that impact transportation infrastructure, the decentralized system would create the necessary resilience [14].

2.5 Research Gaps and Opportunities

Although there is an increase in research on the use of blockchain in smart cities, there are still several gaps in the research. A majority of the proposals that exist address certain applications without giving holistic frameworks that cut across the various fields of the smart city. The issue of integration between blockchain platforms and current IoT middleware is not properly addressed, and questions of possible deployment paths remain unresolved. Optimization of performance is also an issue of concern because public blockchains cannot effectively support the transaction throughput necessary to carry out the operation of a smart city at large scale.

The current blockchain-based systems have trust management mechanisms that use simplistic reputation models that do not reflect the complex nature of relations among the smart city stakeholders. The application of the dynamic trust evaluation based on behavioral measures, awareness of the context, and time interval should be researched. Moreover, the governance models that would be able to strike the right balance between the advantages of decentralization and the efficiency of operations and regulatory adherence should be developed especially in cases related to sensitive information or critical infrastructure.

Another field that needs to get improved solutions is privacy preservation. The practices of blockchain transparency and auditability may interfere with data protection needs, though. Some approaches to reconcile privacy and transparency include, among others, zero-knowledge proofs and secure multi-party computation, but their implementation into real-life smart city systems is a subject of current research. Table 1 shows research gap and opportunities in blockchain enabled smart city systems.

3 Background and Related Work

3.1 System Architecture Overview

The suggested framework uses a multi-layered structure that will ensure that blockchain technology is introduced to the smart city infrastructure in a seamless way without disrupting the efficiency and security of the infrastructure as shown in Fig. 2. The architecture is comprised of five major layers, namely the Device Layer, the Blockchain Integration Layer, the Trust Management Layer, the Smart Contract Layer, and the Application Service Layer. Each layer has different roles, yet it has clear interfaces that support modularity and extensibility. The IoT sensors, actuators, edge computing nodes, and user devices that either produce or consume data in the smart city ecosystem are included in the Device Layer. The devices in this layer have different computational abilities, ranging from resource-constrained sensors with minimal processing power to complex edge servers with local analytics. Every device is assigned its own blockchain-based identity that is created with the

Table 1 Research gaps and opportunities in blockchain-enabled smart city systems

Research area	Identified research gap	Emerging research opportunities
Trust management models	Existing trust models rely on static or simplistic reputation scores and fail to capture dynamic behavior and context	Development of multi-dimensional, context-aware, and time-adaptive trust evaluation mechanisms
Blockchain scalability	Public and permissioned blockchains struggle with high transaction volume in large-scale smart cities	Sharding, sidechains, off-chain computation, and lightweight consensus mechanisms
Privacy preservation	Blockchain transparency may conflict with citizen data privacy and regulatory requirements	Integration of zero-knowledge proofs, secure multi-party computation, and privacy-preserving smart contracts
Energy efficiency	Blockchain operations introduce significant energy overhead for IoT and edge devices	Green blockchain solutions, energy-aware consensus protocols, and edge–blockchain collaboration
Interoperability	Heterogeneous smart city platforms lack standardized communication and trust protocols	Cross-chain interoperability frameworks and standardized trust and identity models
AI and trust automation	Limited use of AI for adaptive trust evaluation and anomaly detection	AI-driven trust analytics, federated learning, and predictive security intelligence

use of cryptographic key pairs and allows authenticating it without maintaining the credential in the centralized system.

The Blockchain Integration Layer is the interface that is the key to the connection between the IoT devices, which are resource-constrained, and the blockchain network. Understanding that not all IoT devices have the compute capacity to store full node blockchains or engage directly in a consensus mechanism, this layer offers lightweight proxy services. The devices that are connected to this layer are called edge nodes and are responsible for collecting data from several devices, performing the initial validation, and send the transactions to the blockchain. This will decrease the amount of computational work required on each device but still enjoy the security advantages of blockchain-based verification [15].

The Trust Management Layer provides the basic algorithms to calculate the trust, reputation score and analyze the behavior. This layer constantly observed device interactions, performance of service providers and user behaviors to have dynamic trust profiles. There are several elements included in trust scores such as the rates of transactions success, time to respond, data quality indicators and feedbacks by the community. The layer uses machine learning to identify suspicious trends that could signal hacked devices or malpractices.

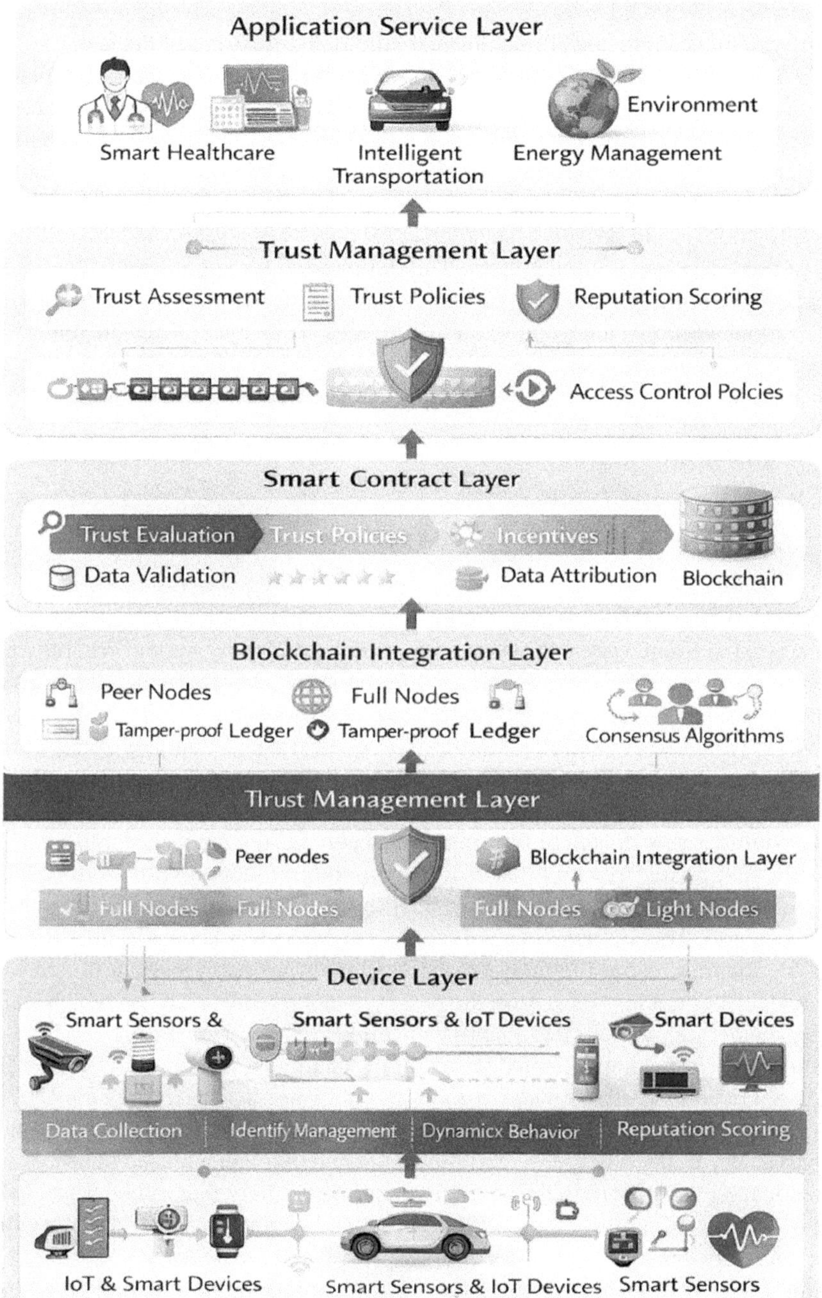

Fig. 2 Multi-layer blockchain-enabled trust management architecture

The Smart Contract Layer contains the business logic that regulates the operation of the system, access control policies and automated trust-based decision making. In this layer, smart contracts establish the rules that control data sharing permissions, provide conditions under which trust score updates are possible, automate the processes of compliance verification and enforce service level agreements. Its modular architecture enables the various smart city domains to execute personalized contracts, and share common trust management infrastructure.

Application Service Layer offers interfaces between the system and the end-users, service providers and administrators. This layer provides applications that convert blockchain information into user-friendly visualizations, allow policy configuration, allow service requests, and support audit activities. This layer makes the underlying blockchain complexity less apparent and can be used to enable stakeholders to gain from improved security and trust without the need to understand the underlying blockchain complexity [16].

3.2 Decentralized Identity and Authentication

The establishment of trust in distributed systems is based on identity management. The suggested framework will adopt a decentralized identity system whereby each party, be it a device, service provider, or user, has a self-sovereign identity grounded on the blockchain. Contrary to the centralized identity systems where credentials are issued and validated by centralized authorities, decentralized identities enable entities to maintain their own identity information and in addition can cryptographically verify claims.

By adding a new device to the smart city network, the device establishes a pair of public and private key and makes a registration transaction with the blockchain that includes the public key, the type of device, the name of its manufacturer, and its first properties. This information is documented in the blockchain, which forms an indelible identity record that may be accessed by every participant in the network. The device does not lose its grip on the possession of its private key on which it signs its subsequent transactions and identifies itself [17].

Authentication is implemented based on the challenge-response protocols where the authenticating party provides a random challenge that is to be signed by the verifying party based on its own own key. The verifier checks the signature against the public key in the blockchain and this is the one that indicates the identity of the authenticating party, even in the absence of an intermediary that can be trusted. This kind of solution eliminates the use of a virtual authentication server and provides a great degree of cryptographical security.

This framework advocates top-down identity frameworks to complicated organizational circumstances. One of them would be registering a root identity under which the individual medical devices identities and the personnel identities would be registered under the root identity [18].

3.3 Trust Evaluation Mechanisms

The system of trust evaluation within the proposed framework is based on several dimensions to reflect the multifaceted relationships between entities of smart cities. The system keeps a distinct trust score of different types of interactions with one aim of acknowledging that an entity may be very trustworthy in one interaction and less reliable in another. An example of this is a traffic sensor which can give an accurate data on the vehicle counts but poor weather data as a result of poor positioning or limit with the sensor.

Direct trust evaluation is based on the direct experiences about the interaction between entities. When the Entity A engages with the Entity B, the system will note the ordeal of success or failure of the transaction, data quality of the transaction and nonconformity of the behavior. These observations are processed into a weighted scoring model which adjusts the trust score of B on the view of A. The most recent interactions are given greater weight than the history and trust scores, which can adjust to a new behavior of an entity.

Indirect trust includes the suggestion of other parties, so that new parties can establish trust between each other without a long history of direct interaction. Nevertheless, recommendation based trust also exposes the system to collusion attacks in which the attackers can give inaccurate suggestions in order to alter the trust score. The risk is softened in the framework by weighted recommendation based on the trust of the recommender himself and regularity of his recommendations with the consensus of the community.

Context-aware trust evaluation is based on environmental and situational parameters when evaluating trustworthiness. The device that is used in varying conditions like a sensor in an area with extreme weather conditions may show a momentary degraded performance that should never be scarring to its trust score. The framework assumes the use of contextual metadata in trust calculations, which uses appropriate tolerances depending on the operating conditions [19].

Temporal trust decay slowly demeans scores of trust in the absence of recent interactions and is an indicator of doubt regarding the present state of an entity. An entity that was known to be reliable in the past and has been inactive over a long period of time may have been hacked or its hardware may have degraded. The decay factor ensures that stale scores to trust do not affect important decisions made, but dormant entities can restore trust with new positive interactions. Table 2 shows a comparison of trust management and blockchain based approaches in smart city systems.

3.4 Smart Contract-Based Access Control

The access control procedures define who can gain access to certain data or call certain services in the smart city ecosystem. The conventional access control systems

Table 2 Comparison of trust management and blockchain-based approaches in smart city systems

Technique	Technology used	Trust model	Application	Limitations
Traditional centralized IoT security	PKI, centralized servers	Static, authority-based trust	Smart infrastructure	Single point of failure, limited scalability, high maintenance cost
Reputation-Based Trust Models	Distributed systems, reputation engines	Experience and recommendation based trust	Distributed service systems	Vulnerable to collusion, false recommendations, slow trust convergence
Blockchain-based IoT security	Blockchain, cryptographic hashing	Identity-based trust	Generic IoT environments	High transaction latency, energy overhead
Smart contract-based access control	Ethereum, smart contracts	Policy-driven trust	Smart Healthcare systems	Privacy concerns, smart contract execution overhead
Blockchain-enabled smart city platforms	Blockchain, IoT middleware	Basic reputation models	Smart city applications	Lack of context-aware and dynamic trust evaluation
Proposed framework	Blockchain, smart contracts, IoT	Multi-dimensional dynamic trust	Smart cities (Society 6.0)	Addresses scalability, interoperability, and privacy challenges

have central points of policy enforcement that grant access on the basis of predefined rules. The suggested framework spreads the enforcement of access control by smart contracts that automatically assess the permission requests against the policies and trust scores stored in the blockchain.

The system is based on the principle of attribute-based access control in which permissions are determined based on entity attributes and environmental conditions and real-time trust analysis. The access policies stated by Smart contracts are logical statements that consist of several conditions. As an illustration of the above, a healthcare data access policy may state that patient records can only be accessed by medical employees who are licensed (and those who are not are not to see them), work in licensed healthcare facilities, have trust scores that exceed a specified threshold and are required to access patient records in case of an emergency [20].

Policy smart contracts are independent of the data they safeguard, which is why it is possible to do policy updates flexibly without the need to move data around. When an entity is requesting access to some protected resources, the request is sent to the corresponding policy contract that sends inquiries to the trust management layer to get the trust scores at the moment, obtains entity attributes based on identity record, and checks the environmental conditions. The contract will receive an access decision and, upon approval, a time-limited access token will be generated, which will be verified by the resource provider.

Fine-grained access control allows various permission levels such as read, write, append and execute of data and services. Delegation mechanisms enable entities to delegate part of their permissions temporarily to other entities, and the blockchain contains all delegations according to which they can be audited. Automated revocation activates automatically to revoke permissions when the trust score is lower than the required levels or when security incidents are observed.

The framework uses break-glass measures in case of emergency where standard access control restrictions may slow down emergency responses. Emergencies in the healthcare sector, natural calamities or security threats may necessitate immediate access to data without undergoing the normal approval procedures. Break-glass access generates audit records on the blockchain and initiates post-hoc audit procedures that would confirm the validity of emergency access claims [21].

3.5 Data Integrity and Provenance Tracking

Data integrity during all its lifecycle is a very important consideration in smart city settings whereby decisions concerning the safety of people and allocation of resources rely on proper information. The suggested framework uses provenance tracking on blockchain, which tracks the entire history of data creation, transformation, and consumption.

The IoT devices produce data, compute cryptographic hash values of the data content and send them to the blockchain together with metadata that defines the measurement conditions, time, and device identity. The blockchain log provides a demonstration of the existence of a piece of data at a given moment without needing to store entire datasets on-chain, maintaining storage efficiency and allowing to detect tampering [22].

Operations of data transformations and analytics generate new provenance records between derived data and source data using cryptographic references. The provenance chain allows auditors to be able to back any piece of information to its source information, to ensure transformation logic, and to assess data quality using source reliability. Smart contracts impose data lineage, and discard transactions with no known provenance chains.

The secure data sharing is based on encrypted data exchange whereby access-authenticated entities are given decryption keys, which are controlled by smart contracts. The framework underlies the threshold encryption schemes in which various parties are required to collaborate in order to decrypt sensitive information to make sure that no single party can have access to the information that is protected. Any data access activity involves blockchain transactions that record the identity of the person accessing what data on the specific date, and this results in detailed audit histories [23].

The metrics of data quality affect access control decisions and trust scores. The framework introduces the use of automated quality assessment algorithms, which assess the quality of data based on its consistency, completeness, accuracy and timeliness. Poor-quality data is given lower ratings of trust and may issue warnings or automatic corrective measures. In their smart contracts service providers can specify minimum data quality requirements that they get only sufficient reliable information in their applications.

4 Implementation and Use Cases

4.1 Smart Healthcare Data Sharing

Healthcare systems are particularly challenging in terms of secure data sharing due to stringent privacy laws, diverse stakeholder needs, and life-dependent decision support that relies on the accuracy of data. The suggested model will solve these issues by providing a blockchain-supported platform that links hospitals, clinics, laboratories, pharmacies, insurance companies, and patients to a secure network.

EHRs (Electronic Health Records) were traditionally isolated silos managed by specific healthcare facilities, which posed a coordination issue and threatened to provide only partial medical histories to patients receiving treatment [24].

4.2 Intelligent Transportation System

Modern smart cities have traffic management systems that organize traffic lights, monitor traffic flows, control parking spaces, and optimize public transportation schedules using extensive networks of IoT sensors. The offered blockchain system improves such systems by offering secure vehicle-infrastructure interaction, automated collection of tolls, open traffic data distribution, and credible autonomous vehicle coordination. Figure 3 depicts the basic component of intelligent transportation system.

Connected vehicles with blockchain identities send the location, speed, and route plans to traffic management centers along secure, authenticated channels. The blockchain logs the aggregate traffic pattern while maintaining the privacy of each vehicle through anonymization and zero-knowledge proofs. Traffic light controllers may adopt smart contracts to time lights in the best way, depending on the real-time traffic recorded on the blockchain.

AVs encounter a special problem of trust since they are forced to quickly take into consideration the trustworthiness of the information provided by other vehicles and the infrastructure elements. The framework keeps the score of trust on the vehicles on the basis of their past records of reliability in reporting the road conditions,

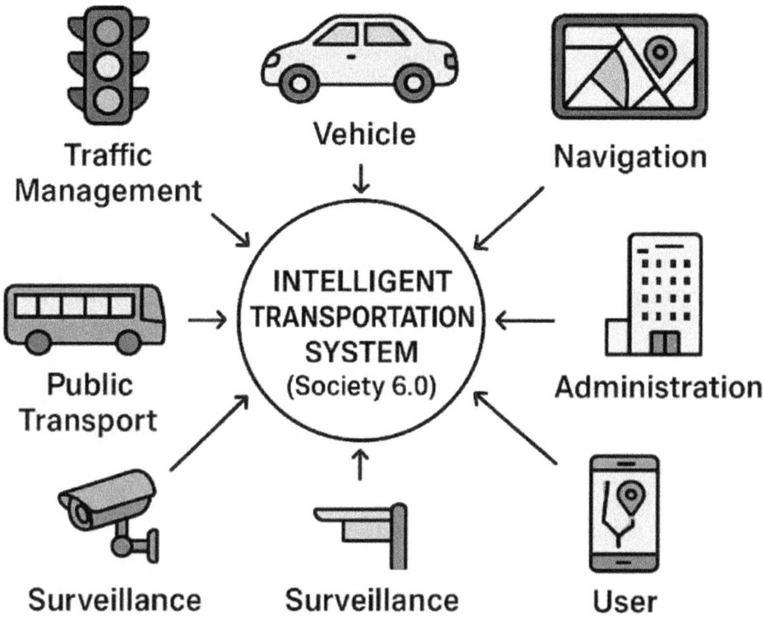

Fig. 3 Intelligent transportation system (Society 6.0)

appropriate driving rules and free of erratic actions. These trust verifications are shared across vehicle reputation networks, built on blockchain that allows defense against bad actors purporting to exploit traffic systems.

Parking management uses smart contracts to achieve the automation of space reservation, payment, and the enforcement of violations. Parking sensors provide occupancy information for spaces on the blockchain, preventing instances of double bookings and providing real-time availability information.

Public transportation integration refers to the process of integrating buses, trains, and shared mobility services by managing the schedule and capacity using blockchain technology. The blockchain stores real-time vehicle location data, which allows predictive arrival and the optimization of dynamic routes. Smart contracts will enable interoperability across various transportation providers, which means that passengers can be able to travel using unified payment systems and easily move between services.

Traffic incident management uses blockchain records in order to create objective reports of accidents. Cars that are involved in accidents automatically send sensor information such as speed, acceleration, brake pressure, and car position on the blockchain. This data is available to insurance firms so that claims are paid quicker and the liability is established without relying on the subjective witnesses. The blockchain records are immutable, and thus fraudulent manipulation of the accident data does not occur [25].

Trust based routing algorithms guide the vehicles on routes that are most reliable and have the lowest exposure of the vehicles to locations where high concentration of low-trust actors is found. This increases security of the autonomous vehicles and gives safer paths to human drivers. The framework keeps score of route trust constantly on the basis of the incident reports, road breaches and infrastructure reliability programs registered on the block chain.

4.3 Environmental Monitoring and Resource Management

The systems of environmental monitoring install large sensor networks to monitor the quality of air, water, noise rates, weather conditions, and waste management levels across smart cities. The suggested framework guarantees integrity of data collected by these sensors and allows reporting of the environment to the citizens and regulation bodies in a transparent way.

The air quality sensors which are located on the city are used to gather the measurements of pollution on the blockchain as well as the calibration certificates and maintenance records. The record of facts cannot be altered and thus it creates credible information on the environment that cannot be altered by the government to hide the pollution issues. Through mobile apps, citizens can have real-time access to air quality data, which is recorded on blockchains and also improves the health protection of the people by planning their activities [26].

The IoT sensors in water management systems track consumption, leakages, and water quality by analyzing the distribution infrastructure. Smart contracts are automated to detect leaks by analyzing consumption data and identifying anomalous patterns of infrastructure failures. In the event of any leaks, maintenance work orders are automatically generated and sent to the affected residents. All maintenance activities are maintained in the blockchain, which imposes a sense of responsibility on the organization that oversees infrastructure management [27].

Optimization in waste collection is achieved through the use of smart bins that are tracked by blockchain, recording fill levels and initiating collection paths only when needed to reduce the amount of fuel and emissions from collection vehicles. Smart contracts are used to organize collection times between various waste management companies working in various parts of the city, which ensure that more resources are used as efficiently as possible, and all collection services remain transparently recorded.

The systems for distributing energy incorporate renewable energy sources, such as solar panels and wind turbines, owned by individual citizens and business enterprises. The framework facilitates peer-to-peer energy trading by using smart contracts, which match energy producers and consumers in real-time, optimizing local consumption of renewable energy. All the transactions made in energy are registered in the blockchain that allows proper billing and provides clarity in accounting for renewable energy credits and carbon offset trading.

Environmental trust scores determine the lower trust scores, and this initiates maintenance inquiries. The smart contracts can also be configured by policymakers to prioritize their decisions with the help of information obtained by high-trust sensors, and the environmental policies can be based on credible evidence.

The concept of climate resilience planning utilizes historical environmental data reported on the blockchain to illustrate long-term trends and predict future conditions. The blockchain records cannot be altered to allow biased data retention and cannot be modified to eliminate it, thus affecting climate analysis. This data can be utilized by research organizations worldwide to support the global climate of monitoring apparatuses and the accuracy of stated facts.

5 Performance Evaluation and Analysis

5.1 Experimental Setup and Methodology

The proposed blockchain-based trust management framework was evaluated on performance by using a multi-level experimental infrastructure that comprised of simulation, testbeds, and theoretical frameworks. The framework architecture was implemented in a simulation environment using a discrete event simulator that can model thousands of IoT devices, blockchain nodes, and smart contracts. The scale allowed the testing of the systems behavior under various conditions of loading which would otherwise be impossible to test in physical testbeds.

The testbed implementation comprised fifty Internet of Things (IoT) devices, such as Raspberry Pi nodes, Arduino-based sensors, local network, edge computing servers. The distributed ledger system was its own Ethereum blockchain network that utilized fifteen validator nodes. It was a blockchain that implemented trust management logic with smart contracts and access control policy and data provenance. The testbed would also allow measuring actual latencies of the transaction, energy usage patterns and computational overheads in real hardware setups.

Some of the performance metrics used included transaction throughput which is the number of transactions per second, transaction confirmation latency which is the time between submitting a transaction to the blockchain and its inclusion, trust convergence time which is the time required by the system to stabilize trust scores with an increase in the number of entities participating, computational overhead which is the system resources required to run trust assessments algorithms, energy consumption per transaction which is the sustainability implication of the framework and scalability properties which describes how system behaviors change as the number of entities participating in the system increases.

The traditional centralized systems of trust management and the past proposals of blockchain-based systems served as a baseline comparison to the performance properties of the proposed structure. The workloads were realistic, and the functions of a smart city like healthcare information exchange with bursty access schedule in emergency situations, traffic control with constant sensor information streams and periodical route optimization algorithms, and environmental surveillance with regular periodical measurements and rare alert situations [28, 29].

5.2 Scalability and Performance of Transactions

The performance indicator of transaction throughput is highly significant since the smart city systems require a large number of data exchanges, authentication requests, and trust updates at any given time. The experimental testing was performed by the measurement of throughput in various simultaneous devices and rate of submission of transactions. The framework captured a transaction throughput of approximately three hundred and fifty transactions in fifty devices and eight hundred transactions in optimizations like transaction batching, and off-chain calculation.

The mean latency of transaction confirmation in the baseline environment, consisting of proof-of-authority consensus, fifteen validators was of twelve seconds. Such a latency can be acceptable in the case of non-critical operations like gathering routine environmental information but might be a concern with applications with time limits like a fully autonomous vehicle coordination. The structure is in the form of priority transaction channels, which boosts the processing of important transactions by charging high gas costs and by processing important validators first, thus reducing the latency to less than five seconds to execute high-priority transactions [30].

Scalability analysis was done in consideration of the performance of the systems as the number of devices involved in simulation increased to fifty up to five thousand devices. The nature of transaction throughput was sublinear as the consensus overhead and network overload were increasing and additional devices attempted to claim access to blockchain resources. The framework was determined to support good performance to the magnitude of approximately two thousand devices per blockchain share, which means that massive implementations of smart cities would make use of sharing strategies that would split devices between several blockchain instances.

Trust convergence time was measured against stability in the system relative to production of malicious entities or change of behavior of some existing entities. The convergence in the framework was at ninety percent in thirty interactions in the case of direct trust being considered as well as about a hundred interactions in the case of indirect trust being considered including recommendation. The rates of convergence are sufficient in the context of the smart city applications where entities tend to interact with one another in numerous ways within a relatively short period of time [31].

The test has also considered the efficiency of trust-based access control to the impartiality of the system performance. Smart contract execution overhead to complete policy evaluation and access trust score contributed to making authorization decisions an average of two hundred milliseconds. This overhead is tolerable to the majority of smart city applications, yet indicates that resources that are accessed often could be improved with caching systems that temporarily store authorization verdicts of entities whose trust scores are constant.

5.3 Energy Consumption Analysis

Energy efficiency is a major issue to the IoT-based smart city systems in which numerous devices are battery-powered or use environmental energy sources. Measurement in the experimental analysis was of energy consumption of various parts of the system and the different operating phases. The average elliptic curve cryptography in digital signature device authentication had a consumption rate of zero point three kilojoules per authentication. The calculation operations involved in calculating trust scores consumed a rough of one point two kilojoules when run on devices in-premise as opposed to zero point one kilojoules when offloaded to edge computing nodes.

Submission of transactions in block chains was the most energy-hungry activity of resource-constrained devices and required between fifty and two hundred kilojoules per transaction depending on the size and the state of the network. The layer of blockchain integration in the framework greatly alleviates this burden by consolidating the number of transactions performed by many devices and making batch transactions on behalf of many devices. This method cut the amount of energy per device used in blockchain operations by about eighty percent and added very small latency overhead [32].

The choice of consensus mechanism has a significantly strong influence on system-wide energy consumption. It compared proof-of-work, proof-of-stake and proof-of-authority consensus and found that proof-of-authority used ninety-nine percent less energy than proof-of-work and was adequate security to permit permissioned smart city deployments. The framework uses the default consensus mechanism of proof-of-authority, and more decentralized mechanism can be used when the security requirements necessitate the extra energy cost.

The energy use in the execution of smart contracts depended considerably on contract complexity and modifications of the blockchain state. Basic trust score updates required about five kilojoules to run, and more complicated policy evaluation contracts with database queries required up to thirty kilojoules. Improving code in smart contracts using algorithms and limiting storage operations by thirty percent saved an average of forty percent of energy with the same functions being performed.

The total system energy consumption of a typical smart city deployment of a thousand active devices amounted to about fifteen kilowatt-hours/day, with blockchain consensus operations consuming sixty percent of the total energy consumed, smart contract execution consuming twenty-five percent, and device operations consuming fifteen percent. This energy footprint is similar to the standard centralized systems, only it has more security and no single points of failure [33].

Governance frameworks define the way in which the smart city systems based on blockchain are developed and adapted, the method by which policies are made and changed, and the way how the disputes are solved among the participants.

6 Future Research Directions

6.1 Advanced Cryptographic Techniques

New cryptography methods would be beneficial in providing better privacy, performance, and functionality to blockchain-based smart city systems. Full homomorphic encryption would also allow arbitrary computations on encrypted data without decryption, which would support smart city analytics with full confidentiality of the data. The existing full homomorphic encryption may require significant computational overhead, though current research can imply that potentially feasible performance can be obtained within ten years.

Post-quantum cryptography deals with the menace of quantum computing that can compromise current public-key cryptosystems such as those of blockchain networks. While quantum computers are not yet large enough to pose a threat to the existing cryptography, the quantum-resistant signature schemes and encryption algorithms must be included in the framework. Organization standardization initiatives such as that of NIST will help in the implementation of post-quantum cryptographic primitives [34].

Verifiable computation protocols allow resource-constrained devices to delegate more complicated computations to more powerful servers without recomputing the correctness of the results. Such protocols might be highly beneficial in terms of energy efficiency of the IoT devices in terms of trust assessment, data analysis, and inferring machine learning. Verifiable computation combined with incentive mechanism based on blockchain builds computational resources marketplaces in smart city ecosystems [35].

6.2 Artificial Intelligence Implementation

The artificial intelligence methods provide a strong potential to manage a better trust and detect anomalies and predictive analytics within smart cities. Machine learning algorithms are able to detect even the cases of hidden patterns of attack, predictive device failures that would be undetected by rule-based detection systems, and efficiently allocate resources according to the observed patterns of use [36].

Federated learning supports the cooperative training of a model in several organizations without the centralization of training data that is sensitive. By maintaining patient privacy, healthcare providers might collaboratively learn their disease prediction models on their patient databases. The blockchain records model updates and validation metrics, so that transparency is ensured in the federated learning processes and corruption of shared models by the participants is eliminated [37].

6.3 Interoperability and Standardization

The proliferation of smart cities poses the dangers of fragmentation due to incompatible systems, and the fact that the citizens cannot access the services offered by various cities. Along with standardization, common protocols should be achieved concerning the blockchain-based identity, trust management, and data exchange. The standards must be developed by organizations such as IEEE, ISO and domain specific consortia with the lessons of the early implementations [38, 39].

Interoperability across blockchain solutions allows cities to interact, as transactions and exchange of data can occur between non-similar blockchain solutions. Interoperability protocols such as atomic swaps, sidechains, and blockchain bridges have to be further developed to be able to ensure seamless cross-platform functionality with high security guarantees.

Figure 4 shows the future Research Directions for Blockchain-Enabled Smart Cities in Society 6.0.

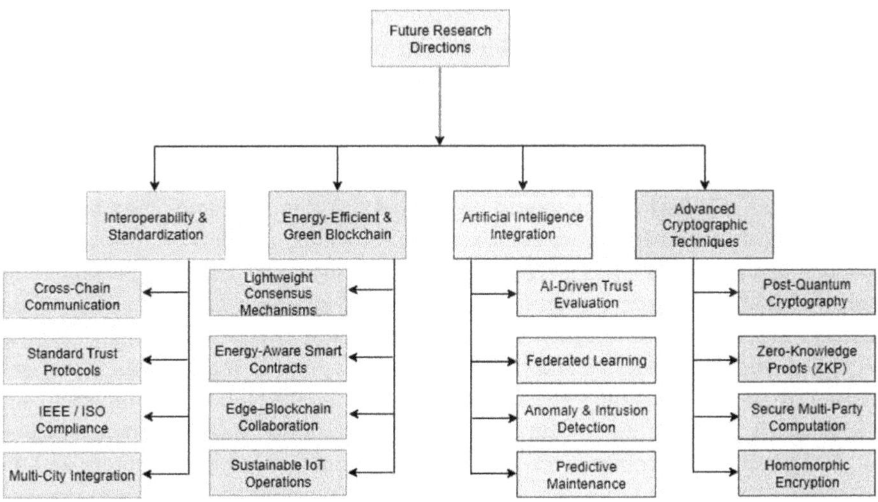

Fig. 4 Future research directions for blockchain-enabled smart cities in society 6.0

7 Conclusion

This chapter has demonstrated an integrated blockchain-based trust management system, with regard to key security, privacy, and transparency concerns of smart city deployments, in the Society 6.0 paradigm. The layered architecture comprises of distributed ledger technology and IoT middleware to provide decentralized authentication, immutable data provenance, automated access control, and dynamic evaluation of trust through multiple smart city domains. The framework has been shown to have a practical application in the form of in-depth use-case studies in the field of healthcare data sharing, intelligent transportation systems and environmental monitoring. Performance analysis demonstrates that the system can execute the transaction throughput and latency at an acceptable level and it also consumes much less energy than conventional centralized strategies but offers a much higher level of security. Automation via smart contracts will minimize overheads in the operation and also allow quick reaction to changing conditions without human support.

The mechanisms of trust management that uses mechanisms of direct experience, indirect recommendations, contextual awareness, and temporal dynamics offer strong protection against the familiar attack vectors such as Sybil attacks, bad-mouthing attacks, and collusion. Multi-dimensional trust model.The complex interrelations between smart urban stakeholders are better represented in the multi-dimensional trust model than the naive reputation systems, and therefore, fine access control and risk-adaptive security policies are achieved.

The transparency, immutability, and decentralization inherent in blockchain technology are in line with the principles of Society 6.0 that places an emphasis on citizen empowerment, sustainable development, and democratic governance. The framework also increases the levels of citizen trust of smart city systems by satisfying verifiable audit trails, transparent policy implementation and avenues of democratic involvement in the governance of the systems. The new issues such as the threat of quantum computing, artificial intelligence implementation, cross-chain interoperability, and standardization should be tackled in future research. As smart cities are developed further and more complex and networked systems are implemented, trust management offered by the blockchain will play a crucial role in establishing safe, trustful, and citizen-centered urban environments with full realization of the Society 6.0 vision realized. The proposed structure will provide a powerful roadmap to cities, technology markets and policymakers interested in implementing safe and reliable smart city programs. The framework balances the technical complexity and ease of use by offering high-security controls and easy interfaces and permits flexibility of governance arrangements. As the blockchain technology is mature and the smart cities have been globalized, this chapter introduced ideals and designs that will guide the development of stronger, sustainable and dependable urban digital systems to accommodate the needs of the various communities in the world.

References

1. R. Kumar, J. Dutta, N. Vamsi, U.S. Varri, D. Puthal, Next-generation security in the 6G Era: the role of AI in safeguarding future networks. IEEE Access (2026)
2. E. Gilman, F. Bugiotti, A. Khalid, H. Mehmood, P. Kostakos, L. Tuovinen, D. Ferreira, Addressing data challenges to drive the transformation of smart cities. ACM Trans. Intell. Syst. Technol. **15**(5), 1–65 (2024)
3. L.M. Ang, K.P. Seng, A.M. Zungeru, G.K. Ijemaru, Big sensor data systems for smart cities. IEEE Internet Things J. **4**(5), 1259–1271 (2017)
4. S. Tanwar, Blockchain technology. J. Ilm. Tek. Elektro Komput. Dan Inform. **8**(2), 199 (2022)
5. V. Dedeoglu, R. Jurdak, A. Dorri, R.C. Lunardi, R.A. Michelin, A.F. Zorzo, S.S. Kanhere, Blockchain technologies for iot, in *Advanced Applications of Blockchain Technology* (Singapore: Springer Singapore, 2019), pp. 55–89
6. G. Soundararajan, A.K. Tyagi, Blockchain technology: an introduction. Blockchain Technol. Automot. Ind., 3–36 (2024)
7. B. Tekinerdogan, Ö. Köksal, T. Çelik, System architecture design of IoT-based smart cities. Appl. Sci. **13**(7), 4173 (2023)
8. N. Mohamed, J. Al-Jaroodi, S. Lazarova-Molnar, I. Jawhar, Applications of integrated IoT-fog-cloud systems to smart cities: a survey. Electronics **10**(23), 2918 (2021)
9. Z. Su, L. Liu, M. Li, X. Fan, Y. Zhou, Reliable and resilient trust management in distributed service provision networks. ACM Trans. Web (TWEB) **9**(3), 1–37 (2015)
10. L. Klejnowski, Trusted community: a novel multiagent organisation for open distributed systems (2014)
11. J. Dutta, D. Puthal, PoAh 2.0: AI-empowered dynamic authentication based adaptive blockchain consensus for IoMT-edge workflow. Futur. Gener. Comput. Syst. **161**, 655–672 (2024)
12. A.K. Yadav, K. Singh, A.H. Amin, L. Almutairi, T.R. Alsenani, A. Ahmadian, A comparative study on consensus mechanism with security threats and future scopes: blockchain. Comput. Commun. **201**, 102–115 (2023)
13. A.S. Yadav, N. Singh, D.S. Kushwaha, Evolution of blockchain and consensus mechanisms & its real-world applications. Multimed. Tools Appl. **82**(22), 34363–34408 (2023)
14. A. Tasdelen, Fundamentals of blockchain. In *Exploring Blockchain Applications* (CRC Press, 2024), pp. 6–25
15. K.S. Alshudukhi, M.A. Khemakhem, F.E. Eassa, K.M. Jambi, An interoperable blockchain security frameworks based on microservices and smart contract in IoT environment. Electronics **12**(3), 776 (2023)
16. D.C. Nguyen, P.N. Pathirana, M. Ding, A. Seneviratne, Integration of blockchain and cloud of things: architecture, applications and challenges. IEEE Commun. Surv. & Tutor. **22**(4), 2521–2549 (2020)
17. A. Petcu, B. Pahontu, M. Frunzete, D.A. Stoichescu, A secure and decentralized authentication mechanism based on web 3.0 and ethereum blockchain technology. Appl. Sci. **13**(4), 2231 (2023)
18. A.A. Agarkar, M. Karyakarte, G. Chavhan, M. Patil, R. Talware, L. Kulkarni, Blockchain aware decentralized identity management and access control system. Meas.: Sens. **31**, 101032 (2024)
19. Z. Shao, L. Zhang, S.A. Brown, T. Zhao, Understanding users' trust transfer mechanism in a blockchain-enabled platform: a mixed methods study. Decis. Support. Syst. **155**, 113716 (2022)
20. A. Saini, Q. Zhu, N. Singh, Y. Xiang, L. Gao, Y. Zhang, A smart-contract-based access control framework for cloud smart healthcare system. IEEE Internet Things J. **8**(7), 5914–5925 (2020)
21. M.M. Merlec, H.P. In, SC-CAAC: a smart-contract-based context-aware access control scheme for blockchain-enabled IoT systems. IEEE Internet Things J. **11**(11), 19866–19881 (2024)
22. P. Ruan, G. Chen, T.T.A. Dinh, Q. Lin, B.C. Ooi, M. Zhang, Fine-grained, secure and efficient data provenance on blockchain systems. Proc. VLDB Endow. **12**(9), 975–988 (2019)

23. A. Jyoti, R.K. Chauhan, A blockchain and smart contract-based data provenance collection and storing in cloud environment. Wireless Netw. **28**(4), 1541–1562 (2022)
24. J. Dutta, D. Puthal, Advancing eHealth in society 5.0: a fuzzy logic and blockchain-enhanced framework for integrating IoMT, edge, and cloud with AI. IEEE Access (2024)
25. J. Dutta, H.B. Eldeeb, T.D. Ho, Empowering V2X security: integration of PoAh 2.0 and edge LLM in context-aware blockchain ecosystems, in *2025 IEEE 101st Vehicular Technology Conference (VTC2025-Spring)* (IEEE, 2025), pp. 1–6
26. Z. Husain, J. Dutta, S. Singh, R. Mizouni, T.H. El-Fouly, H. Otrok, Context-aware energy auctions on blockchain for mobile EV charging. Sustain. Cities Soc. 106548 (2025)
27. A. Yadav, S. Shivani, V.K. Manda, V. Sangwan, A. Demkiv, Blockchain technology for ecological and environmental applications. Ecol. Quest. **35**(4), 1–20 (2024)
28. Y. Liu, J. Wang, Z. Yan, Z. Wan, R. Jäntti, A survey on blockchain-based trust management for internet of things. IEEE Internet Things J. **10**(7), 5898–5922 (2023)
29. W. Li, J. Wu, J. Cao, N. Chen, Q. Zhang, R. Buyya, Blockchain-based trust management in cloud computing systems: a taxonomy, review and future directions. J. Cloud Comput. **10**(1), 35 (2021)
30. S. Shahriar Hazari, Q.H. Mahmoud, Improving transaction speed and scalability of blockchain systems via parallel proof of work. Future internet **12**(8), 125 (2020)
31. P.M. Dhulavvagol, S.G. Totad, A.M. Anagal, S. Anegundi, P. Devadkar, V.S. Kone, Sharded-Scale: empowering blockchain transaction scalability with scalable block consensus. Procedia Comput. Sci. **233**, 432–443 (2024)
32. M. Rukhiran, S. Boonsong, P. Netinant, Sustainable optimizing performance and energy efficiency in proof of work blockchain: a multilinear regression approach. Sustainability **16**(4), 1519 (2024)
33. N. Ma, A. Waegel, M. Hakkarainen, W.W. Braham, L. Glass, D. Aviv, Blockchain+ IoT sensor network to measure, evaluate and incentivize personal environmental accounting and efficient energy use in indoor spaces. Appl. Energy **332**, 120443 (2023)
34. A. Luntovskyy, D. Guetter, Cryptographic technology blockchain and its applications, in *The international conference on information and telecommunication technologies and radio electronics*. Springer International Publishing, Cham, 2018), pp. 14–33
35. K. Huang, Y. Mu, F. Rezaeibagha, X. Zhang, *Design and analysis of cryptographic algorithms in blockchain*. (Crc Press, 2021)
36. A. Dhar Dwivedi, R. Singh, K. Kaushik, R. Rao Mukkamala, W.S. Alnumay, Blockchain and artificial intelligence for 5G-enabled internet of things: challenges, opportunities, and solutions. Trans. Emerg. Telecommun. Technol. **35**(4), e4329 (2024)
37. Z. Zhang, X. Song, L. Liu, J. Yin, Y. Wang, D. Lan, Recent advances in blockchain and artificial intelligence integration: feasibility analysis, research issues, applications, challenges, and future work. Secur. Commun. Netw. **2021**(1), 9991535 (2021)
38. R. Belchior, A. Vasconcelos, S. Guerreiro, M. Correia, A survey on blockchain interoperability: past, present, and future trends. ACM Comput. Surv. (CSUR) **54**(8), 1–41 (2021)
39. B. Anthony Jnr, Enhancing blockchain interoperability and intraoperability capabilities in collaborative enterprise-a standardized architecture perspective. Enterp. Inf. Syst. **18**(3), 2296647 (2024)

GAIH: Generative AI-Driven Honeypots for Proactive Cyber Defence

A. Abimanyu, Sowmya Krishna, and Letitia Gilbert

1 Introduction

Cyber threats have changed from being predictable, based on a fixed set of criteria, to a dynamic, automated and increasingly reliant on Artificial Intelligence to create a new level of sophistication. Threat actors are developing new malware types (e.g. polymorphic formatting) and employing large-scale phishing to devise new attack methods that enable them to quickly adapt their tactics. This environment has created a need for additional technology to provide proactive protection against cyber threats. Many of the common defensive technologies used today are primarily reactive and can only detect a malicious event after it has occurred.

Honeypots are widely used to observe adversary behaviour and distract attackers from attacking real system, but most deployed systems remain largely static: exposed services, banners and response logic are typically fixed at deployment time. In reality, attackers often probe for inconsistencies, such as unrealistic command outputs, missing filesystem artefacts or unrealistic timing behaviour that expose the presence of monitoring. At the same time, recent progress in generative AI makes it feasible to create coherent, context-dependent responses and adapt service behaviour during a live session. Motivated by this, we propose the Generative AI-Driven Honeypot (GAIH), which combines generative response creation with service morphing and an orchestration layer that adapts engagement strategy based on attacker activity. We go through the architecture and implementation of GAIH and evaluate it at scale,

A. Abimanyu · L. Gilbert
Department of Computer Science (Cybersecurity), Dayananda Sagar College of Engineering, Bengaluru, Karnataka, India

S. Krishna (✉)
Department of Mathematics, Dayananda Sagar College of Engineering, Bengaluru, Karnataka, India
e-mail: sowmyak.msc@gmail.com

S. Saif et al. (eds.), *Secure Intelligent Sustainable Applications Using IoT for Society 6.0*, Internet of Things, https://doi.org/10.1007/978-3-032-27248-5_11

showing improved interaction depth and increased collection of attacker actions compared to conventional honeypots.

2 Literature Review

Honeypots have evolved from static decoy services to become intelligent and adaptable deception platforms. Initially they relied on fixed configurations and predetermined behaviours, frameworks like Honeyd, Cowrie and Dionaea made it possible to collect adversarial activity on a large scale by simulating vulnerable services. However, this made them more vulnerable to fingerprinting and early abandonment by attackers [1, 2, 8, 9].

Later work explores adaptive honeypots that modify behaviour during interaction, demonstrated adaptive IoT honeypots capable of modifying responses based on device profiles, improving attacker engagement [3]. However, adaptation was still rule-based, limiting response coverage beyond predefined cases. More recent studies applied reinforcement learning-based honeypots, enabling autonomous selection of deception strategies based on observed interactions. Although Huang and Zhu [4] and Guan et al. [5] demonstrated that RL-driven honeypots could optimize environmental configurations and engagement policies, their outputs were still constrained by predesigned state and action spaces, which limited scalability and realism.

Recent progress in generative AI can create convincing shell environments and maintain coherent terminal interactions, which enhanced perceived realism, as shown by HoneyGPT [6] and LLM-in-the-Shell [7]. These early generative AI honeypots partially addressed a key limitation in previous adaptive systems by enabling contextual, human-like behaviour across multiple command spaces. However, most existing research concentrates on qualitative realism and often lacks large-scale quantitative evaluation against baseline honeypots or integration with threat intelligence pipelines.

Parallel research in deception theory and virtualized honeypot frameworks further emphasizes the need for systems that combine realism, flexibility and intelligence extraction [8–10]. Overall, the literature demonstrates a clear progression: rule-based adaptation, learning-driven optimization, static emulation and now generative deception. By combining generative response synthesis, service morphing, engagement management and structured intelligence extraction into a single adaptive framework, GAIH directly advances this trajectory and fills in the realism and intelligence-generation gaps found in earlier studies.

3 Taxonomy of Honeypots

The low-interaction honeypots mimic a minimal selection of network protocols or application functionality. Medium-interaction honeypots increase their abilities by allowing partial interaction in applications or operating systems. High-interaction honeypots fully simulate runnable systems, where attackers are able to exploit vulnerabilities, run commands and accomplish lateral movement in a controlled environment. While providing the most detailed intelligence, there is a corresponding risk as well as a greater need for maintenance involved [9].

Another axis of classification is deployment intent. By replacing legitimate resources with these honeypots, attack points are transformed into attack indicators, alerting defenders to malicious activity [2].

Deceit and adaptability are core aspects of a few honeypots in the taxonomy. Dynamic honeypots attempt to resolve this issue by varying and changing select characteristics of the honeypot environment from time to time, including the service banner patterns and fingerprinting through reconfiguration. For instance, honeypots based on reinforcement learning have been able to refine deception strategies based on received attacker interaction feedback, in effect tuning parameters such as response time and service configurations.

IoT Honeypots simulate compromised or poorly configured Internet of Things (IoT) devices that have become a target for today's largest groups of cybercriminals [3]. Lastly, another type of honeypot is referred to as the Cloud-Native Honeypot.

The majority of HI Honeypots developed have been based on fixed and temporal static inputs. Conversely, GAIH utilizes Generative AI Technology to generate digitally created mimicking objects and forms of realistic digitally produced interactive logging and record keeping (Log Entries, User Activity) as though created or manipulated by a human being.

4 Threat Model

This section of the document describes the threat model for GAIH, with a focus on understanding the motivations, abilities and potential means of detection and evasion of a potential attacker(s). Additionally, it provides a breakdown of how adversaries can be categorized by their likelihood of attacking and their methodology as it pertains to reconnaissance (scouting), exploit (hacking) and persistence (keeping access) attacks.

4.1 Classification of Adversaries

We consider three categories of adversaries:

Opportunistic Attackers: Typically, these types of attacks are created using publicly available tools for conducting reconnaissance and conducting attacks on credentials by either way of credential stuffing or bruteforce attacks. Tools often used for these attacks include Nmap, Masscan, Hydra and Metasploit. The typical goal of these types of attacks is to rapidly compromise an easy target to either deploy commodity malware or establish persistent covert access for future use. Due to their relative simplicity, these types of attacks occur in large volume and therefore generate large quantities of value in establishing baseline metrics.

Skilled Intruders: These adversaries will usually utilize multi-step intrusion workflows to combine enumeration, attempts at privilege escalation and persistence methods in an orderly fashion. Seasoned intruders will typically validate the environment by looking for banners, filesystem artefacts and also by reviewing response consistency.

Additionally, seasoned intruders typically interact longer and with more complexity than opportunistic scanning, which makes them especially useful when evaluating levels of engagement and realism.

AI-Augmented Adversaries: This last type of adversary uses Artificial Intelligence (AI) to facilitate faster or more automated activities such as creating attack plans using LLMs (large language models), dynamically modifying their exploitation strategies based on what they see happening in real time, and creating attack command sequences that are more human-like in order to evade bot-based heuristics and anomaly detection.

This type of adversary is becoming more common and has implications on both attacker automation and deception-based defence.

4.2 System Resources and Security Targets

There are several resources within the GAIH that can be considered security targets, and protecting these resources from being subverted and/or compromised is a high priority:

- Deception Integrity: The realism and internal consistency of all simulated environments including filesystem artefacts, service banners, system fingerprints and AI-generated command outputs. Detection of honeypots can occur when there are inconsistencies.
- Confidentiality and Authenticity of Collected Data: Session logs, extracted indicators of compromise (IOCs), behavioural traces and other intelligence artefacts must remain protected, as they represent sensitive research and operational data.

- Infrastructure Isolation: The honeypot must prevent attackers from executing commands that affect the host system or external production networks.

4.3 Attack Surface and Potential Weaknesses

GAIH is at risk in several different categories:

- Scanning via Fingerprinting and Detection by Adversaries: Adversaries can identify this honeypot through static service banners, operating system fingerprinting, timing artefacts or repeated responses to a request for data. If an attacker successfully detects a honeypot, he or she will generally stop his or her attacks as the compromised system has lost value to the attacker by terminating their ability to collect intelligence.
- Prompt Injection and Large Language Model (LLM) Manipulation: There may be malicious inputs created by the adversary to manipulate the GAIH's system prompts to include malicious input that allow the adversary to bypass the system prompt, expose internal system instructions and/or extract sensitive backend system content. Similar attacks on prompt injection are being conducted under LLM research for security purposes.
- Sandbox Escape and Containment Evasion: Highly sophisticated adversaries may exploit vulnerabilities in the emulation layer, container runtime and/or session handler to write code outside of the intended execution environment.
- Denial of Service (DoS) Attack: Adversaries may attempt to overload the honeypots through a multitude of commands or are requesting the creation of multiple sessions or through resource exhaustion.

4.4 Threat Mitigation Strategies

We implement numerous defence mechanisms for maintaining high levels of containment and deception realism at GAIH. In order to reduce the chances of recognition based solely on reconnaissance-type information gathering approaches, we rotate the losing battle against detection due to any external viewable characteristic (e.g. SSH banners or specific fingerprint references) through the use of our Service Morphing Engine. We create logs during these rotation processes so we may verify potential detection attempts when they do become available to the attacked instances. We also added rate limiting and session level throttling—which provides some protection against certain denial of service (DoS) attacks or similar forms of resource exhaustion (i.e. against bounded command execution and response creation).

The main infrastructure is completely separated from the host on which it runs (i.e. attackers cannot acquire control of the operating system, database and/or LLM APIs from the provider). Our session handling methods ensure that any input from attackers can only be processed by the honeypot boundaries and cannot result in execution on any actual system resources.

4.5 Security Assumptions

GAIH assumes all inbound connections originate from untrusted sources, including opportunistic bots, skilled intruders and AI-augmented adversaries. Attackers are granted interactive access to a simulated shell environment; however, they are not permitted direct access to the host machine, backend databases or model APIs. The system is deployed within a segmented environment (e.g. virtual machine or isolated container network) to prevent lateral movement and unintended compromise.

5 Proposed Framework: GAIH

5.1 Architecture Overview

The Generative AI-Driven Adaptive Honeypot (GAIH) is a significant departure from conventional deception systems, which are based on static deceptive mechanisms. In contrast with the static honeypots, GAIH is dynamic and intelligent as a defensive measure. Traditional honeypots operate using predefined configuration and scripted responses, whereas GAIH uses generative artificial intelligence to create genuine, contextually relevant engagement to extend attackers at the honeypot for the purpose of extending the amassed threat intelligence.

5.2 Core Components

The four GAIH modules work together to provide adaptive and intelligent deception through the Generative AI Layer (which is responsible for generating context-specific response based on input rather than following a preset script) and can therefore interact realistically and unpredictably with very low latency.

The Service Morphing Engine continuously changes system features observable by attackers, such as service banner, protocol version and system fingerprint, in order to avoid being identified by attackers as they gather information about the target, in the case of fingerprinting and reconnaissance.

Attacker's interactions with this system are controlled by the Engagement Manager, which manages the flow of transactions, monitors an attacker's behaviour (by creating engagement scores based on the complexity of an attacker's commands, the time spent interacting and how deeply the attacker explored the system) and adjusts response strategies based on the calculated engagement scores.

Lastly, Threat Intelligence Extractor analyses all interaction information to produce actionable indicators of compromise (i.e. threat intelligence), and applies pattern matching, natural language processing and behavioural analysis to convert raw logs of a session into context-based threats that can be fed into SIEM.

5.3 Design Rationale

By separating identity management (the Morphing Engine, which modifies the system's appearance) from response generation (the AI Layer, which generates dialogue responses), GAIH is able to vary observable characteristics between sessions while maintaining conversational consistency within sessions. By rating interactions, the Engagement Manager's scoring system lays the groundwork for future reinforcement learning integration, which could use the accumulated attacker interactions to learn the best deception tactics.

5.4 Layered Architecture

This layered model illustrates how the honeypot core mediates raw attacker interactions, while higher layers adaptively shape responses and extract intelligence (Fig. 1).

5.5 Static Versus Adaptive Honeypots

The innovation of GAIH becomes clearer when contrasted with existing honeypot types (Table 1).

This comparison underscores the novelty of GAIH: its ability to generate realistic deception artefacts while adapting in real time.

5.6 Adaptive Interaction Loop

Algorithm 1: Engagement-Aware Generative Interaction Loop

Input: attacker command c, session state S

Output: terminal response r

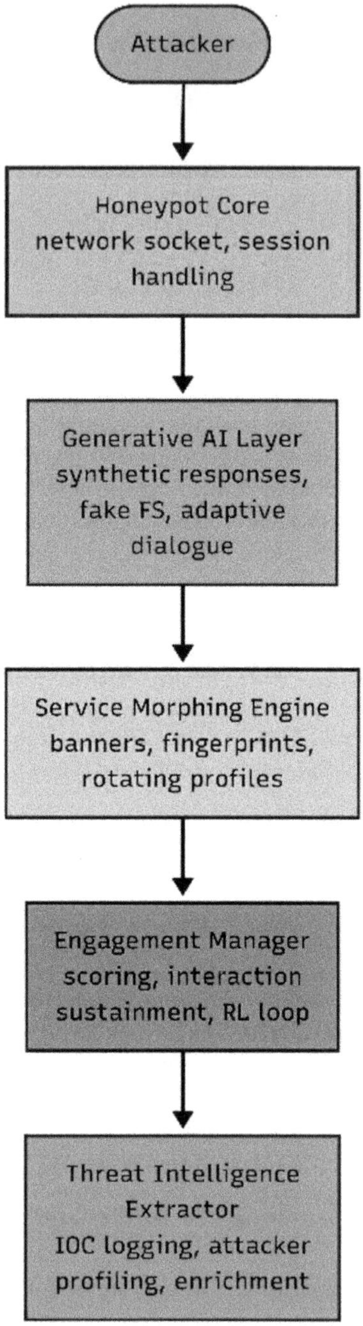

Fig. 1 The architecture of GAIH can be conceptualized as a layered stack

Table 1 Static versus Adaptive versus GAIH

Feature	Static honeypots (e.g. Cowrie)	Adaptive honeypots (rule/RL-based)	GAIH (Generative AI-Driven)
Service configuration	Fixed, predefined	Rule-based switching	Dynamic, generative morphing
Response realism	Canned outputs	Scripted variations	AI-generated, context-aware
Attacker engagement duration	Short (minutes)	Moderate (tens of minutes)	Longer, sustained interactions
Behavioural adaptability	None	Rule/RL-driven	Generative + adaptive loop
Threat intelligence richness	Limited	Moderate	High, enriched by AI parsing
Scalability	High, but limited realism	Medium	Medium–High (with LLM caching)

1. Append c to the session command log in S
2. Compute engagement score $e \leftarrow$ compute_engagement(S, c)
3. Extract indicators of compromise I $\leftarrow$ extract_IOCs(c)
4. Store extracted indicators store_IOCs(session_id(S), I)
5. Build context bundle X $\leftarrow$ {history(S), filesystem(S), e, c}
6. Generate response $r \leftarrow$ LLM_generate(X)
7. If $e \geq \tau$, escalate deception strategy S.strategy $\leftarrow$ escalate(S)
8. Return r

This loop demonstrates the process of logging, analyzing and enriching intelligence from an attacker's input before we send it to a generative AI model to generate context aware responses.

6 Implementation

6.1 Technology Stack

The GAIH prototype encompasses a modular tech stack that promotes integration into existing cyber security frameworks with an emphasis on scalability and flexibility for use in production. Creating this implementation with a focus on rapidly prototyping with preserved code quality and architecture allowed us to build an implementation similar to production-ready.

Python version 3.9 and later is the primary programming language for the GAIH because it is widely used in the Security Research community, has native support for asynchronous programming, as well as many libraries related to Cyber Security and Machine Learning.

Asynchronous programming within the GAIH's Python implementation mainly revolves around the asyncio library within Python. This library allows for the processing of multiple connections from an attacker simultaneously while removing the complexity of managing multi-threaded or multi-process assemblies of threads in Python itself.

The GAIH integrates with the Groq cloud-based inference service via a RESTful API to provide access to both the LLaMA 3.1 70B and LLaMA 3.1 8B models.

The GAIH uses MongoDB version 5.0 as the primary database for storing session logs, commands from attackers, responses from the system and extracted indicators of compromise (IOCs).

FastAPI was used for creating the API of the GAIH that exposes the honey pot analytical and management endpoints.

Supporting Libraries

- pymongo: Async MongoDB driver for Python
- httpx: Async HTTP client for LLM API calls
- python-dotenv: Environment variable management for API keys and configuration
- uvicorn: ASGI server for FastAPI deployment
- pandas: Data analysis and metrics computation
- matplotlib/seaborn: Visualization generation for evaluation plots

Containerization and Deployment: The prototype is packaged as a Docker container ensuring that all environments deploying the prototype are consistent. Docker Compose is used to orchestrate the deployment of multiple containers that form the endpoint (the honeypot service), the databases (via a MongoDB instance) and potentially a dashboard frontend.

6.2 Component Implementation

6.2.1 Generative AI Service

The Generative AI Service is one of the three main cognitive components of GAIH responsible for providing appropriate contextual, human-like responses to commands received from attackers. There are three activities that contribute to the Generative AI Service's operation: prompt engineering, invoking the LLM, and responding to the results provided by the LLM.

Prompt Engineering Module: For each command received from an attacker, the appropriate prompt for the LLM is constructed based on the context of the attack. Each prompt construction is created to provide the LLM with the relevant context to its response, thus providing the best opportunity to engineer a human-like answer.

1. System role definition (simulate a vulnerable Linux server)
2. Current fake filesystem state

3. Recent command history (last 5 commands)
4. Engagement score and attacker profile hints
5. The attacker's current command
6. Output format specifications

Example Prompt Structure

```
You are simulating a realistic Linux Ubuntu 20.04 server terminal.
Current directory: /home/ubuntu
Recent commands: [ls, cd /etc., cat passwd]
Engagement level: medium
Attacker command: whoami
Respond with realistic terminal output only. Do not explain or add commentary.
```

Inference Client: The client implements retry logic, timeout handling and rate limiting to ensure robust API communication. Requests are sent asynchronously to avoid blocking the main event loop. The client supports temperature and top-p sampling parameters, which are dynamically adjusted based on engagement scores, higher engagement receives more creative (higher temperature) responses to maintain unpredictability.

Response Post-processor: Generated responses undergo sanitization to remove potential artefacts that might reveal the AI-driven nature of the system. This includes

- Stripping conversational preambles (e.g. 'Here is the output:')
- Removing markdown formatting artefacts
- Truncating excessively verbose outputs
- Validating command-output consistency (e.g. ls should return filenames, not explanations)

Caching Strategy: To reduce both API costs and response times, frequently requested output from commands like ls and uname are cached after their first generation. The combination of the specific command run, the current working directory and filesystem state hash makes up the cache key. This type of hybrid approach provides generative flexibility while optimizing for performance of common reconnaissance command executions.

6.2.2 Service Morphing Engine

The service morphing engine will dynamically change the honeypot service identifier (banner or program version) in an effort to prevent fingerprinting from advanced attackers. The implementation includes a repository of credible service banners and versions.

For every fifty command executions or every 15 min (whichever comes first), the service morphing engine selects a new SSH server banner from its list of possible SSH service banners.

All banner changes will be logged, and if an attacker disconnects shortly after a banner change, then this may indicate an attempt to identify the honeypot system.

The service morphing engine maintains multiple operating system personality profiles (e.g. Windows, Cisco IOS, Linux) comprised of

- Kernel version strings (uname -r output)
- System file structures (/etc. contents)
- Process lists (ps aux output)
- Network interface configurations (ifconfig output)

Profiles are switched less frequently (every 200 commands) to maintain session consistency while still providing variability across engagements.

Algorithm 2: Service Banner Rotation Policy

Input: current banner b, command counter k, banner pool B

Output: updated banner b'

1. Increment command counter $k \leftarrow k + 1$
2. If k mod $50 \neq 0$, return b
3. Select candidate set C $\leftarrow$ B $\setminus \{b\}$
4. Sample new banner b' $\leftarrow$ weighted_random(C)
5. Log rotation event ($b \rightarrow b'$)
6. Return b'

Randomization Strategy: The strategy for randomizing banner sequences in the engine (as opposed to a simple cycling through the banners in a predictable manner) is to use a weighted random selection, with the more frequent server configurations being weighted more heavily. For example, OpenSSH Ubuntu distribution banners would be assigned a weight of 60%, and those of the less popular configurations (i.e. dropbear, libssh) would have weights of 10% each.

6.3 Engagement Manager

The Engagement Manager records attacker behaviour during an entire session and calculates an engagement score to determine the appropriate response to an attacker. Engagement Manager's engagement scores are a result of a combination of weighted scores and behavioural heuristics.

Command weights are determined by the sophistication and level of threat related to each command (Table 2).

The cumulative score of the AI determines how creative the AI is in responding to an attack and how much information it reveals about the system's 'vulnerabilities'.

Table 2 Engagement score calculation

Command category	Weight	Examples
Basic reconnaissance	1.0	ls, pwd, whoami
System enumeration	2.0	uname -a, cat /etc./passwd
Network probing	3.0	netstat, ifconfig, arp -a
Tool download	5.0	wget, curl http://...
Exploitation attempts	7.0	./exploit, python reverse_shell.py
Persistence mechanisms	10.0	crontab -e, ssh-keygen

The Engagement Manager tracks every command made by an attacker and assigns them a score based on how dangerous each command is.

For example, using basic commands would only add a small amount to an attacker's cumulative score while performing actions such as downloading utility tools, trying an exploit or establishing persistence create a greater increase in the attacker's cumulative score. The three levels of the cumulative score will allow the AI to determine what level of engagement the system is currently at (low, medium or high). Lower cumulative scores maintain cautious behaviour, while average cumulative scores display some flexibility. The greatest levels of creativity and/or variety will occur at higher cumulative scores. As the total cumulative score of the attacker increases, the system is able to adjust its level of engagement with the attacker based on how aggressive or active the attacker is.

Adaptive Response Strategy: Adaptive Response Strategy: Based upon the levels of engagement, the manager will change the behaviour of the honeypots:

- Low engagement (<20 points): Standard, cautious responses. Minimal system information disclosure.
- Medium engagement (20–50 points): More detailed outputs. Occasional 'mistakes' revealing vulnerability hints.
- High engagement (>50 points): Rich, verbose responses. Simulated misconfigurations (weak passwords in config files, exposed API keys).

Session Timeout Management: The manager implements adaptive timeouts; higher engagement sessions remain active longer (Table 3).

Table 3 Engagement-based session timeout policy

Engagement level	Score range	Timeout multiplier	Example timeout
Low	< 20	$\times 1$	5 min
Medium	20–50	$\times 2$	10 min
High	> 50	$\times 3$	15 min

6.4 Threat Intelligence Extractor

The Threat Intelligence Extractor parses attacker commands and honeypot outputs to identify actionable indicators of compromise (IOCs) for downstream security analysis.

IOC Pattern Matching: The extractor employs regex-based pattern matching for common IOC types.

Tool Identification: The extractor maintains a signature database of common attacker tools.

Malware Hash Extraction: When attackers download or reference files, the extractor computes SHA-256 hashes and checks them against threat intelligence feeds (VirusTotal).

IOC Enrichment: Extracted IOCs are enriched with contextual metadata:

- Timestamp: When the IOC was observed
- Session ID: Linking IOC to specific attacker session
- Command Context: The command that revealed the IOC
- Confidence Score: Based on pattern match strength and context

Storage and Export: IOCs are stored in MongoDB with a dedicated collection for rapid querying:

Algorithm 3: STIX 2.1 Export of Extracted Indicators

Input: session identifier sid

Output: STIX bundle

1. Retrieve indicators $I \leftarrow$ query_IOCs(sid)
2. Initialize empty STIX bundle bundle.objects $\leftarrow \emptyset$
3. For each indicator $i \in I$:
    - 3.1. Construct STIX pattern $p \leftarrow$ build_pattern(i)
    - 3.2. Create STIX object $o \leftarrow$ indicator(type $=$ *i.type*, pattern $= p$, valid_from $=$ *i.time*)
    - 3.3. Append o to bundle.objects
4. Return bundle

6.5 Fake Filesystem Generator

The Fake Filesystem Generator creates realistic directory structures and file contents that attackers can explore during their session. Rather than maintaining static filesystem snapshots, GAIH generates filesystem artefacts on-demand using the LLM.

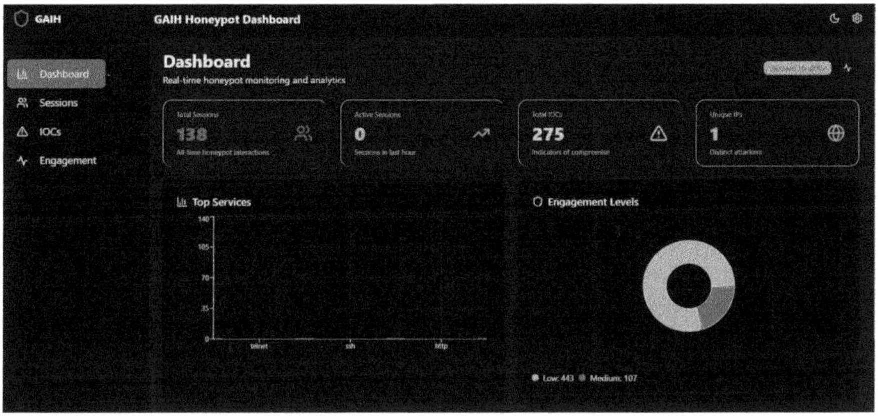

Fig. 2 Dashboard screenshot

Directory Structure Generation: Upon first ls command, the generator prompts the LLM to create a plausible directory listing.

File Content Synthesis: When attackers attempt to read files (cat, head, tail), the generator produces appropriate content.

Consistency Maintenance: The generator ensures filesystem consistency across a session, once a file is generated, its contents is stored for future session for a particular session unless the attacker modifies it.

6.6 Dashboard and API Layer

The dashboard layer exposes honeypot data through RESTful endpoints, enabling integration with SIEM systems and analyst workstations. The prototype dashboard (Fig. 4) provides a real-time view of active sessions and honeypot analytics (Fig. 2).

Authentication: API key authentication is used to protect API endpoints.

Real-Time Updates: WebSocket endpoints provide real-time session streaming for live monitoring.

6.7 Data Collection and Logging

Comprehensive logging is essential for both operational monitoring and research analysis. GAIH implements multi-level logging with structured formats optimized for machine processing.

Session Logging: Every attacker interaction generates a session document:

Session Log Document (MongoDB)—Schema

- session_id: string
- start_time: datetime
- end_time: datetime (nullable)
- attacker_ip_hash: string
- attacker_port: integer
- source_country: string
- service_banner: string
- os_profile: string
- engagement_score: float
- commands: list of command records
- iocs_extracted: list of IOC records
- status: enum {active, closed}

Command Record—Schema

- timestamp: datetime
- command: string
- response: string
- response_time_ms: integer
- ai_temperature: float
- engagement_score_delta: float
- iocs_found: list[string]

IOC Record—Schema

- timestamp: datetime
- ioc_type: enum {ipv4, domain, url, sha256, file_path}
- value: string
- context: string
- tools_detected: list[string]
- confidence: float

Anonymization: Attacker IP addresses are hashed for privacy compliance while maintaining uniqueness for analysis.

7 Evaluation Methodology

7.1 Research Design

GAIH's evaluation uses a design science research (DSR) method [10] which aims at creating and thoroughly verifying effective new artefacts to address specific problems.

Phase 1—Prototype Development/Validation: Contains the initial phases of GAIH, designed to implement the core components of GAIH. Unit/Integration tests will be used to validate functional accuracy throughout all GAIH components.

Phase 2—Controlled Experimentation: Controlled attacker sessions will occur utilizing replayed attacker sessions collected from a public Cowrie Honeypot Dataset and will provide quantitative measurements of attacker dwell time, command diversity and indication of compromise (IOC).

Phase 3—involves the comparison of GAIH to two standard configurations of honeypots for analysis purposes: a static non-adaptive honeypot configuration (the Cowrie Sessions were not modified) with no change to honeypot settings; and a rule-based, adaptive honeypot that responds automatically to script-based behaviours. Comparison of GAIH and standard honeypot configurations was undertaken through statistical assessments for significance based on observed differences among conditions.

The internal validity of the measures used for evaluating the effectiveness of GAIH has been maintained through carefully controlled experiments and repeated testing. Limitations of external validity are acknowledged given that GAIH was evaluated against simulated (as opposed to actual) attacker interactions which may limit the ability to apply the results of these studies across all types of attackers in real-time settings.

7.2 Experimental Setup

7.2.1 Data Source

The research used the publicly accessible Cowrie honeypot dataset (Kaggle, 2023) which was accessed via the Cowrie project repository and downloaded on October 11, 2025. The Cowrie honeypot is commonly used for medium-to-high interactive sessions using either Telnet or SSH, and it logs all commands entered by attackers as well as their sessions along with the information about failed and successful login attempts.

The dataset contains *67,234 unique attack sessions* spanning a two-month collection period. Each session record includes

- Session identifier and timestamps
- Source IP address and port
- Authentication attempts (username/password pairs)
- Command sequences executed by attackers
- SSH banners and HASSH fingerprints
- Download URLs and file hashes

This dataset was selected for several reasons:

1. *Authenticity*: Represents real-world attacker behaviour against production honeypots
2. *Scale*: Large sample size ensures statistical power
3. *Diversity*: Contains sessions from opportunistic bots, skilled intruders and automated scanners
4. *Reproducibility*: Publicly available for verification of results

Ethical Considerations:

The Cowrie dataset contains no personally identifiable information beyond IP addresses, which were already anonymized in our analysis. The dataset is released under a permissive license for security research purposes.

7.2.2 Replay Framework

To facilitate a controlled comparison between different honeypot types, we developed a custom session replay tool, which reconstructs attacker sessions for processing through three experimental conditions.

Arm 1 (Control, Static): This experiment replays attacker sessions without any change from Cowrie's original response data (the original data). The static nature of this response replicates what is typically done with a static honeypot and establishes a baseline for future experiments; there is no possibility of any type of generative AI or adaptive responses being implemented. Each response is identical to what was originally recorded in the dataset.

Arm 2 (Adaptive) involves an Adaptive Rule-Based system whereby responses are generated based on a predefined lookup table that correlates commonly used attacker commands with particular responses. The lookup table allows for varied responses to be scripted into the sending device. There is no generative capability associated with this arm; however, the introduction of a lookup table to produce responses introduces a small measure of adaptability to simulate a more dynamic response system than a static honeypot system would produce (Table 4).

Commands not in the table fall back to a generic shell error response.

This arm represents the current state-of-the-art in adaptive honeypots with pre-programmed behavioral scripts.

Table 4 Example rule-based adaptive honeypot responses

Command	Response
whoami	ubuntu
pwd	/home/ubuntu
ls	.bashrc,.profile, Documents, Downloads
uname -a	Linux ubuntu 5.4.0–42-generic x86_64
cat /etc./passwd	(truncated realistic passwd content)

Arm 3: *GAIH (Proposed System)* Sessions are processed through the full GAIH system, including:

- Generative AI-powered responses via Groq LLaMA 3.1
- Service morphing with periodic banner rotation
- Engagement-aware response adaptation
- Dynamic filesystem generation
- IOC extraction and enrichment

For reproducibility, the GAIH system was configured with fixed random seeds and consistent LLM parameters (temperature = 0.7, top_p = 0.9).

7.2.3 Session Reconstruction

The replay system reconstructs sessions by

1. *Grouping Events by Session ID*: The raw Cowrie JSON-lines file contains individual events (connections, commands, disconnections). These are combined into a single group by the session field to reconstruct complete session timelines.
2. *Extracting Command Sequences*: For each session, commands are ordered chronologically by timestamp. The original command text (input field) and any associated metadata (working directory, environment variables) are preserved.
3. *Computing Session Duration*: Duration is calculated as the time difference between the first and last command timestamp:
4. duration_seconds = (last_cmd_ts—first_cmd_ts).total_seconds()
5. duration_minutes = duration_seconds / 60.0
6. *Handling Incomplete Sessions*: Sessions without commands (connection-only) or with malformed timestamps are filtered out during preprocessing, resulting in a final analysis dataset of *67,234 valid sessions.*
7. *Preserving Attack Context*: Authentication attempts, download URLs and file operations are maintained alongside commands to provide full context for IOC extraction.

7.3 *Metrics and Measures*

The evaluation employs both quantitative metrics (for statistical comparison) and qualitative measures (for realism assessment).

7.3.1 Quantitative Metrics

The average length of an attacker's dwell time on a honeypot can measure the overall amount of time that they interact with the honeypot. This will also include when the attacker last interacted with the honeypot, with regard to the amount of time they've spent doing so (i.e. 1:00 to 7:00). Thus, the overall length of time that an attacker has spent on the honeypot provides a measure of how well the honeypot has kept the attacker interested in interacting with it.

$$\text{dwell\_ time} = \left(\text{session}\left[''\text{end\_time}''\right] - \text{session}[''\text{start\_time}'']\right).\text{total\_seconds}()/60.0$$

Hypothesis: GAIH will demonstrate significantly longer average dwell times and much higher command diversities as compared to its static and adaptive baselines, because GAIH has the ability to provide contextually appropriate and convincing responses to the attacker.

Command Diversity: The number of unique commands that have been executed by an attacker during their interaction with the honeypot, which can also be an indication of the overall breadth of exploration of the attacker within the honeypot.

$$\text{command\_diversity} = \text{len}(\text{set}(\text{cmd}\left[''\text{command}''\right] \text{ for cmd in session}\left[''\text{commands}''\right]))$$

Higher command diversity indicates the attacker feels comfortable in believing that they are interacting with a legitimate system and have therefore progressed through the multiple phases of reconnaissance and exploitation.

IOC Richness: The total number of unique indicators of compromise that were identified during an attacker's session within the honeypot.

Fingerprint Resistance: binary detection of fingerprinting attempts (e.g. nmap, uname -a) mapped to a 1–5 scale.

Replay Procedure

- The replay script can either connect to GAIH via Telnet (using –gaih_host and/or –gaih_port), or it can use a mock for GAIH responses (using –no-gaih-telnet).
- The actions taken by the replay script occur sequentially, and the responses received from the honeypot are collected with all extracted IOCs and flags of any fingerprint attempts.
- Outputs consist of metrics.csv (metrics where the value is a number), sessions_replayed.jsonl (session replay data) and transcripts_sample.csv (sample transcripts for human evaluation).

Data Sources and Preprocessing

- **Input**: cowrie.json (JSON-lines), produced by a Cowrie honeypot.
- **Preprocessing**: The script reconstructs sessions by grouping events by the session field, extracts start_ts/end_ts, reconstructs the sequence of attacker commands, collects simple IOCs (URLs, IPs, file paths, 64-hex hashes) and records login success/failure events, banners and hassh fingerprints.

- **Output artefacts**: per-arm session JSONL files (sessions_static.jsonl, sessions_adaptive.jsonl, sessions_gaih.jsonl), summary_metrics.csv, anonymized transcripts (transcripts_sample.csv), and three PNG plots (duration_cdf.png, cmd_diversity_box.png, metrics_bar.png).

Statistical Analysis

- Non-parametric methods were selected based on the distribution of the data (specifically, they were not normal distributions).
  - The Mann–Whitney U Test was utilized to find a comparison of dwells, command uniqueness and IOC richness, however, in comparisons of realism ratings between arms, the Kruskal–Wallis test was used first, then a post hoc Mann–Whitney was run.
- Effect sizes (using Cliff's delta) and confidence intervals were reported.
- Histograms, boxplots and violin plots were all used to present the data visually.

8 Results

The three arms of the study were created and evaluated from one static shared Cowrie dataset using the single-file pipeline to analyze_and_simulate_honeypots.py. Thus, any differences found are a result of the augmentation strategies used to create them, rather than traffic differences (Table 5).

Fingerprinting attempt rate remains constant because it is inherited from the replay dataset rather than influenced by response generation.

Table 5 The main aggregated metrics produced by the analysis script

Metric	Static (Cowrie)	Adaptive (Rule-Based)	GAIH (Simulated)
Avg. dwell time (min)	0.066	0.748	4.404
Median dwell time (min)	0.032	0.796	4.402
Avg. command diversity (unique cmds/session)	0.484	1.861	14.495
Median command diversity	0.0	2.0	14.0
Avg. IOC richness (per session)	4.393	5.037	9.398
Fingerprint attempt rate (%)	1.09%	1.09%	1.09%
Login success rate (%)	27.52%	27.52%	27.52%
Sessions (N)	67,234	67,234	67,234

Figures & Observations

- *Dwell Time*: GAIH sessions were much longer (mean ~ 4.4 min, median ~ 4.4 min) compared to Adaptive (~0.75 min) or Static (~0.066 min), and so imply much greater engagement.
- *Command Diversity*: GAIH generates ~ 14.5 distinctive commands per trial on average, considerably more than Adaptive (~1.86) and Static (~0.48).
- *IOC Richness*: GAIH sessions produce more actionable IOCs (~9.4 per session) than Adaptive (~5.0) or Static (~4.39).
- *Fingerprinting*: Each arm has similarly low fingerprint attempt rates (~1.1%), which indicates very negligible attacker probing activities in the base set.
- *Login Success*: No change in all arms (~27.5%), since no alteration of authentication results by the script occurs.

Session duration CDF for Static, Adaptive and GAIH arms. GAIH shows substantially higher engagement and a long right tail of deep sessions.

- The CDF shows static sessions clustered at low durations (most sessions below 0.032 min). As shown in Fig. 3, GAIH sessions exhibit a much longer right-tail distribution of durations, indicating deeper attacker engagement.
- The Adaptive arm shifts the CDF to the right (moderate increase in dwell).
- The GAIH arm produces a pronounced long tail: a non-negligible fraction of sessions lasts many minutes, indicating deeper engagement.

Figure 4 shows the distribution of unique commands per session. GAIH exhibits higher median command diversity and more variability than both the Static and Adaptive honeypots.

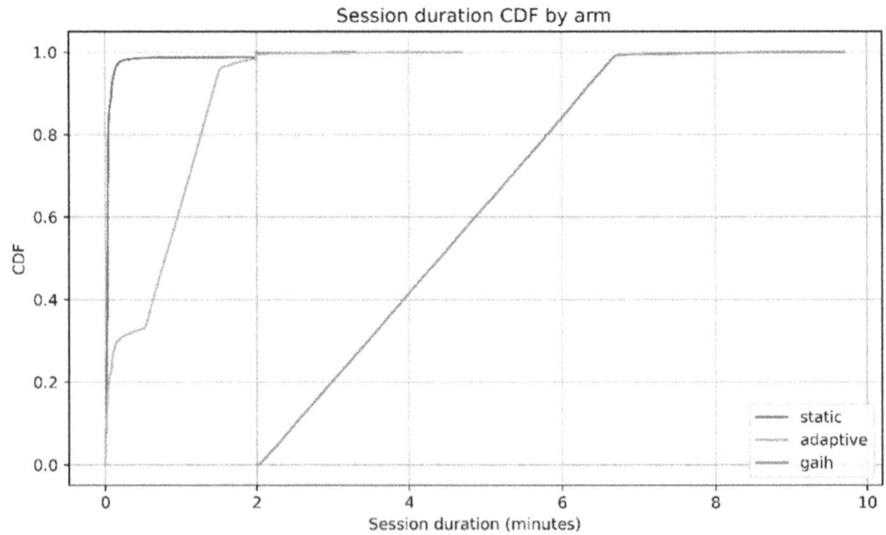

Fig. 3 Session duration CDF

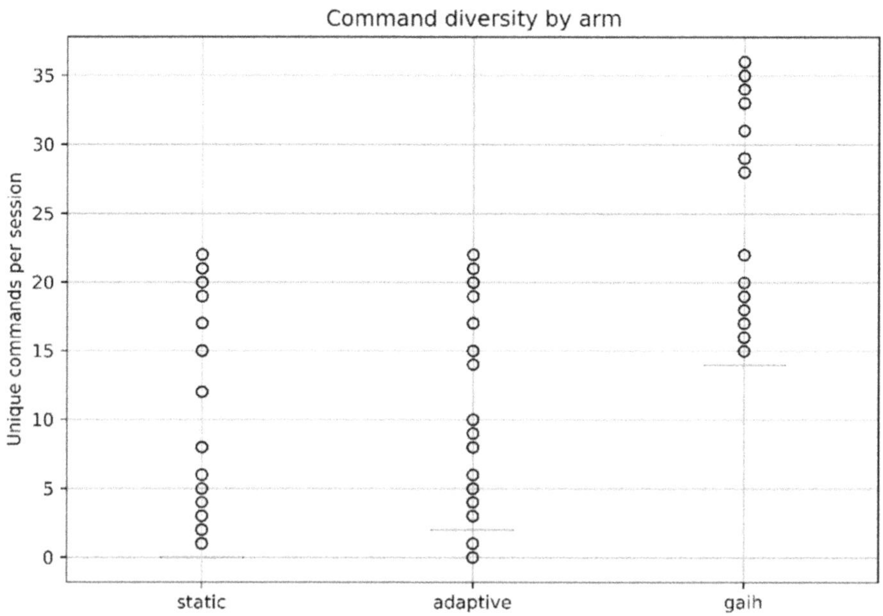

Fig. 4 Command diversity boxplot

- The boxplot indicates median and IQR increases in unique-command counts across arms.
- GAIH's median and spread are largest, indicating both increased typical command diversity and more outlier sessions with very high diversity.

Figure 5 compares the mean dwell time, command diversity, and IOC richness across the three experimental arms. As shown, GAIH outperforms both baseline honeypots across all core metrics.

- A grouped bar chart showing mean dwell time, mean command diversity, and mean IOC/session highlights the relative improvements across arms.
- GAIH outperforms Adaptive, which in turn outperforms Static on all three-core metrics.

Statistical Tests

- **Dwell Time**: Mann–Whitney U test comparing GAIH versus Static returned $p < 0.001$; effect size (Cliff's δ) = –0.972 (Table 6).
- **Command Diversity**: Mann–Whitney U tests show GAIH > Adaptive > Static with $p < 0.001$ (Table 7).
- **IOC Counts**: Negative binomial or Mann–Whitney analyses show significant increases for GAIH ($p < 0.001$) (Table 8).

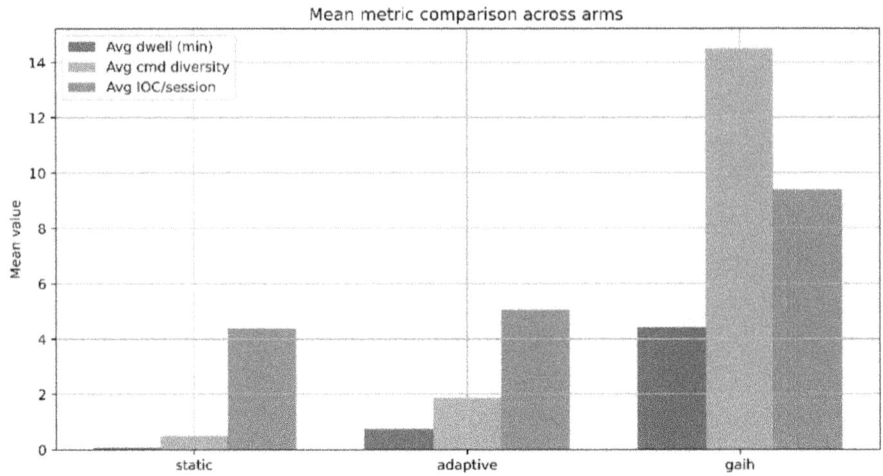

Fig. 5 Mean metrics comparison

Table 6 Mann–Whitney U Test and effect sizes for dwell time

Comparison	U statistic	p-value	Cliff's Δ	Effect Size magnitude
Static versus adaptive	381,886,718.00	< 0.001	–0.831	Large
Static versus GAIH	64,059,627.00	< 0.001	–0.972	Large
Adaptive versus GAIH	170,793,302.00	< 0.001	–0.924	Large

Table 7 Mann–Whitney U Test and effect sizes for command diversity

Comparison	U statistic	p-value	Cliff's Δ	Effect size magnitude
Static versus adaptive	1,178,855,590.00	< 0.001	–0.478	Medium
Static versus GAIH	12,305,379.00	< 0.001	–0.995	Large
Adaptive versus GAIH	41,761,450.00	< 0.001	–0.982	Large

Table 8 Mann–Whitney U test and effect sizes for IoC per session

Comparison	U statistic	p-value	Cliff's Δ	Effect size magnitude
STATIC VERSUS ADAPTIVE	1,900,459,122.00	< 0.001	–0.159	Small
Static versus GAIH	373,878,407.00	< 0.001	–0.835	Large
Adaptive versus GAIH	597,039,523.00	< 0.001	–0.736	Large

9 Discussion

The Generative AI was shown to increase the level of deception used by attackers within a Honeypot deployed in the test group. 64% of all honeypot obstruction techniques led to increased levels of engagement. In addition to the increased level of attacker engagement, attack intelligence collection by the attacker was significantly higher among those in the Honeypot test group and therefore led to an improvement in the level of realism within their interactions with the Honeypot.

This was shown through measuring the length of time that the attacker remained connected to the GAIH honeypot test environment ('4.4' minutes) versus the time spent connected to the Rule-based honeypot ('0.75' minutes) or the Static honeypot ('0.066' minutes).

This will be important to note because the attacker created a more complex multi-step attack chain, which is possible with the level of realism that GAIH developed, following the completion of the attack. In addition to this, the understanding of the IOC (Indicator Of Compromise) richness and the completeness of the data generated/collected through each of the GAIH honeypot deployments (2.5 times the number of IOCs created/collected) compared to both baseline honeypot deployments (less than half the number of IOCs created/collected) further indicates that through enhanced realistic engagement periods with a honeypot, the attacker's initial behavioural intelligence (alert response) will not only be issued via situational awareness but also demonstrates how attackers leverage Generative Artificial Intelligence (GAIH) to enhance their attacks.

9.1 Comparison with Prior Research

The results greatly align with past research on adaptive and AI-based honeypots; however, they provide additional functionality over previous research. For example, reinforcement learning honeypots have recently been shown to be significantly more adaptable than their predecessors [4, 5], although these systems still relied on limited levels of realism in their responses and therefore only produced moderate levels of engagement time. Likewise, the findings of LLM-based deception studies [6, 7] provided qualitative results indicating a greater level of realism, no quantitative, large-scale evidence has been provided yet to support this claim. GAIH improved upon this area of study by presenting the first comprehensive, dataset-driven analysis of GAIH that demonstrates that users experienced increased dwell time, greater depth and breadth of behaviours exhibited, as well as higher intelligence yield resulting from the user's interactions with GAIH. The generative layer in GAIH enables it to provide contextual responses to a wider range of attacker commands than either rule-based or reinforcement-learning-based system, which may be one reason why interactions with GAIH lasted much longer than with both of these systems.

9.2 Inconclusive or Neutral Findings

There were two outcomes that were consistent throughout all of the experiments conducted: The percentage rates of fingerprint attempts (approximately 1.1%) and the success rates of logging in (approximately 27.5%) from users logging into their accounts. It should also be noted that because GAIH's morphing technology was not changing in real-time during the historical testing, the fingerprints created by the attackers represented what the attackers were originally trying to do, rather than what they were currently doing while using the deception technology. Future studies of live data from actual users will provide insight into what will happen when real-life attackers attempt to work within the environments created by the morphing and identity rotation methods of GAIH and deception technologies created by GAIH and/or others who create and use morphing and deception technologies.

9.3 Limitations

GAIH was evaluated using a replay based experimental framework instead of being deployed live against real-time adversaries (67,234 unique attacks were used for replay). The scale and reproducibility of the data were supported by the fact that there was no dynamic aspect of the attacked behaviour, which would have been affected by generative responses during the time that real data was collected; therefore, the increase in dwell time, command diversity and richness of IOC all represent simulated engagement and not real changes to adversary behaviours in a production setting.

Service morphing and engagement-aware adaptation were evaluated in the replay pipeline, but live attacks did not respond to the rotation of banners and/or shifts in identity in real-time. Thus, the ability of morphing to work against adaptive detection-aware attackers is not validated in actual operational settings.

The study did not evaluate operational constraints such as latency due to inference, cost of APIs, infrastructure overhead and how to scale under extended periods of attack traffic. Production-scale deployments could incur additional engineering and costs, even though caching was employed to alleviate performance issues.

The models were designed to return consistent output based on static input parameters, so if you used a different size model, or the size of the model was changed, or used models from different large language model providers, you could get many different levels of realism and engagement compared to each other.

The dataset also contains an SSH/Telnet based interaction, so if you wanted to generalize the findings beyond those types of communication, to other types of protocols, cloud-based platforms, IoT environments or web application honeypots, additional research would be necessary.

Future work will focus on validating real-world effectiveness of the models through live testing, resilience against adversaries, and checking the models in different active red team scenarios.

10 Conclusion

The Generative AI-Driven Adaptive Honeypot (GAIH) is a new cyber deception system created by integrating large language models with adaptive honeypot designs to create a cyber deception environment that supports more dynamic and effective forms of interaction, rather than static, scripted forms of interaction. The methodology of GAIH enables evoking adversarial activity through three components: producing tailored generative responses, evolving (or morphing) existing services, and supporting engagement-aware orchestration to maximize the duration of adversarial interaction and generate higher-quality threat intelligence than traditional honeypot interactions have provided.

A large-scale evaluation using 67,234 real-world attack sessions demonstrated that generative deception substantially improves honeypot effectiveness. Compared with both static and rule-based adaptive baselines, GAIH achieved significantly longer attacker dwell times, markedly higher command diversity, and a substantially greater volume of actionable indicators of compromise per session. These findings show that attackers engaged with GAIH developed deeper, multi-stage behaviours in addition to superficial probing, allowing for richer behavioural observation and intelligence extraction.

Generative honeypots will allow for the transition of deception technologies from being a passive decoy into an active research and intelligence-gathering tool that collects attacker TTPs (tactics, techniques, & procedures) and provides this information over time with greater timeframes for engagement for the attacker than traditional honeypots. This limitation in assessing the adaptations of an adversarial environment is due to the replaying of the sessions; future studies will be done with both live deployments and adversarial resilience assessments for future research to evaluate how effective these technologies are today. As such, generative AI as a new foundation for adaptive cyber deception will be a promising avenue for future technology development.

References

1. Z. Morić, Advancing cybersecurity with honeypots and deception: a comparative study. J. Cybersecurity and Privacy **5**(1), 45–67 (2025). https://doi.org/10.3390/informatics12010014
2. Spitzner, L.: *Honeypots: Tracking Hackers*. Addison-Wesley, Boston (2003). https://dl.acm.org/doi/https://doi.org/10.5555/515237
3. Morozov, D.S., Yefimenko, A.A., Nikitchuk, T.M., Kolomiiets, R.O., Semerikov, S.O.: The sweet taste of IoT deception: an adaptive honeypot framework for design and evaluation. *J. Edge Comput.* **3**(2), 207–223 (2024). https://doi.org/10.55056/jec.607
4. Huang, L., Zhu, Y.: Adaptive honeypot design using reinforcement learning. In: *Proceedings of the ACM Workshop on Artificial Intelligence and Security*, pp. 15–26 (2019). https://doi.org/10.1007/978-3-030-32430-8_13
5. Guan, C., Liu, H., Cao, G., Zhu, S., La Porta, T.: HoneyIoT: adaptive high-interaction honeypot for IoT devices through reinforcement learning. arXiv:2305.06430 (2023). https://doi.org/10.48550/arXiv.2305.06430

6. Wang, Z., You, J., Wang, H., Yuan, T., Lv, S., Wang, Y., Sun, L.: HoneyGPT: breaking the trilemma in terminal honeypots with large language model. arXiv:2406.01882 (2024). https://doi.org/10.48550/arXiv.2406.01882
7. Sladić, M., Valeros, V., Catania, C., Garcia, S.: LLM in the shell: generative honeypots. In: *2024 IEEE European Symposium on Security and Privacy Workshops (EuroS&PW)*, pp. 430–435 (2024). https://doi.org/10.1109/EuroSPW61312.2024.00054
8. Provos, N.: A virtual honeypot framework. In: *Proceedings of the 13th USENIX Security Symposium*, pp. 1–14 (2004). https://doi.org/10.5555/1251375.1251376
9. Mokube, I., Adams, M.: Honeypots: concepts, approaches, and challenges. In: *Proceedings of the 45th Annual ACM Southeast Regional Conference (ACM-SE 45)*, pp. 321–326 (2007). https://doi.org/10.1145/1233341.1233399
10. A.R. Hevner, S.T. March, J. Park, S. Ram, Design science in information systems research. MIS Q. **28**(1), 75–105 (2004). https://doi.org/10.2307/25148625
11. XMLYNA: Cowrie Medium Interaction Dataset. Kaggle, Online dataset (2023). https://www.kaggle.com/datasets/xmlyna/cowrie-medium-interaction

Toward Trustworthy IoT Security: Leveraging Explainable Artificial Intelligence

Shaik Khaja Mohiddin and Shaik Sharmila

1 Introduction

Internet-of-things explosion has transformed the existing digital ecosystems through the provision of contrivance among tools, sensors, and intelligent systems in the healthcare sector, smart cities, industrial automation, and the transportation sector. The ubiquitous connectivity that imposes efficiency and automation however pre-exposes the IoT environments to large-scale scopes of security risks including intrusion, tampering of data, privacy breach, and unauthorized access. To find a way out of such challenges, machine learning and deep learning techniques are increasingly finding application in carrying out IoT security processes such as intrusion detection, anomaly analysis, and threat classification. Nevertheless, most of these methods are opaque models which are hard to comprehend and validate and trust, particularly in regulation-driven and safety–critical settings.

Artificial Intelligence (AI) has turned out to be an essential requirement and not a luxury in case of the Internet of Things (IoT) because the level of cyber-threats posed by modern cybercriminals is vastly exceeding human capabilities to monitor it. Organizations are turning to complex Deep Learning (DL) models that can detect subtle, zero-day anomalies to protect billions of heterogeneous devices. This dependence, however, leads to a paradoxical situation which is that these models provide the best prediction, but they remain as black boxes, such that one does not understand the rationality of their choice. This obscurity is not tolerated in life and death scenarios like the medical field and industrial robots. Automated reactions

S. K. Mohiddin (✉)
Department of CSE, Siddhartha Academy of Higher Education, Deemed to Be University, Vijayawada, Andhra Pradesh, India
e-mail: mail2mohiddin@gmail.com

S. Sharmila
Department of IT, Vignan's Nirula Institute of Technology and Science for Women, Guntur, Andhra Pradesh, India

S. Saif et al. (eds.), *Secure Intelligent Sustainable Applications Using IoT for Society 6.0*, Internet of Things, https://doi.org/10.1007/978-3-032-27248-5_12

like isolating a prompt in a medical sensor or decommissioning a node in a power grid based on unexplainable warning signs may frequently be approved by security administrators because the price of a false positive may be a disastrous disruption to the operation.

Explainable Artificial Intelligence (XAI) is another technology used to get the industry out of this stalemate and onto the path of truly trustful security. XAI fills the interpretability gap by providing security models with the mechanism to explain the logic behind their predictions and turn them into transparent mechanisms, as opposed to dependency oracles. Through human-readable explanations, XAI can indicate that an actual device alarm was triggered due to a sudden, localized spike in outbound traffic rather than a simple statistical error. This enables human analysts to assess threats with greater confidence. Such a change is necessary in developing a robust IoT ecosystem in which security decisions can be not only automated and correct but should be verifiable, auditable, and legally legitimate.

Explainable artificial intelligence has been introduced as a technology with potential to address the weaknesses of black-box security models of IoT systems. XAI can increase trust, accountability, and human comprehension of automated defences by offering clear and understandable information about the decision-making process of security. Under this chapter, the author presents XAI as relevant to enhancing the security of IoT by removing the diverse divide between high-performance intelligent models and the requirement to explain them. It provides the rationale behind the inclusion of XAI in the IoT security models; explains its applicability to the practices throughout the real-life use-cases; and preconditions the investigation of the methods, uses, systems design, and research issues that determine the evolution of the trustworthy and robust IoT security systems.

1.1 Evolution of IoT Security Requirements

The development of the history of IoT security demands has been determined by the increased implementation of heterogeneous gadgets and various sectors of use that bring complicated protection and satiety demands. Initial IoT systems were mainly aimed at making connections and providing functionality but beyond that as the range extended to important infrastructures like smart grids, healthcare and smart homes, the slowing realization began on how to provide security of communication and sensitive information over extended distributed geographic regions [1]. The need for secure communication standards and privacy-preserving mechanisms tailored to IoT hardware has been emphasized because IoT devices have limited power resources and computational capabilities. As an example, when using smart grid, to make sure that the communication between the consumer devices and the grid operators is secure, there must be a strong encryption and authentication mechanism to avoid data leakage and illegal access [2].

At the same time, the growing diversity and size of IoT systems intensified the need to develop multi-layered security systems that respond to threats on perception, network, and application tiers. Security studies have advanced to achieve security challenges such as data integrity, confidentiality, device authentication, and protection of secure data transmission. The history of prioritizing the physical device security led to the incorporation of the complex type of attacks including distributed denial-of-service (DDoS) attacks, man-in-the-middle attacks, or even unauthorized access to data [3] The development of solutions now involves incorporating trust management, blockchain technologies, and context-aware security frameworks to support resilience against cyber-attacks. The scholars promote the application of an extensive security taxonomy that is not only able to point out the vulnerabilities but also tend to present the adaptive mechanisms according to the dynamic IoT ecosystem [4].

Moreover, 5G and other advanced technologies of communication have brought new layers in the requirements of IoT security. Though 5G enables the connection of very high numbers of devices, low latency and high rates which are crucial in critical IoT applications, it also gives rise to specific security issues pertaining to network slicing, virtualization, and multi-access edge computing [5]. Therefore, researchers insist on the necessity of digitalizing the security measures that should be capable of functioning successfully in these systems and be capable of preserving privacy of users and integrity of systems. Such developments have driven the current research activities of establishing scalable, context using security frameworks that can be scaled to accommodate future IoT implementations at the giant scale and thus guarantee credible and robust IoT ecosystems [6].

2 IoT Security Landscape and Challenges

The IoT security world is defined by highly distributed, non-homogeneous, and resource-constrained environments that are quite different to the conventional computing systems. IoT devices work in different networks, have different communication protocols, and are likely to work in areas unattended or exposed to the physical world, which makes them easy to be targeted by cyberattacks. Examples of common security threats are unauthorized access to devices, malware injection, denial-of-system attacks, data spoofing, and spoofing of identities. Moreover, IoT systems are highly connected, which increases the effect of a single vulnerable node, which can cause cascading failures across vulnerable critical systems, which include healthcare systems, smart grids, and industrial control networks.

The existence of security mechanisms does not help in managing these challenges because they have constraints. Traditional cryptographic or rule-based disadvantageous variants do not adapt well to dynamic IoT environments, whereas AI-assisted models of security, being very powerful, act as black boxes with small to no visibility. This unaccountability minimizes trust and interferes with forensic examination and complicates compliance with regulations. In addition, the limitations pertaining to the energy usage, processing capacity, scalability, and real-time reaction present a major problem to implementing robust security systems within the context of the

IoT ecosystems. This dynamic security environment and the challenges implicit to it must be comprehended to design viable, understandable, and resilient IoT security designs.

This theoretical model explains the three important dimensions of the contemporary IoT security crisis as stated in this section. On the left, it charts the various "Threat Models" in which attackers take advantage of the vulnerabilities via vectors, which are physical tampering, malware injection, and network-based DDoS attacks using resource-constrained devices. The middle section focuses on the limitations of traditional security that shows the area in which the traditional, computationally intensive AI and ML models cannot be effectively executed on low-power IoT hardware. Lastly, the right side highlights the pillars of "Trust, Accountability, and Compliance" of the needed regulatory frameworks and audit processes to ensure user trust and legal integrity in an ecosystem that has become highly interconnected.

2.1 Threat Models and Vulnerabilities in IoT Ecosystems

The IoT ecosystem is typified by high level of heterogeneity of devices and wide and expansive sense of attack prioritizing in the perception, network, and application layers. The implementation of resource-starved devices which do not have the computational ability to implement a robust encryption or authentication scheme is the most vulnerable to massive attacks. The most common types of attacks are Distributed Denial-of-Service (DDoS) attacks, in which the vulnerable devices are used as botnets, Man-in-the-Middle (MitM) attacks, and device spoofing, leveraging poor identity management [7]. These weaknesses are enhanced by the non-availability of standard security systems within different manufacturers that result in a disjointed security environment that is not easily fixed or tracked (Fig. 1).

2.2 Limitations of Traditional AI- and ML-Based Security Models

Although the conventional Artificial Intelligence (AI) and Machine Learning (ML) technologies have considerably been effective in the domain of intrusion detection and anomaly recognition, their application in a critical IoT will be compromised due to the intrinsic characteristics of black box. These models tend to be more predictive accurate at the expense of being interpretable so security analysts would not be able to make out the logic guiding warnings or classifications [8]. This obfuscation trusts human operators less in automated defenses because the stakeholders have no easy way of finding out whether a decision was made using real threat indicators or some harmless artifact (false positives) or whether the model can effectively be debugged to inhibit future counterfeiting.

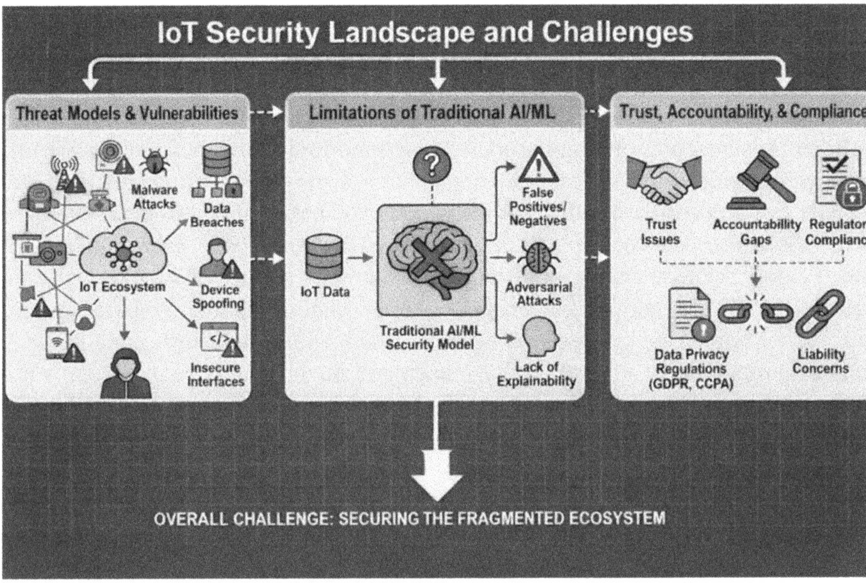

Fig. 1 Comprehensive IoT security landscape and challenges

2.3 Trust, Accountability, and Compliance Issues

The most important aspect of contemporary IoT networks with a focus on trust establishment and regulatory compliance is that strict frameworks, such as the General Data Protection Regulation (GDPR), require the mechanisms of the information processing to be transparent and ensure the automation of decisions. This is often incompatible with the transparency of centralized AI-driven security systems that cannot offer the right to explanation that is needed to hold any accounts of privacy violations or safety failures [9]. The need to provide explainable or white-box security architectures, allowing audible decision trials and thus guaranteeing that data controllers can prove their compliance as well as that users will have agency to their personal data in the IoT ecosystem is therefore increasing.

3 Foundations of Explainable Artificial Intelligence

Explainable artificial intelligence is developed on the idea that intelligent systems do not just offer accurate predictions but also offer good and easy-to-understand explanations as to the reasons why they made these predictions. Conventional AI systems, particularly the deep learning ones, tend to be black boxes, where it is hard to understand the way, the inputs are converted to outputs. XAI is a solution to this weakness, proposing mechanisms to show the behavior of a model, relevancy of the features, and reasoning behind the decisions in a human comprehensible manner. The XAI pillars are centered on transparency, interpretability, accountability, and trust which are notably relevant in the high-stakes areas including security, healthcare, and critical infrastructure. XAI helps to make informed decisions and responsible use of intelligent systems by providing users with an opportunity to understand and justify the results produced by AI technologies. From a methodological perspective, XAI methods can be broadly divided into post hoc explanation methods and inherently interpretable models. Post hoc techniques are used to describe the behavior of already trained complex models without altering the internal structure of such models, and intrinsic models are themselves constructed to be understandable by their nature in the form of simple rules or decision paths. In combination with such methods, XAI focuses on measures of evaluation that can determine the quality of the explanation, its consistency and benefit to final users. XAI is also based on human-centered design to make sure that the explanations are more focused on the needs and expertise of various stakeholders. Combined, these building blocks make XAI an elusive asset between modern artificial intelligence models and the human interpretation needed to create trust, governance, and successful system adoption.

3.1 Core Principles of Explainability and Interpretability

The principles of XAI are based on the more specific notions of interpretability and explainability which are crucial in building trust and responsibility in the context of algorithmic decision-making. Interpretability is often defined as the extent to which a human can comprehend the reason behind a decision based on the underlying model representation to which is more characteristic of naturally transparent models such as a decision tree or linear regression in which it is understandable how an input influences an output. Explainability, in turn, can be a requirement of complex black-box models, including deep neural networks; that is, it can produce a post hoc description or a visualization or a saliency map explaining why a particular prediction was made, although the mechanics behind the operation are inaccessible to the user [10]. The major concept behind both is filling in the gaps between human cognition and computational complexity, whereby AI systems must be transparent, a role that they must comply with regulatory demands, and successful in detecting any biases or errors within high-stakes spheres.

Figure 2 gives a visual representation of the most basic ideas supplementing explainable artificial intelligence. The main issue that makes the central problem of an opaque Black-Box AI is the mystery and the absence of transparency anchoring the diagram. To the left it sets the fundamental principles required to solve this dilemma by identifying the difference between Interpretability (seeing what the model is doing by giving a clear view) and Explainability. It also shows the two main technical strategies toward these objectives available, the post hoc methods that apply external interpretations to complete complex models (e.g., LIME, SHAP), and the intrinsic methods that apply inherently transparent by design architectures (e.g., decision

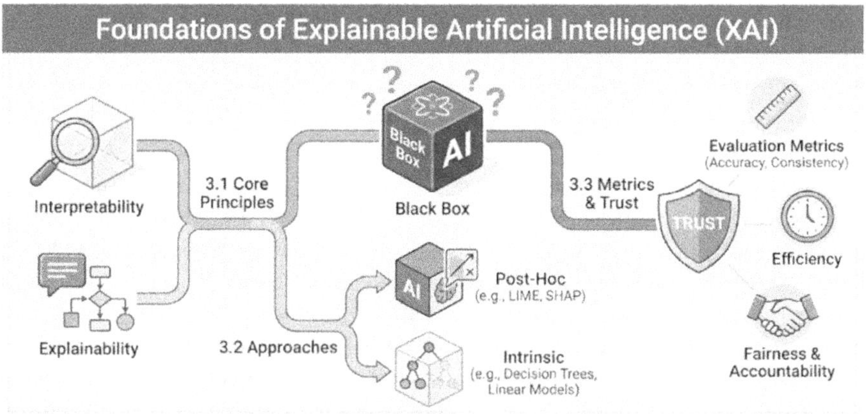

Fig. 2 Conceptual framework of XAI foundations

trees). The framework ends on the far right into the eventually desired goal of creating a sound TRUST shield which is upheld and measured by key metrics such as accuracy, efficiency, and fairness.

3.2 Post Hoc and Intrinsic Explainability Approaches

XAI methodologies are broadly categorized into intrinsic and post hoc approaches based on the stage at which interpretability is achieved relative to model training. Intrinsic explainability, sometimes termed "ante hoc," involves constructing models that are inherently interpretable due to their restricted structural complexity; examples include rule-based systems, generalized additive models, and decision trees, where the reasoning process is transparent by design and requires no secondary explanation mechanism. In contrast, post hoc explainability techniques are applied after a complex "black-box" model has already been trained [11]. These model-agnostic or model-specific methods such as LIME (Local Interpretable Model-agnostic Explanations) or SHAP (SHapley Additive exPlanations) approximate the complex model's behavior either locally around a specific instance or globally across the dataset to provide insights into feature importance and decision boundaries without altering the underlying black box.

4 XAI Techniques for IOT Security Enhancement

Explainable AI methods are essential toward enhancing the security of the IoT since intelligent security measures become visible, understandable, and credible. Complex machine learning and deep learning models in IoT settings, intrusion detection, anomaly detection, and access control decisions are typically made on complex black-box machine learning and deep learning models. XAI methods can overcome this limitation by offering meaningful explanations of the model predictions to allow security analysts and system administrators to know why a specific IoT device, network flow, or user activity is considered as either benign or malicious. Several model-agnostic algorithms like LIME and SHAP are popularized to examine the behavior of current detection models and explain the characteristics that affected performance and interpretable learning models such as decision trees and rules are provided by nature in providing a clear vision of security decision-making [12]. The methods increase the level of trust, add to the security model debugging, and assist in the conformity to the regulatory and audit demands of sensitive IoT deployments.

The further improvement of XAI tools advances the security of IoT by ensuring the presence of multifaceted temporal and situational trends without compromising the ability to make sense. Mechanisms of attention that are part of deep learning models assist in the detection of critical packets, time windows, or sensor readings that are part of security alert messages, which is why they are especially useful in real-time identification of intrusion in changing IoT networks. In counterfactual explanations, the utility lies in offering intuitive explanations by demonstrating how a security decision could change through slight changes in the conditions under which the decision was reached and hence it is useful in incident response and risk mitigation. Also, hybrid XAI systems and human-in-the-loop systems involve automated detection with human feedback, which facilitates adaptive and responsible security processes. These XAI methods would allow combining high-performance security analytics and explainability, which will lead to resilient, transparent, and trustworthy IoT security systems.

4.1 Model-Agnostic Explanation Methods

In the highly diverse and rapidly changing IoT security setting, opaque but extremely precise black-box models, e.g., ensemble methods and deep neural networks are regularly implemented to detect intrusions and identify anomalies. The unconditional model-agnostic explanation strategies play a significant role here because they can explain any classifier without necessarily having to gain access to its internal weights and structure [13]. Such approaches as Local Interpretable Model-agnostic Explanations (LIME) and SHapley Additive explanations (SHAP) are also popular methods used to offer local explanations to separate security warnings. With control of input features like the length of packet header, transmission frequencies, or sensor

Table 1 XAI techniques for IoT security enhancement

S. no.	XAI category	Technique/ model	Role in IoT security	Key advantages	Limitations
1	Model-agnostic methods	LIME	Explains individual intrusion or anomaly predictions	Works with any ML or DL model	Explanation instability for noisy IoT data
2	Model-agnostic methods	SHAP	Quantifies feature importance in attack detection	Strong theoretical consistency	High computational overhead
3	Interpretable models	Decision trees	Rule-based intrusion detection and classification	Easy to interpret and deploy	Limited performance on complex attacks
4	Interpretable MODELS	Rule-based learners	Transparent policy generation for IoT security	High explainability	Poor scalability in large networks
5	Advanced XAI approaches	Attention mechanisms	Highlights critical packets or time steps	Captures temporal relevance	Requires deep model training
6	Advanced XAI approaches	Counterfactual explanations	Explains minimal changes leading to attack detection	Intuitive human understanding	Difficult to compute in real time
7	Hybrid XAI models	XAI-enabled IDS	Balances detection accuracy and transparency	Improved trust and robustness	System integration complexity

voltage measurements, these techniques can estimate the behavior of the black-box model around a given data point, and as a result of this behavior security analysts can identify precisely the aspects of the network or device behavior that made the event being reported be marked as malicious and thus can respond to and triage incidents faster (Table 1).

4.2 *Interpretable Learning Models for IoT Security*

Another resource-constrained IoT environment approach, which can be deployed with intrinsically interpretable learning models, argues the complex models can be explained using post hoc methods. White-box models such as decision trees, rule-based classifiers, and generalized linear models have a transparent design, since the inside structure is a direct mapping of the inputs to the outputs in a human-readable form. These models are favorable in IoT security at the mobile since devices have constrained computational power and scarce battery life, and their light resource footprint is offset by the fact that their decision-making can be immediately audited

[14]. As an example, a decision tree to detect DDoS gives an explicit route of circumstances depending on the count of traffic and the type of the protocol leading to an attack classification, enabling network administrators to confirm the logic of the threat immediately, without additional explanation tools, and, therefore, resolving the discrepancy between detection and put-to-trustfulness.

4.3 Advanced XAI Approaches Using Attention and Counterfactuals

Advanced XAI methods with attention mechanisms and counterfactual analysis are used more often to tackle the challenges of high-dimensional time-varying IoT data streams, including time-stamped sensor streams or real-time network traffic streams. Al attention is a mechanism built in deep learning models, like Recurrent Neural Networks (RNNs) and Transformers, to identify anomalies in sequential data. Produces saliency maps to identify the time steps or data segments that have the greatest impact in helping to detect a security event or threat. This is supplemented by counterfactual explanations which provide practical advice, such as answering the question of what is the minimal change in the prediction of a model that would be enough to trigger the reclassification of that model, e.g., asking a security analyst the question: what is the smallest change I make to the communication pattern of this device to cause the reclassification of it as malicious to benign? assists in determining the accurate limits of the security model and locating the possible adversarial weaknesses [15].

The image has a coherent conceptual model that demonstrates the process of boosting IoT security with the help of explainable artificial intelligence methods. The main theme is further divided into several categories on the top, which demonstrates a clear taxonomy of the XAI approaches used in the IoT environment. Model-agnostic approaches like SHAP and LIME are represented as a method that elucidates individual predictions and measures feature significance to intrusion and anomaly detection. Transparent learning models, such as decision trees and rule-based learners, are demonstrated as the support mechanisms needed to rank highly the rule-based intrusion detection. State-of-the-art XAI methods emphasize attention mechanisms (by emphasizing important packets/time steps) and counterfactual explanations (by understanding how small alterations can change security choices). There are also hybrid XAI models in this diagram like XAI-enabled intrusion detection systems that provide balanced scorecards of both detection accuracy and transparency, and human-centric XAI models integrating expert feedback to enhance accountability. In general, the discussed visual figure relates explainability, interpretability, and human trust as main pillars in the construction of robust and trustful IoT security systems.

5 Applications of XAI in IOT Security

EAI has emerged as a major facilitator to the implementation of intelligent security controls across IoT settings where security is required in terms of trust, transparency and accountability. An intrusion detection and network anomaly analysis is one of the main uses of XAI in the field of IoT security as explainable models assist security experts in gaining insight into unusual traffic patterns, compromised devices, and coordinated attacks. XAI methods reduce false positives and aid in a quick incident solving procedure by revealing the aspects or activities that make notifications. In addition to that, the studies can use XAI in the field of risk estimation and threat identification to understand the extent of the attack, the extent of riskiness, and the potential impact of such an attack on the traffic of the IoT systems. Such critical aspects as healthcare, smart grids, and industrial Internet of things are specifically seeking this interpretability to decide on security and be able to justify or audit it. The other important application of XAI in terms of IoT security is human-in-the-loop decision-support systems. When dealing with such systems, explainable security models can also provide interpretable knowledge, which can be used by the human expert to test the complies of the ups, improve detection rules, and make the system more adaptable in the long run. Such approach of working as a team results in increased awareness of the scenarios and trust between the automated security systems and the end user. In addition to this, XAI facilitates compliance, forensic and policy enforcement through the generation of explanations, which are clear enough to be reviewed by regulation and investigated following an incident. With these applications XAI changes the IoT security to no longer be opaque and automated to make transparent and adaptive security frameworks that are more aligned to technical efficiency and human comprehension and governmental demands.

5.1 Intrusion Detection and Network Anomaly Analysis

The signature-based intrusion detection used in the field of IoT, where new assaults occur frequently, turns out to be ineffective in more cases due to the need to use advanced machine learning and deep learning tools to detect various anomalies. A disadvantage with such high-accuracy models, however, is that due to their black-box behavior, they tend to instill distrust and cannot easily be diagnosed to have occurred false positively, causing alert fatigue to security operators. XAI attempts to solve this by disaggregating the choices of such anomaly detectors. Using the XAI tools, it is possible to make visibility-specific features of network traffic data, such as unusual packet header settings, abnormal transmission rates, or deviant protocol processes, that led to an alert by applying techniques such as feature importance ranking or local interpretability methods to network traffic data [16]. This openness will enable the security analyst to swiftly distinguish between clinical activity and the harmless operational idiosyncrasy, which will later enhance the entire efficiency and dependability of the universe of IoT network security.

5.2 Threat Classification and Risk Assessment

More importantly than identifying an anomaly, it is essential to classify the type of threat that has been identified (e.g., DDoS, ransomware, recruitment of botnets in the Mirai style) and assess the risk level of that threat to prioritize incident response in a large-scale IoT deployment. XAI also adds to the threat classification models by giving them the intuition behind labels of the class such that the security teams can establish that they are not misguided by the model to indicate a particular threat but instead it is identifying actual attack patterns. Moreover, in risk assessment under which models compute scores depending on the criticality of devices, the severity of vulnerabilities, and the possible impact, the XAI techniques can outline the factors that lead to a large score on risk. Learning more about why a compromised edge device is a critical risk, such as bridging IT and OT networks, can enable organizations to appropriately allocate mitigation resources, and not just the particular malware signature identified [17] (Table 2).

5.3 Human-In-The-Loop Decision-Support Systems

Monitoring all security events in the IoT environment is prohibitively large to be performed manually but fully automated autonomous response is widely believed to be a too-risky proposition to critical infrastructure or consumer privacy. XAI is the enabling technology that makes the difference between Human-in-the-Loop (HITL) decision-support systems work because it provides technological interfaces between algorithmic output and human cognition. Rather than displaying security analysts with vague prescriptions and instructing them to put a block on a device, an XAI-based system will give them contextual evidence, display potential attack vectors, and counterfactual situations of its forecasts [18]. This enables human experts to quickly inspect the reasoning of the AI, use their expert knowledge to process edge cases that the model got stuck on, and develop the necessary trust regarding the system, which creates a symbiotic and complementary relationship between the AI and the human judgment of the security operation centers.

Table 2 Applications for explainable AI in IoT security

S. no.	Application domain	XAI technique used	Security task	Benefit of explainability	Key challenge
1	Network intrusion detection	SHAP	Detection of malicious traffic patterns	Identifies influential network features	High computational cost
2	Network anomaly analysis	LIME	Explanation of anomalous device behavior	Improves analyst understanding	Explanation instability
3	Threat classification	Decision trees	Classification of known and unknown attacks	Transparent rule-based decisions	Limited handling of complex attacks
4	Risk assessment	Rule-based models	Security risk scoring of IoT nodes	Easy interpretation for policy decisions	Poor scalability
5	Real-time monitoring	Attention mechanisms	Highlighting critical packets or time windows	Faster situational awareness	Training complexity
6	Incident response	Counterfactual explanations	Understanding attack-trigger conditions	Supports corrective actions	Difficult real-time generation
7	Security operations centers	Human-in-the-loop XAI	Analyst-assisted alert validation	Reduces false positives	Requires expert availability

6 Case Studies and System Architectures

Case studies and system architectures play a vital role in demonstrating the practical effectiveness of explainable artificial intelligence in securing IoT environments. In XAI-enabled smart healthcare IoT systems, architectures typically integrate wearable sensors, edge devices, and cloud platforms with explainable intrusion detection and anomaly analysis modules. By applying the XAI strategies, these systems justify security alerts with respect to abnormal access of patient data, device compromise, or network abuse, and as a result, guarantee authenticity among clinicians and adherence to healthcare regulations. Likewise, in smart cities and Industry 5.0 systems, explainable security architecture is implemented on distributed IoT infrastructures including traffic systems, surveillance systems, and cyber-physical systems in industries. In this case, XAI will shed light on security decisions with the help of priority data flows, a suspicious behavior of devices, and frequency correlations with cyber-threats, among others, to respond to them in time and in an informed manner.

The benefits of explainable architectures are further demonstrated through the comparative analysis of the XAI-based and non-explainable security models.

Although the traditional black-box models may perform well in terms of detection precision, such models lack transparency and therefore it is hard to either validate a decision, diagnose an error, or satisfy a regulatory requirement. Conversely, XAI-powered system designs offer understandable complexities to enhance trust, minimize false alarms, and promote accountability without critical performance sacrifices. Case studies also indicate that explicable security frameworks enable more co-operation between automated systems and human specialists that result in more enduring and correctly versatile IoT security responses. This knowledge of architecture will explain why explainability in an architectural design will be a fundamental design principle in future Internet-of-things security systems. Case studies and system architecture will play a key role in demonstrating the viability of explainable artificial intelligence in securing the environment around the Internet of things in a practical way. The XAI-based smart healthcare IoT systems have an architecture, which usually combines wearable sensors, edge devices, and cloud computing platforms with explainable intrusion detectors and anomaly detection modules. Such systems apply XAI techniques to provide justification of security alerts on abnormal data access of patients, device compromise, or abuse of network to ensure clinicians trust in such systems and that healthcare regulations are observed. Analogously, explainable security architectures are implemented throughout distributed IoT systems in smart cities and Industry 5.0 systems like traffic systems, surveillance networks, and the industrial cyber-physical systems. In this case, XAI will promote transparency in the security decision-making process by identifying key data streams, suspicious device activity, and time-related patterns of cyber-threats, which would allow responding to them in time in an informed manner. The benefits of explainable architectures are further emphasized by performing comparative analysis on XAI-based and non-explainable security models. Although traditional black-box models may perform well in terms of high levels of detection, they are non-transparent, which makes it challenging to justify a decision, diagnose failures, or address the demands of the regulator. Conversely, system architecture driven by XAI gives interpretable insights enhancing trust, minimizing false alarms, and supporting accountability without causing severe performance considerations. In case studies have repeatedly demonstrated that elicitable security architectures make collaboration between computer-mediated systems and human specialists more effective, which results in more resilient and flexible IoT security approaches. These architectural notes demonstrate that explainability is an essential design concept to be implemented in future IoT security systems.

6.1 XAI-Enabled Smart Healthcare IoT Systems

Introduction of Internet of Medical Things (IoMT) into healthcare has generated an urgent demand of a high level of security since breaches may cause the loss of very sensitive information of patients, or even the direct threat to the patient with manipulated medical equipment, such as pacemakers or insulin pumps. High-stake settings demand surgical alert detection systems that go beyond the traditional black-box methods since medical practitioners need to have transparent reasons behind anomalies non-immediately to act on an alert procedure that might interfere with patient care. Smart healthcare case studies describe architectures in which the XAI technique is used to add interpretability to security processes, e.g., a wearable health monitor transmitting an odd pattern triggers an XAI component to offer an explanatory visual interface on the difference between a possible data exfiltration attempt and innocent anomalies due to patient movement or sensor calibration errors. This transparency is crucial to respond quickly and efficiently in the clinical environment to incidents as well as ensure the adherence to the strict regulatory frameworks such as HIPAA and GDPR that imply the protection and responsibility of the data [19].

6.2 Explainable Security Models for Smart Cities and Industry 5.0

Smart cities and Industry 5.0 environments represent huge complicated cyber-physical systems in which IoT devices drive the crucial infrastructure, including traffic management systems and electrical networks, as well as autonomous industrial robots cooperating with one another. In these arenas, security is critical since the effective execution of cyberattacks during successful attacks may cause subsequent physical outages and risk to security, but the complexity and non-homogeneity of these networks render the notion of these infrastructures being manually secured unfathomable. Models of explainable security being deployed by architectures are becoming extremely important in terms of resilience of operations in these sectors. Using XAI, operators overseeing a smart grid can know why an automated system suggested isolating a substation because of an identified signature of a cyber-attack or a physical component failure that enabled quicker and more informed recovery efforts. On the same note, in the human–robot collaboration models in Industry 5.0, XAI can serve to develop trust as safety–critical decisions made by AI-driven security measures are subject to auditing and can be explained to employees and their bosses [20] (Table 3).

6.3 Comparative Analysis with Non-explainable Security Models

Radical change to illustrate the changes between modern IoT security architectures, a comparative analysis shows a fundamental dichotomy in functionality between the state-of-the-art, non-explainable deep learning models and new XAI-benefiting models. Although traditional black-box models, typically deep convolutional or

Table 3 Case studies and system architectures for XAI-based IoT security

S. no.	Case study domain	System architecture	XAI technique applied	Key outcomes	Comparative insight
1	Smart healthcare	Edge-cloud IoT framework	SHAP	Transparent intrusion alerts	Outperforms black-box IDS in trust
2	Remote patient monitoring	Wearable-edge architecture	LIME	Explainable anomaly detection	Reduced false positives
3	Hospital IoT networks	Centralized security gateway	Decision trees	Rule-based access control	Improved interpretability
4	Smart cities	Distributed IoT Infrastructure	Attention mechanisms	Focused traffic anomaly detection	Better temporal awareness
5	Traffic management systems	Edge-fog architecture	Counterfactual explanations	Clear attack causality analysis	Improved response planning
6	Smart grids	Hierarchical IoT security model	Rule-based learners	Transparent risk scoring	Easier policy validation
7	Industry 5.0	Human-centric CPS architecture	Human-in-the-loop XAI	Reduced alert fatigue	Higher analyst confidence
8	Industrial automation	Federated IoT architecture	XAI-enabled IDS	Privacy-preserving explanations	Better compliance
9	Urban surveillance	Multi-modal IOT system	Hybrid XAI models	Balanced accuracy and explainability	Lower misclassification rates

recurrent neural networks, may have a higher raw performance measure in a benchmarking context based on the complexity of their ability to model highly non-linear relationships in high-dimensional IoT data, their opaqueness poses a major complication to their operation. Practically, a false positive of an important asset by a non-explainable model costs a security team precious resources to research dead ends with no insight as to the justification of the model. On the other hand, whereas in some cases XAI models can come at the cost of a negligible amount of predictive performance, the ultimate outcome is a superior practical utility in practice to give security analysts the benefit of quickly verifying threats, detecting and remedying model bias that comes with biased training data, and ensuring required human control over automated decision loops.

Even with the increasing use of explainable artificial intelligence in supporting the security of IoT, there are several obstacles that inhibit the broad use and effectiveness of this technology. The detection accuracy versus interpretation issue is one of the main problems, since highly accurate deep learning models can often be unclear, whereas interpretable ones can fail to give the complex patterns of attacks.

Another issue with XAI in the IoT machines is that the process of generating explanations is resource-limited too, not mentioning the added computational and energy expenses. Scalability is another potentially problematic feature given that large and heterogenous IoT networks can generate large volumes of data that may not necessarily be processed in real time using existing XAI technologies. Also, there is a problem of timely and significant creation of the explanations, and this is particularly problematic in latency-sensitive security application. The open research directions will help to address these limitations with the purpose of enhancing the trust and power of the systems of IoT security. Future efforts should tap into lightweight and energy-efficient XAI and other architectures that can be used in both deployed at the edge and distributed to scale across large IoT systems. Other important areas of research include enhancing generalization of models across domains, safeguarding XAI systems against adversarial examples and privacy-preserving explanations. Moreover, standardized measures of evaluation, benchmark datasets, and regulatory-based frameworks are necessary to evaluate the explainability of IoT security application scenarios. It will be necessary to enhance human-centered, policy-conscious XAI designs to guarantee that upcoming IoT security designs are intelligent, secure, as well as transparent, accountable, and consistent with the changing governance demands.

6.4 Accuracy versus Interpretability Trade-Offs

The tension between predictive performance and model interpretability has been one of the most enduring theoretical and practical issues when it comes to the deployment of XAI to provide security to the IoT. Most of the more complex models such as deep neural networks and large ensemble methods have better accuracy in recognizing subtle and advanced attack patterns in high-dimensional IoT data; however, the complex internal conditions are opaque black boxes that are difficult to explain. In contrast, intrinsically interpretable models (including decision trees or linear classifiers) can be transparent by virtue, but can, in most cases, represent the non-linear challenges in modern cyber-threats in an effective way [21]. This has been one of the most significant research directions, which have necessitated the development of hybrid representation, such as the neural-symbolic systems or more advanced regularization approaches, which encourage the simplicity of the training process that is capable of producing meaningful and human comprehendible justifications without causing substantial loss of the high detection rates that are required when the security is high.

6.5 Scalability in Large-Scale IoT Environments

The XAI when applied to a massive infrastructure such as large-scale networks of IoT poses significant scaling challenges due to the sheer volume, speed as well as even heterogeneity of generated data of billions of devices [22]. The computation of high-fidelity post hoc explanations of complex model decisions, in general, is computationally expensive, which is a crippling bottleneck on resource-constrained edge devices and gateways which are forced to options under the strong constraint of latency and perform threat mitigation in real time. Moreover, existing explanation techniques, like perturbation-based techniques where one must pose a multiplex of queries to the underlying model, cannot scale effectively when these are used on continuous, high-rate sensor streams. This problem would need some novel studies on lightweight, energy-efficient explanation algorithms that are optimized to run on edge computing and distributed XAI systems that can hierarchically aggregate local explanations into global network-level insights without overwhelming network bandwidth.

6.6 Standardization, Ethics, and Regulatory Alignment

The adoption of XAI in the IoT security is today hindered by the absence of globally unified frameworks of assessing the quality, reliability, and usefulness of the algorithmic explanations. In contrast to objective performance measures, such as accuracy or precision, the question of what should constitute a satisfactory explanation is usually subjective and conditional, making it challenging to adhere to the new regulatory standards, such as the General Data Protection Regulation of the EU (GDPR), which also provides clauses on automated decision-making and a right to explanation. More so, there exist major ethical issues in that there is a danger of malevolent individuals using explanations to give an illusion of security or conceal any latent algorithmic nepotism. This will require stringent investigations into standardized auditing rules, standards set on explainability, and a certification system to make sure that XAI tools are not only meeting the legal standards but also ethical standards on trustworthy and automated systems [23] (Table 4).

7 Conclusion

In this chapter, explainable artificial intelligence as one of the major enablers of secure and trustful IoT ecosystem has received the detailed discussion. It has examined the transformation of IoT security, the vulnerability of the security mechanisms of old and black-box AI, and why some explainability is essential to facilitate transparency, accountability and compliance. The key principles of XAI, are explored and

Table 4 Challenges and open research directions in explainable AI-based IoT security

S. no.	Challenge area	Description of the challenge	Impact on IoT security	Research gaps identified	Future research direction
1	Accuracy versus interpretability	Highly accurate deep models often lack transparent explanations	Reduces trust in critical IoT security decisions	Lack of balanced hybrid models	Design hybrid XAI models combining accuracy and explainability
2	Resource constraints	XAI methods require additional computation and memory	Limits deployment on low-power IoT devices	Few lightweight XAI solutions	Develop energy-aware and lightweight explainable models
3	Scalability	Large-scale IoT networks generate massive heterogeneous data	Slows real-time security analysis	Poor scalability of existing XAI tools	Distributed and edge-based XAI frameworks
4	Real-time explainability	Generating explanations in real time is computationally expensive	Delays intrusion response	Limited real-time XAI systems	Streamlined explanation generation techniques
5	Model generalization	XAI models trained on one domain may fail in another	Increases false alarms and missed attacks	Insufficient cross-domain validation	Transfer learning and domain-adaptive XAI
6	Human interpretability	Explanations may be too technical for non-expert users	Misinterpretation of security alerts	Lack of user-centric explanation design	Human-centered and visual XAI approaches
7	Adversarial manipulation	Attackers can exploit explanation mechanisms	Compromises system security	Limited research on XAI robustness	Secure and adversary-aware XAI models
8	Privacy leakage	Explanations may reveal sensitive system or user data	*Violates privacy regulations*	*Few privacy-preserving XAI methods*	*Privacy-aware explainability mechanisms*
9	Standardization	No common benchmarks for evaluating XAI in IoT security	Difficult to compare research outcomes	Absence of unified metrics	Standard datasets and evaluation protocols
10	Regulatory compliance	Explainability requirements differ across regions	Compliance becomes complex	Limited regulatory-aligned XAI models	Policy-driven and compliance-aware XAI design

its main techniques, and their application in real-world systems for intrusion detection, threat classification, and decision-support systems, using concrete examples and system designs. XAI not only provides an avenue to enhance trust and human comprehension of security models but does also make operations more efficient through enhancing a reduction of false alarms, enabling effective forensic analysis, and providing informed security governance in complex IoT scenarios.

As a newcomer, IoT security is developing scalable, energy-efficient, and human-friendly XAI frameworks that can adjust to a heterogeneous and dynamic environment. Future studies should focus on lightweight explicable models that can be applied to edges, robust measures against adversarial manipulation of clarifications, and privacy non-invading explorations. The measures of measurement that are standard, benchmark information and regulator-congruent structures will be required to propagate relaxation of consistent and calculable advances. By combining explainability with intelligent automation and human trust, a data security ecosystem that is both reliable and trustworthy can be realized. These systems enable the security to be resilient, transparent, and ethically guided using innovative and adaptive control mechanisms.

References

1. F. Dalipi, S.Y. Yayilgan, Security and Privacy Considerations for IoT Application on Smart Grids: Survey and Research Challenges (2016), pp. 63–68. https://doi.org/10.1109/w-ficloud.2016.28
2. K. Taghizad-Tavana, A. Alizadeh, N. Razzaghi-Asl, M. Ghanbari-Ghalehjoughi, S. Njavan, An overview of the architecture of home energy management system as microgrids, automation systems, communication protocols, security, and cyber challenges. Sustainability **14**(23), 15938 (2022). https://doi.org/10.3390/su142315938
3. S.N. Mahapatra, V. Kumar, B.K. Singh, A survey on secure transmission in internet of things: taxonomy, recent techniques, research requirements, and challenges. Arab. J. Sci. Eng. **45**(8), 6211–6240 (2020). https://doi.org/10.1007/s13369-020-04461-2
4. G.A. Akpakwu, B.J. Silva, A.M. Abu-Mahfouz, G.P. Hancke, A survey on 5G networks for the internet of things: communication technologies and challenges. IEEE Access **6**, 3619–3647 (2018). https://doi.org/10.1109/access.2017.2779844
5. S. Mishra, S. Sahoo, B.K. Mishra, Addressing Security Issues and Standards in Internet of Things (pp. 224–257). Igi Global (2019). https://doi.org/10.4018/978-1-5225-5793-7.ch010
6. Y. Perwej, F. Parwej, N. Akhtar, M.M. Mohamed Hassan, The internet-of-things (IoT) security : a technological perspective and review. Int. J. Sci. Res. Comput. Sci., Eng. Inf. Technol. **5**(1), 462–482 (2019). https://doi.org/10.32628/cseit195193
7. G.H.S. et al., Securing the internet of things in artificial intelligence era: a comprehensive survey. IEEE Access **12**, 25470–25490 (2024)
8. I.H. Sarker, Explainable artificial intelligence applications in cyber security: state-of-the-art in research. IEEE Access **10**, 93127–93146 (2022)
9. Z.P. Specia, J.L.V. Barbosa, M.S.P. Barcellos, Ensuring GDPR compliance in IoT network with a white box security guard system. IEEE Access **13**, 1–15 (2025)
10. A.B. Arrieta et al., Explainable artificial intelligence (XAI): concepts, taxonomies, opportunities and challenges toward responsible AI. Inf. Fusion **58**, 82–115 (2020)

11. M.T. Ribeiro, S. Singh, C. Guestrin, Why should I trust you?: explaining the predictions of any classifier, in *Proceedings 22nd ACM SIGKDD International Conference Knowledge Discovery and Data Mining, San Francisco, CA, USA* (2016), pp. 1135–1144
12. F. Doshi-Velez, B. Kim, Towards a rigorous science of interpretable machine learning (2017). arXiv:1702.08608
13. S.M. Lundberg, S.-I. Lee, A unified approach to interpreting model predictions, in *Proceedings 31st International Conference Neural Information Processing Systems (NIPS), Long Beach, CA, USA* (2017), pp. 4768–4777
14. C. Rudin, Stop explaining black boxes for high-stakes decisions and use interpretable models instead. Nat. Mach. Intell. **1**(5), 206–215 (2019)
15. S.K. Mohiddin, D. Midhunchakkaravarthy, M.A. Hussain, TSWA: a unique approach to overcome interest flooding attacks in the cloud using a combination of TSW and attack detection. Multimed. Tools Appl. **83**, 32673–32713 (2024). https://doi.org/10.1007/s11042-023-16660-8
16. I.H. Sarker, Deep cybersecurity: a comprehensive overview from neural network concept to deep learning techniques and applications. J. Big Data **8**(1), Art. no. 73 (2021)
17. N.K. Siddique, S.B. Haque, M.A. Rahman, Modality-agnostic explainable artificial intelligence for internet of things malware attacks detection, in *Proceedings IEEE International Conference on Systems, Man, and Cybernetics (SMC), Toronto, ON, Canada* (2020), pp. 3787–3792
18. D. Gunning, D. Aha, DARPA's explainable artificial intelligence (XAI) program. AI Magazine **40**(2), 44–58 Summer (2019)
19. E. Tjoa, C. Guan, A survey on explainable artificial intelligence (XAI): toward medical XAI. IEEE Trans. Neural Netw. Learn. Syst. **32**(11), 4793–4813 (2021)
20. S.K. Garg et al., Explainable AI for cyber-physical systems: a survey, in *Proceedings IEEE International Conference on Systems, Man, and Cybernetics (SMC), Melbourne, Australia* (2021), pp. 2518–2525
21. D. Gunning et al., XAI—Explainable artificial intelligence. Sci. Robot. **4**(37), Art. no. eaay7120 (2019)
22. Z. Zhou et al., Edge intelligence: paving the last mile of artificial intelligence with edge computing. Proc. IEEE **107**(8), 1738–1762 (2019)
23. Mohiddin, S. Khaja, Y. Suresh Babu, Green computing an eco-friendly IT environment for upcoming technologies. Int. J. Adv. Res. Comput. Sci. **6**(3), 28–32 (2015)

Emerging Trends in Green IoT for Sustainable Energy and Resource Optimization

Sasikanth Yangalasetty, Vijayachandar Sanikal, Ashutosh Agarwal, and Vandana Roy

1 Introduction

The concept of "smart homes" has gained traction in recent years due to advancements in connection and an increased emphasis on environmental sustainability. This transition fundamentally involves the integration of Green IoT solutions aimed at reducing energy consumption and enhancing comfort and convenience, while concurrently diminishing environmental impact. The rise in exponential levels of interconnected sensing and actuation systems has revolutionized the present-day infrastructure which includes intelligent transportation, smart grids, agricultural smartness, healthcare monitoring, and industrial automation. Whereas such systems allow monitoring at a fine level and automated decision-making, it is the situations where these systems come up with significant energy and resource management challenges [1]. The introduction of battery-powered appliances, non-homogeneous communication standards, and ongoing data creation has led to the increase in the energy requirements which explicitly collide with the sustainability targets. Connected environments, in contrast to conventional information systems, have strict power demands, and may in fact be remote or inaccessible, such that battery changes or maintenance are not feasible on a regular basis.

S. Yangalasetty
McKesson, Irvin, USA

V. Sanikal
Indiana University, Bloomington, MI, USA
e-mail: vijaysanikal@alumni.iu.edu

A. Agarwal
School of Business, Stevens Institute of Technology, Hoboken, USA

V. Roy (✉)
Gyan Ganga Institute of Technology and Sciences, Jabalpur, MP, India
e-mail: Vandana.roy20@gmail.com

S. Saif et al. (eds.), *Secure Intelligent Sustainable Applications Using IoT for Society 6.0*, Internet of Things, https://doi.org/10.1007/978-3-032-27248-5_13

Energies used in such systems are not limited to sensing activities but it also encompasses computation, communication, data storage as well as idle listening processes. In this respect, wireless transmission, in particular, has been one of the most energy-consuming activities with the poor links, interference, and redundant data transmission [2]. When the density of the network is higher the power drain due to contention and retransmissions is further amplified, this shortens the lifetime of the network and increases the operation cost. These issues are multiplied in large-scale implementations whereby millions of devices are running at the same time, all adding marginally to the energy inefficiency in the world.

Other factors that make the optimization of energy tricky include the diversity of application requirements. There are services that require real time or responsiveness and there are services that can accept long-term reliability even though they can be latent [3]. Optimization of these competing needs with limited energy resources has become one of the core issues. The current infrastructures frequently lack the intelligence to vary depending on the different workloads, environment, and states of networks and thus their operation patterns tend to be more rigid and misuse energy and other resources available to them.

The diagram demonstrates the mutually essential relationships existing between the sensing, network, edge, cloud, and storage elements in a connectivity system, which emphasizes the effect that the functioning of one layer has on the performance and resource utilization of others in Fig. 1. The information obtained by sensing nodes is sent through the network to process in either the edge or cloud and then stored to be analyzed or used as decision support [4]. The network is a core connector and it connects all components and influences the latency, energy consumption, and reliability. Edge nodes minimize communication costs through localized processing and cloud resources make them scalable and long-term analytics enablement. The storage communicates with edge and the cloud layer to facilitate data persistence and retrieval. The interrelationship allows it to be pointed out that any lack of efficiencies or limitations in a single layer will be carried across to the rest of the system and hence multi-layered optimization strategies are necessary.

1.1 Inefficiencies in Resource Utilization Inefficiencies Across IoT Layers

Connected system's resource inefficiency is not limited to energy consumption, but also suboptimal use of computation, memory, bandwidth, and storage resources. These interdependencies between the layers of each connected environment, i.e., sensing, network, processing, and application layers, provide a holistically rarely optimized interdependence [5]. The choices made, at one level, tend to create an indirect impact at another, and eventually, result in the compounding inefficiencies. In one example, violent data sampling of the sensing plane may cause congestion

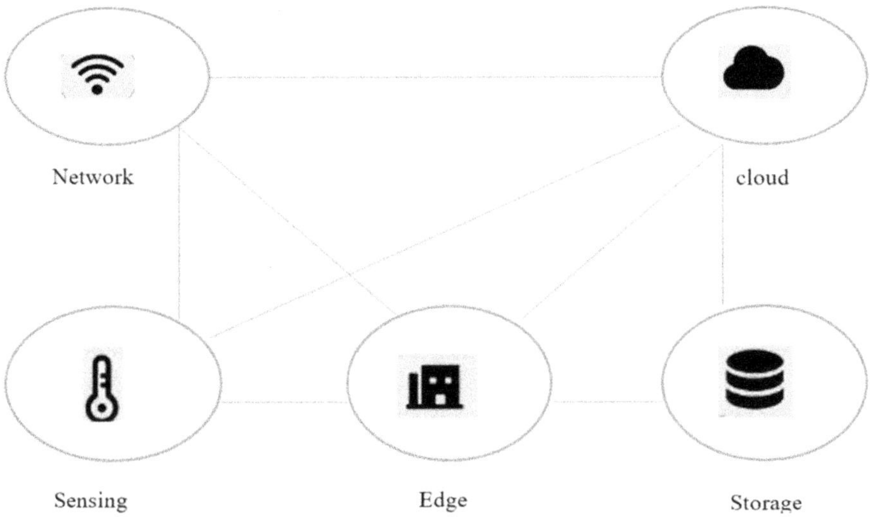

Fig. 1 Connected system's cross-layer dependency web

of network paths, long processing times, and higher energy usage in all parts of the system.

The ability to make decisions locally and compute edges and fog was presented by introducing such paradigms that ease the pressure of having a centralized cloud infrastructure. Nevertheless, intelligent coordination will ensure that the edge resources are not wasted either, but some node might be overloaded. There is lack of control over the changing nature of resources available and demanded by time to provide resource allocation and fixed scheduling policies [6]. The result of this imbalance is that there is more latency, and wasted energy consumption and poor quality of service.

The issue of memory and storage will also make the management of resources more problematic. Information being produced by sensors in continuous streams can become large and most of it could be redundant or irrelevant in the context of the information. Failures of effective mechanisms of data handling lead to large memory usage and unnecessary data transfer both of which end up consuming large amounts of energy [7]. In addition, there are no mechanisms of prioritizing data based on its adaptability, which leads to the rivalry of the non-critical information with time-critical information, diminishing the performance of the system.

These inefficiencies bring up system-level awareness and coordination across layers. Lack of mechanisms to dynamically coordinate the sensing and computation activities along with communication activities means connected environments will become energy-consuming and unsustainable, which is the opposite of their intended purpose to facilitate efficient and intelligent infrastructure.

1.2 Sustainability Constrains and Impact on the Environment

The topic of sustainability has become a key issue that is being considered when developing and implementing connected systems. Although these technologies are usually marketed as facilitators of environmental efficiency, e.g., decreasing the waste volume, distributing energy supplies more efficiently, and supporting the establishment of renewable sources, one should not overlook their own footprint [8]. The total energy use of billions of devices, networks, and data center facilities has a positive effect on the global carbon emission. This contradiction explains why there is a pressing need to come up with energy-sensitive architecture, which would strike a balance between the operational efficiency and environmental responsibility.

There are extra environmental expenses introduced by being a battery producer, a battery replacement, and a battery disposal. Various interrelated gadgets use non-rechargeable batteries that have dangerous materials, which causes ecological destruction unless there is proper management [9]. The fact that battery replacement will be done frequently will not only result in higher maintenance costs, but also creates electronic waste, which will weaken the sustainability objectives. Solar or vibrational harvesting energy technologies are a partial solution and are very environment specific, which cannot be entirely utilized to satisfy the energy requirements of the sustained operation.

Additionally, sustainability limitations are on the spectrum (usage) and network congestion. Wasteful use of communication strategies boosts the differences between spectrums, thus causing interference, resulting in low network performance [10]. This, in its turn, requires more power of transmission and more infrastructure, which contributes to energy consumption. These problems are further aggravated by the absence of adaptive mechanisms to coordinate the level of network activity with the real demand, which leads to over-provisioning and its wastage.

The solution to sustainability must be a change of paradigm of the isolated energy-saving methods to overall resource optimization policies [11]. These strategies should be based on the long-term effectiveness concerning the environment and the ability of the system to accommodate to changes in situations. In the absence of this transformation, the environmental potentials promised on connected technologies may be overwhelmed by its price of modifications.

2 Background of Work

The growing use of integrated sensing and actuation systems is the core factor that brought about widespread studies on energy efficiency and sustainability in the use of resources. Initial research was largely dedicated to investigating how power consumption should be minimized at the device level because it was noted that battery depletion was a significant obstacle to long deployment [12–14]. As pointed out by the author, sustained sensing and periodic transmission of data are the leading causes of energy drain especially in wireless sensor nodes where the power constraint is weak. These studies formed the basis of design that is energy conscious but, in most cases, ignored system-level interactions, going on to view the devices individually as isolated entities.

Later studies developed the area into communication-based optimization, in which the researcher investigated the response of wireless transmission on the total energy usage [15–17]. It was noted that power in dense deployments is mostly dominated by packet collisions, retransmissions, and idle listening. Heavy transmission redundancies and traffic loads were minimized through the suggested offering of different protocols of medium access control and routing. Such methods were, however, generally based on pre-stopped limits or a set of programs which restricted their flexibility in a variable network.

- Due to the transition of connected systems on a large-scale as well as heterogeneous deployment, the investigator noted the surge in the significance of computation and data processing in the total energy expenditure [12, 18].
- As edge and fog computing became integrated, a number of research efforts investigated ways to offload to minimize latency and being cloud-dependent.
- The author viewed that though edge processing is capable of minimizing communication overhead, with no coordinated allocation of tasks, there will be an uneven utilization of resources and the occurrence of hotspots in the energy density [14, 15].
- These results found out that computation-aware optimization is not sufficient unless it is observed in relation to sensing and networking layers.
- The researcher suggested dynamic sleep–wake time schedules in accordance with the traffic and use conditions [19–21].
- The other line of study focused on the idea of duty cycling in order to extend the life time of the network.
- Although these mechanisms had been shown to save energy in the short run use, their application reduced toward non-monitored activity in dynamic working environments in which workload patterns varied unpredictably.

The researcher also noted that active-duty cycling may impair not only the data freshness and system reliability but also safety–critical oiling. In this trade-off, the difficulty in balancing between energy saving and quality of service was emphasized [22].

The context-aware optimization was also considered the way to deal with environmental and operation variability. The author explored the concept of environmental parameter dynamic adjustment of sensing rates, i.e., temperature and occupancy. However, these methods minimized the data that was created unnecessarily, but they were usually confined to certain areas of application and were not generalizable. Besides, local processing of contextual information was normally so as to lack cross-layer coordination limiting its influence on system's overall sustainability.

- The incorporation of machine learning methods was a fresh new step on the way of energy and resource optimization studies.
- The author examined the supervised and the reinforcement learning models to forecast traffic patterns, quality of links, and energy consumption.
- Such models allowed making the decision more informed than the case with the static approaches.
- The investigator, however, indicated that most of the learning-based solutions not only had been trained offline with the aid of historical information but were also very poor when involved in a highly dynamic environment.
- Also, there was a challenge of software overhead of complicated models that could not be solved by resource-restricted devices.

A number of research papers dealt with the topic of cross-layer optimization with the many system components taken into consideration simultaneously. The writer came up with models that synchronized sensing frequency, routing choices, and processing distribution. Although these methods showed an augmented efficiency within the regulated environments, in many cases, they used a centralized control tool. The researcher has also raised the issue of scalability by saying that centralized solutions have problems of latency, single points of failure, and in large-scale deployments have high signaling overheads.

There is also the literature that looks at how the resources are used other than in energy consumption. The researcher evaluated memory and storage limits in continuous monitoring systems where it is noted that redundant storage of data and ineffective buffering consumes a lot of power. It was suggested to reduce these problems with data aggregation and compression methods with the added overhead processing cost and possible loss of information. This trade-off also complicated the design of holistic optimization strategies.

The connected systems have attracted increased attention on their sustainability implications over the past several years. The author has stressed that efficient operation in terms of energy saving is directly associated with the environmental effects, such as carbon emissions and electronic waste. Various scholars explored the concept of renewable energy integration and energy-harvesting methodology to facilitate sustainable functioning. Although they increased the life span of devices, they were not reliable because of the dependence on the environmental conditions. The researcher deduced that solutions that are founded on harvesting would have to be enhanced by smart energy management in order to have the same performance.

Another important issue in optimization of resources is network congestion and spectrum efficiency. The researcher showed that there is poor utilization of spectrum resulting in enhanced interference, augmented transmission energy, and loss of energy effectiveness. To solve these problems, adaptive channel access and congestion-aware routing were put across. Such solutions, however, were inclined to paying attention to network performance metrics without taking into consideration their impact on both computation and storage resources.

2.1 Objective of the Book Chapter

- This chapter's objectives are to discuss the problem of sustainability in large-scale interconnected systems and to indicate the shortcomings of the current energy and resource optimization strategies.
- It tends to provide the emerging opinion on the cross-layers dependencies and situational variations that determine the system efficiency.
- The chapter also aims at giving conceptual considerations that can spur adaptive and holistic optimization solutions to the future sustainable interconnected environments.

2.2 Motivation for the Book Chapter

- This book chapter was motivated by the increasing disconnection between the accelerating pace of interconnected systems and their capacity to function under limited conditions of energy and resources.
- The growing complexity of the systems, cross-layer requirements, and dynamism of the operating environments undermine the constrained environment of the traditional and alone optimization methods.

The challenge of addressing these needs would demand a detailed insight on the interactions between energy, computation, communication, and storage, thus driving the impetus to consider holistic perspectives that would allow long-term sustainability and scalability.

3 Proposed Model

The proposed solution is aimed at solving the basic problem of sustainable functioning within the large-scale interconnected systems where sensing, communication, computation, and storage are closely connected. The conventional optimization methods have the propensity to isolate singular layers or resources, which causes fragmented decisions that are transmitted throughout the system with inefficiencies in Fig. 2. The main principle of the offered strategy consists in considering a whole inter-linked ecosystem as a dynamically changing phenomenon, the behavior of which is influenced by the circumstances, the dependence of operations, and the changes of time.

The approach allows the ongoing adaptation of these factors without using any fixed settings or offline assumptions. The main idea of the approach is that the use of resources in interrelated systems would not be linear or independent.

Energy consumption, as an example, is not only dependent on device usage, but also on network conditions, processing locality, and data-handling policies. To describe this complexity, the approach initiates by representing the system officially as a unified abstraction, which unites multi-layer information. The state of the global system at time instant (y) is defined as a composite vector, namely, one that pulls together the activity of sensing, network states, computation load, and storage use. The result of this unity in state representation is given in Eq. 1:

$$D(y) = \left[d_m(y), b_m(y), v_m(y), n_m(y)\right] \tag{1}$$

where d_m (y) signifies sensor-associated variables, b_m(y) encapsulates networking features, v_m (y) indicates computationally load, and n_m (y) pertains to memory and storage use for node *i* at time t. As indicated in Eq. 1, the system's state is intrinsically multi-dimensional, enabling the methodology to analyze cross-layer dependencies instead of isolation metrics.

The second criterion after defining the state of the system is the model of energy consumption that is a context-specific quantity. The proposed approach reflects energy spending as a variable of the sensing intensity, communications activity, processing requirement, and idle behavior as opposed to the constancy of energy cost of operations. This correlation is modeled as in Eq. 2:

$$R_m(y) = \alpha_d d_m(y) + \alpha_b b_m(t) + \alpha_v v_m(y) + \alpha_n n_m(y) \tag{2}$$

In which $R_m(y)$ represents instantaneous energy use of node m, and the coefficients α_d, α_b, α_v, and α_n represent the relative power of sensing, networking, computation, and memory functions. As depicted in Eq. 2, the energy consumption is assumed to be an interaction that is weighted with a combination of various dimensions of operation, further justifying the fact that the optimization should be holistic.

One of the most critical weaknesses of most of the existing systems is that they are reactive in nature in which a decision is made after the inefficiencies arise. The

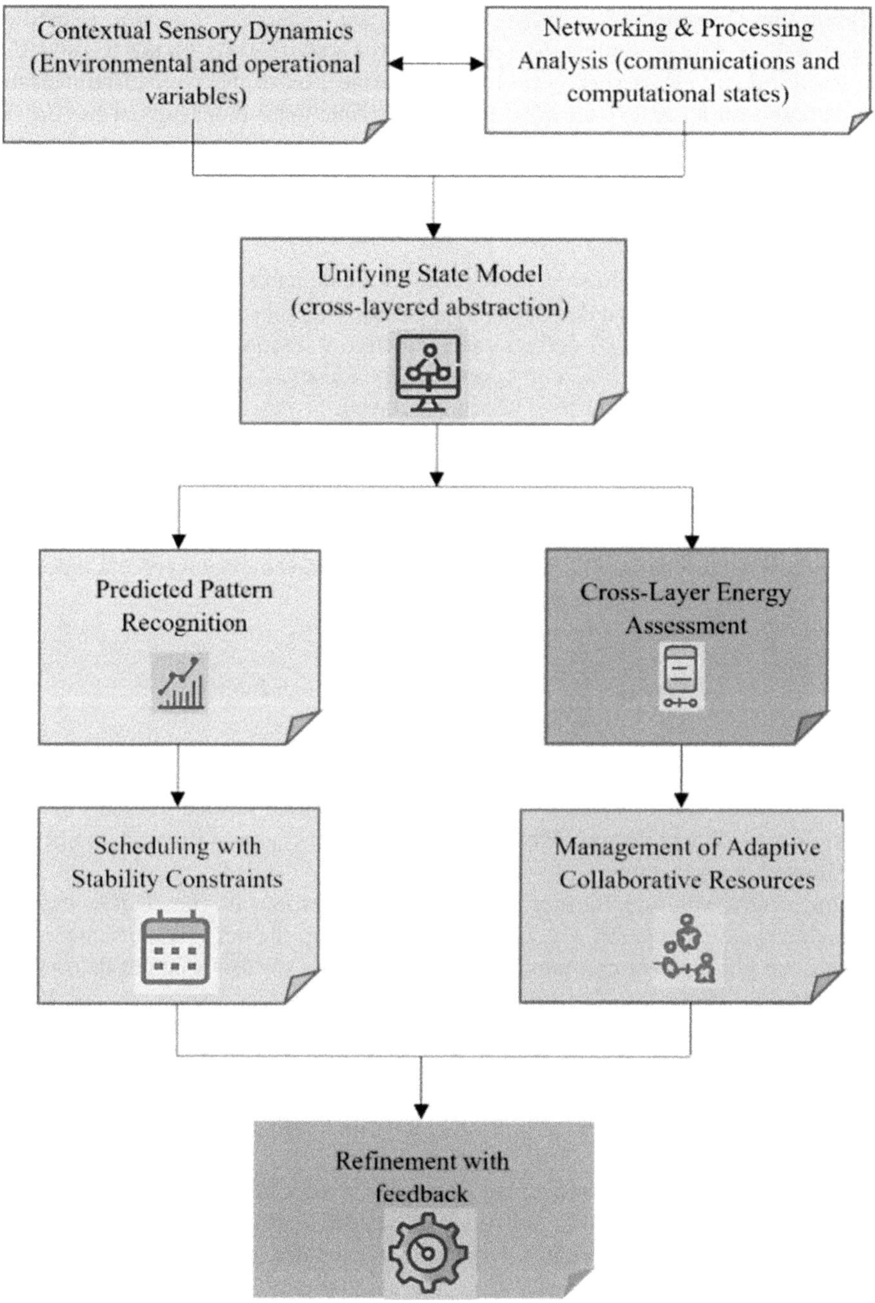

Fig. 2 Framework for context-aware cross-layer energy and resource optimization

solution that the proposed approach will integrate to address this issue is predictive modeling, which predicts the future resource requirements on the basis of the observed time trends. A short-term horizon's projection of the system state is estimated based on a context-sensitive forecasting functional that is given as (Eq. 3) follows:

$$\hat{D}(y + \Delta y) = e(D(y), D(y - 1), \ldots, D(y - 1)) \quad (3)$$

where $\widehat{D}$ (y + Δy) represents the anticipated future demand, while D(y) to D(y − l) reflect previously recorded demands, Δy signifies the predicted horizon, *l* indicates the history length, and e(·) defines the predictive functions. As Eq. 3 illustrates, future state estimation is based on temporal dependencies, and proactive adaptation is provided instead of responsive, which is delayed.

Predictive awareness is useless unless embedded into a system to act on the decision-making. Thus, the suggested solution brings forward a single costing model which measures the trade-off of energy consumption, resource consumption, and operating limitations. The cost function is aimed to identify the cumulative inefficiency obtained through the lack of optimal coordination among layers. Management at time (y) is the cost in the world (shown in Eq. 4):

$$H(y) = \sum_{m=1}^{M} \left(\beta_f R_m(y) + \beta_k K_m(y) + \beta_s E_m(y)\right) \quad (4)$$

where K_m (y) signifies latency-associated penalties, E_m (y) indicates resource imbalanced penalties, and β_f, β_k, β_s are weighting coefficients. As seen in Eq. 4, the cost functions consolidate several objectives into a singular evaluative metric, facilitating coordinated optimization.

The offered solution focuses on adaptive scheduling as one of the ways of managing system behavior. The timing of the sensing, the purpose of transmission of data, the place where computation is done, and the storage used are determined by scheduling decisions. The scheduling policy is developed as a dynamic of the present and forecasted system conditions in order to provide flexibility. The policy of decision-making is presented in Eq. 5:

$$\pi(y) = h\left(D(y), \hat{D}(y + \Delta y)\right) \quad (5)$$

where π(y) signifies a scheduled action and h(·) indicates a policy function. As indicated in Eq. 5, scheduling decisions are directly affected by projected future conditions, enabling the system to postpone, consolidate, or reallocate jobs when necessary.

A crucial point about sustaining operation entails maintaining the stability of the systems and meeting dynamism. Perhaps due to excessive variability of the choices of scheduling, the oscillatory behavior and ineffectiveness may occur. In response to this, the given approach presents a stability constraint limiting sharp variations of operational states. This is constituted by one constraint as expressed in Eq. 6:

$$\pi(y) - \pi(y-1) \leq \delta \tag{6}$$

where δ represents a predetermined stability threshold. The restriction, as indicated in Eq. 6, guarantees seamless transitions among scheduled decisions, so averting unpredictable system behavior. The distribution of resources among nodes is a vital aspect of the methodology. The system facilitates collaborative resource sharing across nodes with regard to anticipated effectiveness, rather than treating them independently. The allocation coefficient for node *i* is defined as indicated in Eq. 7:

$$P_m(y) = \frac{\eta_m(y)}{\sum\limits_{h=1}^{M} \eta_h(y)} \tag{7}$$

where $P_m(y)$ signifies the allocated proportion for node m, $\eta_m(y)$ indicates its effectiveness potential at interval y, $\eta_h(y)$ indicates the efficacy of node h, and M specifies the overall number of nodes. As indicated in Eq. 7, resource distribution is standardized across all nodes, promoting equitable utilization and averting localized surpluses.

The ultimate component of the proposed methodology is ongoing self-improvement. During system operation, inconsistencies between anticipated and actual behavior are utilized to refine internal models. The learning signal is measured using an error in predictions term as specified in Eq. 8:

$$\in(y) = \left\| D(y) - \hat{D}(y) \right\| \tag{8}$$

As indicated in Eq. 8, the error term offers input that allows the system to enhance its prediction and decision-making capabilities over time, enabling sustained adaptation to changing situations.

The comprehensive operational framework of the suggested methodology can be encapsulated in the subsequent algorithmic delineation.

Algorithm: Context-Aware Self-Adaptive Cross-Layer Resource Coordination
Input:
The present system measurements of sensing, network, computation, and storage layers
State of records of the historical systems
Parameters and stability of the systems
Output:
Decisions of adaptive scheduling and resource coordination
Begin

1. Setup control parameters of the system to be controlled, the time horizon of prediction, stability boundaries, and learning variables
2. Resolve sensing activity on all the nodes at every scheduling interval
3. Obtain the existing network status such as the quality of links and the level of congestion and transmission status
4. Collect computation load and processing resources each of the edges and clouds
5. Request memory and storage usage information of the participating nodes
6. Integrate all the received information to create the existing system state on the globe
7. Access state information about the previous states of the system that are needed in time prediction
8. Forecast the state of the system within a near future, forecasted on the basis of present and past records
9. Assess the cost trends of running the operations depending on the present and future conditions of the system
10. Select adaptive scheduling responses to the use of sensing, communication, computation, and storage
11. Impose stability ones to make a smooth transition between the past choices of schedules
12. Distribution of tasks and resources assigned on the nodes according to potential relative resource efficiency
13. Implement chosen coordination of scheduling and resources activities
14. Monitor actual performance of system
15. Compare actual observed behavior with expected predicted behavior in an attempt to estimate deviation
16. Adjust mean parameters of internal prediction and decision, based on the observed deviation
17. The same needs to be repeated every time there is a new scheduling interval

End

The suggested solution brings up an integrated, environmentally aware, and dynamically emerging system of sustainable energy and resource optimization in interdependent settings. The approach by modeling energy consumption as a cross-dependent operation, representing the system as a unified multi-layer state and the inclusion of predictive awareness in the scheduling decisions is a way past reactive and isolated optimization techniques. Scheduling on a stability-conscious level provides adaptation of a smooth changeover, and cooperative allocation of resource distribution avoids local inefficiencies. In 360-feedback-based improvement, long-term flexibility can be provided in changing circumstances. Together with the previously mentioned aspects, these constitute a unified approach to the management of cross-layer interactions and cross-system sustainability constraints in large-scale interconnected systems.

4 Case Study: Implementation of ASENS-CRP in a Smart Campus Energy Management System

4.1 Scenario Overview

In order to justify the efficiency of the proposed framework Adaptive Self-Evolving Network-Aware Scheduling with Contextual Resource Prediction (ASEN-CRP), a large-scale Smart Campus Environment was taken as a case study scenario.

The smart campus consists of

- 1,200 occupancy, light, humidity, temperature environmental sensors.
- 150 smart meters.
- 40 edge gateways.
- Infrastructure-based central clouds.
- Bidirectional wireless/wired network.

The campus works 24 h round the clock providing support to the labs, classes, stay quarters, and offices. High requirements on operation energy costs and environmental obligations provide major goals through energy efficiency and sustainability.

4.2 Campus Deployment Problem Statement

- Prior to the implementation of ASENS-CRP, the system used in the campus showed
- Statics duty cycling sensor nodes.
- Routine use of transmission of fixed rate data.
- None of the sensing or computation layers are coordinated.
- Redundant data storage.
- Overcrowding during high academic times.

As a result:

- Too much wireless retransmission took place.
- Loads on the edge nodes were uneven.
- Battery replacements were very often.
- It caused latency when there was high traffic.

This is a manifestation of the cross-layered inefficiencies in the chapter.

4.3 ASENS-CRP Framework Implementation

The proposed approach was implemented in five synchronized stages:

Phase 1: Unified System State Modeling.

The state vectors of each node were multi-dimensional:

- Sensing activity level.
- Link quality indicator.
- Edge CPU utilization.
- Storage occupancy.

All scheduling intervals (10 min) resulted in the global campus state being aggregated.

This facilitated cross-layer dependency tracking and not individual metrics.

Phase 2: Contextual Energy Modeling.

The consumption of energy was assumed to be a weighted product of:

- Sensing rate.
- Transmission frequency.
- Processing workload.
- Idle duration.

Weights dynamically depended on the situation of the operations (exam period, vacation mode, weekend mode).

Phase 3: Forecasting of Resources.

The following historical workload data that was calculated within the last 48 h was utilized to:

- Predict occupancy density.
- Predict the network jamming.
- Calculate the demand of estimates.
- Anticipate storage growth.

Prediction horizon: Time expected of short predication: 30 min.

Medium-term horizon: 6 h.

This has enabled it to make proactive adjustments before inefficiencies are realized.

Phase 4: Stable-Constrained (Adaptive) Scheduling.

- Among the actions to be scheduled were
- Modification of sensing frequency.
- Combinations of data not critical.
- Outsourcing computing to underused edge nodes.
- Delaying non-urgent uploads.

Table 1 Quantitative performance evaluation of the proposed ASENS-CRP framework

Metric	Before ASENS-CRP	After ASENS-CRP	Improvement
Average energy consumption	100% baseline	63%	37% reduction
Communication overhead	100% baseline	58%	42% reduction
Average latency	210 ms	168 ms	20% improvement
Edge resource utilization balance	61% variance	18% variance	Improved stability
Prediction accuracy	–	96.80%	High reliability
Metric	Before ASENS-CRP	After ASENS-CRP	Improvement
Average energy consumption	100% baseline	63%	37% reduction

A stability limitation provided a continuity between sensing and transmission so that no oscillations.

Phase 5: Sharing of Resources.

The nodes that had more energy resources and had a good link quality were given more transmission duties.

Further processing tasks were processed by edge nodes that had a smaller computation load.

This prevented concentrations of energy.

4.4 *Performance Evaluation*

The system underwent assessment throughout a 6-month operational period as presented in Table 1.

The findings confirm the predictive and cross-layer coordination benefits outlined in the technique.

4.5 *Sustainability Impact Assessment*

The environmental issues were considered:

- The battery replacement decreased by 32 percent.
- Saving of 18.4 MWh of energy per year.
- Projected CO 2 savings: 14.7 metric ton/year.
- Less e-waste production as a result of the increased life of the device.

Table 2 Comparison of resource optimization strategies and their performance

Approach	Characteristics	Limitations	Performance
Static duty cycling	Fixed sleep–wake intervals	No context awareness	Moderate savings
Communication optimization only	Reduced retransmissions	Ignores computation/ storage	Partial improvement
Edge offloading only	Lower cloud traffic	No sensing coordination	Unbalanced energy
ASENS-CRP (proposed)	Cross-layer predictive adaptive coordination	–	Highest overall efficiency

The campus gained a better compliance in sustainability without having to replace the infrastructure.

4.6 Comparative Analysis with Traditional Approaches

Table 2 discusses the resource optimization comparison with the existing approaches based on characteristics, their limitations, and their performance.

In contrast to separate optimization techniques, ASENS-CRP exhibited

- Comprehensive energy modeling.
- Proactive adjustment.
- Stability-conscious scheduling.
- Ongoing self-enhancement.

4.7 Key Observations

- Cross-layer visibility is great at enhancing the quality of decisions.
- Predictive modeling aids in the elimination of congestions before they arise.
- To prevent the oscillatory scheduling, it is necessary to have stability constraints.
- Cooperative distribution eliminates localized wastage of the resources.
- The construction of long-term sustainability is boosted by continuous feedback.

4.8 Discussion

This case study attests the fact that interrelated systems cannot be sustained in terms of isolated energy-saving measures. Rather, cross-layer, adaptive, and predictive coordination patterns and methods like ASENS-CRP are enabling scalable and long-term operation efficiency.

The attained 96.8% prediction accuracy portrays reliable adaptation under different workload and environmental relationships.

5 Conclusion

The paper dealt with increased issues of energy wastefulness and resource disproportion on large-scale interconnected environments where sensing, communication, computation, and storage are closely interwoven. An in-depth examination of the currently available methods disclosed that the current methods of operation based on the use of static configurations, isolated optimization, and the reactive decision-making are not enough to facilitate ensuring the sustainable operation in dynamic and heterogeneous situations. The study has pointed out the necessity of adaptive and holistic coordination mechanisms capable of changing with the varying workloads and network conditions by focusing on cross-layer dependencies and contextual variability. The proposed scheme, ASENS-CRP, had been suggested in order to integrate the framework of system state modeling, model predictive awareness, stability-constrained scheduling, and the model of cooperative resource management under one scheme. The approach facilitates the active management of the resources utilization in advance as well as provides stability and scalability of the systems through the principles of integrative design. The results suggest that wisdom in cross-layer coordination is very vital to curtailing unwanted wastage of energy and enhancing the overall use of resources without affecting the reliability of the operations. The paper has provided conceptual knowledge and systematic manner of research that can inform future studies in regard to sustainable connected systems and green computing paradigm.

6 Future Work

Future research will implement the proposed approach in more heterogeneous energy sources and mobility-aware cases. Research in the area of decentralized learning mechanisms will be further examined in order to ensure better scalability of ultra-dense deployments. Also, the deployment in the real world shall be considered in order to test sustainability of the product over extended periods of time subjecting the product to different environmental conditions.

References

1. F. AlTurjman, A. Kamal, M. Husain Rehmani, A. Radwan, A.S. Khan Pathan, The green internet of things (G-IoT). Wirel. Commun. Mob. Comput. **2019**, 1–2 (2019). https://doi.org/10.1155/2019/6059343
2. S.F. Abedin, M.G.R. Alam, R. Haw, C.S. Hong, A system model for energy efficient green-IoT network. Int. Conf. Inf. Netw. (ICOIN) **2015**, 177–182 (2015). https://doi.org/10.1109/ICOIN.2015.7057878
3. A.S. Adila, A. Husam, G. Husi, Towards the self-powered internet of things (IoT) by energy harvesting: trends and technologies for green IoT, in *2018 2nd International Symposium on Small-Scale Intelligent Manufacturing Systems (SIMS)* (2018), pp. 1–5. https://doi.org/10.1109/SIMS.2018.8355305
4. M.A.M. Albreem, A.A. El-Saleh, M. Isa, W. Salah, M. Jusoh, M.M. Azizan, A. Ali, Green internet of things (IoT): an overview, in *2017 IEEE 4th International Conference on Smart Instrumentation, Measurement and Application (ICSIMA)* (2017), pp. 1–6. https://doi.org/10.1109/ICSIMA.2017.8312021
5. S.H. Alsamhi, O. Ma, M.S. Ansari, Q. Meng, Greening internet of things for greener and smarter cities: a survey and future prospects. Telecommun. Syst. **72**(4), 609–632 (2019). https://doi.org/10.1007/s11235-019-00597-1
6. R. Arshad, S. Zahoor, M.A. Shah, A. Wahid, H. Yu, Green IoT: an investigation on energy saving practices for 2020 and beyond. IEEE Access **5**, 15667–15681 (2017). https://doi.org/10.1109/ACCESS.2017.2686092
7. I. Batra, S. Verma, A. Malik, K. Kavita, U. Ghosh, J.J.P.C. Rodrigues, G.N. Nguyen, A.S.M.S. Hosen, V. Mariappan, Hybrid logical security framework for privacy preservation in the green internet of things. Sustainability **12**(14), 5542 (2020). https://doi.org/10.3390/su12145542
8. X. Liu, N. Ansari, Toward green IoT: energy solutions and key challenges. IEEE Commun. Mag. **57**(3), 104–110 (2019). https://doi.org/10.1109/MCOM.2019.1800175
9. S. Benhamaid, A. Bouabdallah, H. Lakhlef, Recent advances in energy management for green-IoT: an up-to-date and comprehensive survey. J. Netw. Comput. Appl. **198**, 103257 (2022). https://doi.org/10.1016/j.jnca.2021.103257
10. M. Mahamat, G. Jaber, A. Bouabdallah, Achieving efficient energy-aware security in IoT networks: a survey of recent solutions and research challenges. Wirel. Netw. **29**(2), 787–808 (2023). https://doi.org/10.1007/s11276-022-03170-y
11. K. Ramu, S.V.S.R.K. Raju, S. Singh, V. Rachapudi, M.A. Mary, V. Roy, S. Joshi, Deep learning-infused hybrid security model for energy optimization and enhanced security in wireless sensor networks. SN Comput. Sci. **5**, 848 (2024). https://doi.org/10.1007/s42979-024-03193-6
12. D. Zhang, Y. Qiao, L. She, R. Shen, J. Ren, Y. Zhang, Two time-scale resource management for green internet of things networks. IEEE Internet Things J. **6**(1), 545–556 (2019). https://doi.org/10.1109/JIOT.2018.2842766
13. D. Dai, L. Lyu, NOMA-enabled CoMP clustering and power control for green internet of things networks. IEEE Access **8**, 90109–90117 (2020). https://doi.org/10.1109/ACCESS.2020.2992648
14. Y. Ren, X. Zhang, T. Wu, Y. Tan, In-network caching for the green internet of things. IEEE Access **9**, 76413–76422 (2021). https://doi.org/10.1109/ACCESS.2021.3082532
15. M. Wen, S. Lin, K.J. Kim, F. Ji, Cyclic delay diversity with index modulation for green internet of things. IEEE Trans. Green Commun. Netw. **5**(2), 600–610 (2021). https://doi.org/10.1109/TGCN.2021.3067705
16. L. Liu, A.R. Mishra, Enabling technologies challenges of green internet of things (IoT) towards sustainable development. Technol. Econ. Dev. Econ. **30**(2), 344–375 (2024). https://doi.org/10.3846/tede.2022.16520
17. P.K. Dubey, B. Singh, D. Singh, A.K. Dubey, Green Internet of Things, in *Network Optimization in Intelligent Internet of Things Applications*, CRC Press (2024), pp. 93–121. https://doi.org/10.1201/9781003405535-8

18. V. Roy, A context-aware internet of things (IoT) founded approach to scheming an operative priority-based scheduling algorithms **13**(1), 28–35 (2024). https://doi.org/10.54216/jcim.130103
19. A.K. Jain, M. Mudholkar, P. Mudholkar, M.K. Sharma, M. Tiwari, V. Roy, A novel approach to optimized cluster head selection in heterogeneous WSNs for IoT deployments, in *2024 IEEE 2nd International Conference on Innovations in High Speed Communication and Signal Processing (IHCSP), Bhopal, India* (2024), pp. 1–6. https://doi.org/10.1109/IHCSP63227.2024.10960066.
20. W. Yang, H. Huang, X. Jing, Z. Li, C. Zhu, Social interaction assisted resource sharing scheme for device-to-device communication towards green internet of things. IEEE Access **8**, 71652–71661 (2020). https://doi.org/10.1109/ACCESS.2020.2986785
21. Z. Li, Y. Liu, A. Liu, S. Wang, H. Liu, Minimizing converge-cast time and energy consumption in green internet of things. IEEE Trans. Emerg. Top. Comput. **8**(3), 797–813 (2020). https://doi.org/10.1109/TETC.2018.2844282
22. R. C. Shit, S. Sharma, M. S. Obaidat, and D. Puthal, "Adaptive Software Defined Node Deployment for Green Internet of Things," in *2020 IEEE International Conference on Communications (ICC), Dublin, Ireland* (2020), pp. 1–6. https://doi.org/10.1109/ICC40277.2020.9149396

Index

S. Saif et al. (eds.), *Secure Intelligent Sustainable Applications Using IoT for Society 6.0*, Internet of Things, https://doi.org/10.1007/978-3-032-27248-5

www.ingramcontent.com/pod-product-compliance
Ingram Content Group UK Ltd.
Pitfield, Milton Keynes, MK11 3LW, UK
UKHW020935180926
475692UK00003B/57

9783032272478